LIE GROUPS, PHYSICS, AND GEOMETRY

An Introduction for Physicists, Engineers and Chemists

Describing many of the most important aspects of Lie group theory, this book presents the subject in a 'hands on' way. Rather than concentrating on theorems and proofs, the book shows the relation of Lie groups with many branches of mathematics and physics, and illustrates these with concrete computations. Many examples of Lie groups and Lie algebras are given throughout the text, with applications of the material to physical sciences and applied mathematics. The relation between Lie group theory and algorithms for solving ordinary differential equations is presented and shown to be analogous to the relation between Galois groups and algorithms for solving polynomial equations. Other chapters are devoted to differential geometry, relativity, electrodynamics, and the hydrogen atom.

Problems are given at the end of each chapter so readers can monitor their understanding of the materials. This is a fascinating introduction to Lie groups for graduate and undergraduate students in physics, mathematics and electrical engineering, as well as researchers in these fields.

ROBERT GILMORE is a Professor in the Department of Physics at Drexel University, Philadelphia. He is a Fellow of the American Physical Society, and a Member of the Standing Committee for the International Colloquium on Group Theoretical Methods in Physics. His research areas include group theory, catastrophe theory, atomic and nuclear physics, singularity theory, and chaos.

LIE GROUPS, PHYSICS, AND GEOMETRY

An Introduction for Physicists, Engineers
and Chemists

ROBERT GILMORE
Drexel University, Philadelphia

CAMBRIDGE
UNIVERSITY PRESS

University Printing House, Cambridge CB2 8BS, United Kingdom

Published in the United States of America by Cambridge University Press, New York

Cambridge University Press is part of the University of Cambridge.

It furthers the University's mission by disseminating knowledge in the pursuit of education, learning and research at the highest international levels of excellence.

www.cambridge.org
Information on this title: www.cambridge.org/9780521884006

© Robert Gilmore 2008

This publication is in copyright. Subject to statutory exception and to the provisions of relevant collective licensing agreements, no reproduction of any part may take place without the written permission of Cambridge University Press.

First published 2008
Third printing 2011

A catalogue record for this publication is available from the British Library

ISBN 978-0-521-88400-6 Hardback

Cambridge University Press has no responsibility for the persistence or accuracy of URLs for external or third-party internet websites referred to in this publication, and does not guarantee that any content on such websites is, or will remain, accurate or appropriate.

Contents

		Preface	*page* xi
1	Introduction		1
	1.1	The program of Lie	1
	1.2	A result of Galois	2
	1.3	Group theory background	3
	1.4	Approach to solving polynomial equations	8
	1.5	Solution of the quadratic equation	10
	1.6	Solution of the cubic equation	11
	1.7	Solution of the quartic equation	15
	1.8	The quintic cannot be solved	17
	1.9	Example	18
	1.10	Conclusion	21
	1.11	Problems	22
2	Lie groups		24
	2.1	Algebraic properties	24
	2.2	Topological properties	25
	2.3	Unification of algebra and topology	27
	2.4	Unexpected simplification	29
	2.5	Conclusion	29
	2.6	Problems	30
3	Matrix groups		34
	3.1	Preliminaries	34
	3.2	No constraints	35
	3.3	Linear constraints	36
	3.4	Bilinear and quadratic constraints	39
	3.5	Multilinear constraints	42
	3.6	Intersections of groups	43
	3.7	Embedded groups	43

	3.8	Modular groups	44
	3.9	Conclusion	46
	3.10	Problems	47
4	Lie algebras	55
	4.1	Why bother?	55
	4.2	How to linearize a Lie group	56
	4.3	Inversion of the linearization map: EXP	57
	4.4	Properties of a Lie algebra	59
	4.5	Structure constants	61
	4.6	Regular representation	62
	4.7	Structure of a Lie algebra	63
	4.8	Inner product	64
	4.9	Invariant metric and measure on a Lie group	66
	4.10	Conclusion	69
	4.11	Problems	69
5	Matrix algebras	74
	5.1	Preliminaries	74
	5.2	No constraints	74
	5.3	Linear constraints	75
	5.4	Bilinear and quadratic constraints	78
	5.5	Multilinear constraints	80
	5.6	Intersections of groups	80
	5.7	Algebras of embedded groups	81
	5.8	Modular groups	81
	5.9	Basis vectors	81
	5.10	Conclusion	83
	5.11	Problems	83
6	Operator algebras	88
	6.1	Boson operator algebras	88
	6.2	Fermion operator algebras	89
	6.3	First order differential operator algebras	90
	6.4	Conclusion	93
	6.5	Problems	93
7	EXPonentiation	99
	7.1	Preliminaries	99
	7.2	The covering problem	100
	7.3	The isomorphism problem and the covering group	105
	7.4	The parameterization problem and BCH formulas	108
	7.5	EXPonentials and physics	114

7.6	Conclusion	119
7.7	Problems	120

8 Structure theory for Lie algebras 129
8.1	Regular representation	129
8.2	Some standard forms for the regular representation	129
8.3	What these forms mean	133
8.4	How to make this decomposition	135
8.5	An example	136
8.6	Conclusion	136
8.7	Problems	137

9 Structure theory for simple Lie algebras 139
9.1	Objectives of this program	139
9.2	Eigenoperator decomposition – secular equation	140
9.3	Rank	143
9.4	Invariant operators	143
9.5	Regular elements	146
9.6	Semisimple Lie algebras	147
9.7	Canonical commutation relations	151
9.8	Conclusion	153
9.9	Problems	154

10 Root spaces and Dynkin diagrams 159
10.1	Properties of roots	159
10.2	Root space diagrams	160
10.3	Dynkin diagrams	165
10.4	Conclusion	168
10.5	Problems	168

11 Real forms 172
11.1	Preliminaries	172
11.2	Compact and least compact real forms	174
11.3	Cartan's procedure for constructing real forms	176
11.4	Real forms of simple matrix Lie algebras	177
11.5	Results	181
11.6	Conclusion	182
11.7	Problems	183

12 Riemannian symmetric spaces 189
12.1	Brief review	189
12.2	Globally symmetric spaces	190
12.3	Rank	191
12.4	Riemannian symmetric spaces	192

12.5	Metric and measure	193
12.6	Applications and examples	194
12.7	Pseudo-Riemannian symmetric spaces	197
12.8	Conclusion	198
12.9	Problems	198

13 Contraction — 205

13.1	Preliminaries	205
13.2	Inönü–Wigner contractions	206
13.3	Simple examples of Inönü–Wigner contractions	206
13.4	The contraction $U(2) \to H_4$	211
13.5	Conclusion	216
13.6	Problems	217

14 Hydrogenic atoms — 221

14.1	Introduction	221
14.2	Two important principles of physics	222
14.3	The wave equations	223
14.4	Quantization conditions	224
14.5	Geometric symmetry $SO(3)$	227
14.6	Dynamical symmetry $SO(4)$	230
14.7	Relation with dynamics in four dimensions	233
14.8	DeSitter symmetry $SO(4, 1)$	235
14.9	Conformal symmetry $SO(4, 2)$	238
14.10	Spin angular momentum	243
14.11	Spectrum generating group	245
14.12	Conclusion	249
14.13	Problems	250

15 Maxwell's equations — 259

15.1	Introduction	259
15.2	Review of the inhomogeneous Lorentz group	261
15.3	Subgroups and their representations	262
15.4	Representations of the Poincaré group	264
15.5	Transformation properties	270
15.6	Maxwell's equations	273
15.7	Conclusion	275
15.8	Problems	275

16 Lie groups and differential equations — 284

16.1	The simplest case	285
16.2	First order equations	286
16.3	An example	290

16.4	Additional insights	295
16.5	Conclusion	302
16.6	Problems	303
Bibliography		309
Index		313

Preface

Many years ago I wrote the book *Lie Groups, Lie Algebras, and Some of Their Applications* (New York: Wiley, 1974). That was a big book: long and difficult. Over the course of the years I realized that more than 90% of the most useful material in that book could be presented in less than 10% of the space. This realization was accompanied by a promise that some day I would do just that – rewrite and shrink the book to emphasize the most useful aspects in a way that was easy for students to acquire and to assimilate. The present work is the fruit of this promise.

In carrying out the revision I have created a sandwich. Lie group theory has its intellectual underpinnings in Galois theory. In fact, the original purpose of what we now call Lie group theory was to use continuous groups to solve differential (continuous) equations in the spirit that finite groups had been used to solve algebraic (finite) equations. It is rare that a book dedicated to Lie groups begins with Galois groups and includes a chapter dedicated to the applications of Lie group theory to solving differential equations. This book does just that. The first chapter describes Galois theory, and the last chapter shows how to use Lie theory to solve some ordinary differential equations. The fourteen intermediate chapters describe many of the most important aspects of Lie group theory and provide applications of this beautiful subject to several important areas of physics and geometry.

Over the years I have profited from the interaction with many students through comments, criticism, and suggestions for new material or different approaches to old. Three students who have contributed enormously during the past few years are Dr. Jairzinho Ramos-Medina, who worked with me on Chapter 15 (Maxwell's equations), and Daniel J. Cross and Timothy Jones, who aided this computer illiterate with much moral and ebit ether support. Finally, I thank my beautiful wife Claire for her gracious patience and understanding throughout this long creation process.

Robert Gilmore

1
Introduction

Lie groups were initially introduced as a tool to solve or simplify ordinary and partial differential equations. The model for this application was Galois' use of finite groups to solve algebraic equations of degree two, three, and four, and to show that the general polynomial equation of degree greater than four could not be solved by radicals. In this chapter we show how the structure of the finite group that leaves a quadratic, cubic, or quartic equation invariant can be used to develop an algorithm to solve that equation.

1.1 The program of Lie

Marius Sophus Lie (1842–1899) embarked on a program that is still not complete, even after a century of active work. This program attempts to use the power of the tool called group theory to solve, or at least simplify, ordinary differential equations.

Earlier in nineteenth century, Évariste Galois (1811–1832) had used group theory to solve algebraic (polynomial) equations that were quadratic, cubic, and quartic. In fact, he did more. He was able to prove that no closed form solution could be constructed for the general quintic (or any higher degree) equation using only the four standard operations of arithmetic $(+, -, \times, \div)$ as well as extraction of the nth roots of a complex number.

Lie initiated his program on the basis of analogy. If finite groups were required to decide on the solvability of finite-degree polynomial equations, then "infinite groups" (i.e., groups depending continuously on one or more real or complex variables) would probably be involved in the treatment of ordinary and partial differential equations. Further, Lie knew that the structure of the polynomial's invariance (Galois) group not only determined whether the equation was solvable in closed form, but also provided the algorithm for constructing the solution in the case that the equation was solvable. He therefore felt that the structure of an ordinary differential equation's invariance group would determine whether or not the equation could be solved or simplified and, if so, the group's structure would also provide the algorithm for constructing the solution or simplification.

Lie therefore set about the program of computing the invariance group of ordinary differential equations. He also began studying the structure of the children he begat, which we now call Lie groups.

Lie groups come in two basic varieties: the simple and the solvable. Simple groups have the property that they regenerate themselves under commutation. Solvable groups do not, and contain a chain of subgroups, each of which is an invariant subgroup of its predecessor.

Simple and solvable groups are the building blocks for all other Lie groups. Semisimple Lie groups are direct products of simple Lie groups. Nonsemisimple Lie groups are semidirect products of (semi)simple Lie groups with invariant subgroups that are solvable.

Not surprisingly, solvable Lie groups are related to the integrability, or at least simplification, of ordinary differential equations. However, simple Lie groups are more rigidly constrained, and form such a beautiful subject of study in their own right that much of the effort of mathematicians during the last century involved the classification and complete enumeration of all simple Lie groups and the discussion of their properties. Even today, there is no complete classification of solvable Lie groups, and therefore nonsemisimple Lie groups.

Both simple and solvable Lie groups play an important role in the study of differential equations. As in Galois' case of polynomial equations, differential equations can be solved or simplified by quadrature if their invariance group is solvable. On the other hand, most of the classical functions of mathematical physics are matrix elements of simple Lie groups, in particular matrix representations. There is a very rich connection between Lie groups and special functions that is still evolving.

1.2 A result of Galois

In 1830 Galois developed machinery that allowed mathematicians to resolve questions that had eluded answers for 2000 years or longer. These questions included the three famous challenges to ancient Greek geometers: whether by ruler and compasses alone it was possible to

- square a circle,
- trisect an angle,
- double a cube.

His work helped to resolve longstanding questions of an algebraic nature: whether it was possible, using only the operations of arithmetic together with the operation of constructing radicals, to solve

- cubic equations,
- quartic equations,
- quintic equations.

This branch of mathematics, now called Galois theory, continues to provide powerful new results, such as supplying answers and solution methods to the following questions.

- Can an algebraic expression be integrated in closed form?
- Under what conditions can errors in a binary code be corrected?

This beautiful machine, applied to a problem, provides important results. First, it can determine whether a solution is possible or not under the conditions specified. Second, if a solution is possible, it suggests the structure of the algorithm that can be used to construct the solution in a finite number of well-defined steps.

Galois' approach to the study of algebraic (polynomial) equations involved two areas of mathematics, now called field theory and group theory. One useful statement of Galois' result is the following (Lang, 1984; Stewart, 1989).

Theorem A polynomial equation over the complex field is solvable by radicals if and only if its Galois group G contains a chain of subgroups $G = G_0 \supset G_1 \supset \cdots \supset G_\omega = I$ with the properties:

(i) G_{i+1} is an invariant subgroup of G_i;
(ii) each factor group G_i/G_{i+1} is commutative.

In the statement of this theorem the field theory niceties are contained in the term "solvable by radicals." This means that in addition to the four standard arithmetic operations $+, -, \times, \div$ one is allowed the operation of taking nth roots of complex numbers.

The principal result of this theorem is stated in terms of the structure of the group that permutes the roots of the polynomial equation among themselves. Determining the structure of this group is a finite, and in fact very simple, process.

1.3 Group theory background

A group G is defined as follows. It consists of a set of operations $G = \{g_1, g_2, \ldots\}$, called **group operations**, together with a combinatorial operation, $\cdot$, called **group multiplication**, such that the following four axioms are satisfied.

(i) Closure: if $g_i \in G$, $g_j \in G$, then $g_i \cdot g_j \in G$.

(ii) Associativity: for all $g_i \in G$, $g_j \in G$, $g_k \in G$,

$$(g_i \cdot g_j) \cdot g_k = g_i \cdot (g_j \cdot g_k)$$

(iii) Identity: there is a group operation, I (identity operator), with the property that

$$g_i \cdot I = g_i = I \cdot g_i$$

(iv) Inverse: every group operation g_i has an inverse (called g_i^{-1}):

$$g_i \cdot g_i^{-1} = I = g_i^{-1} \cdot g_i$$

The Galois group G of a general polynomial equation

$$(z - z_1)(z - z_2) \cdots (z - z_n) = 0$$
$$z^n - I_1 z^{n-1} + I_2 z^{n-2} + \cdots + (-1)^n I_n = 0 \quad (1.1)$$

is the group that permutes the roots $z_1, z_2, \ldots, z_n$ among themselves and leaves the equation invariant:

$$\begin{bmatrix} z_1 \\ z_2 \\ \vdots \\ z_n \end{bmatrix} \longrightarrow \begin{bmatrix} z_{i_1} \\ z_{i_2} \\ \vdots \\ z_{i_n} \end{bmatrix} \quad (1.2)$$

This group, called the permutation group P_n or the symmetric group S_n, has $n!$ group operations. Each group operation is some permutation of the roots of the polynomial; the group multiplication is composition of successive permutations.

The permutation group S_n has a particularly convenient **representation** in terms of $n \times n$ matrices. These matrices have one nonzero element, $+1$, in each row and each column. For example, the $6 = 3!$ 3×3 matrices for the permutation representation of S_3 are

$$I \to \begin{bmatrix} 1 & 0 & 0 \\ 0 & 1 & 0 \\ 0 & 0 & 1 \end{bmatrix} \quad (123) \to \begin{bmatrix} 0 & 1 & 0 \\ 0 & 0 & 1 \\ 1 & 0 & 0 \end{bmatrix} \quad (321) \to \begin{bmatrix} 0 & 0 & 1 \\ 1 & 0 & 0 \\ 0 & 1 & 0 \end{bmatrix}$$

$$(12) \to \begin{bmatrix} 0 & 1 & 0 \\ 1 & 0 & 0 \\ 0 & 0 & 1 \end{bmatrix} \quad (23) \to \begin{bmatrix} 1 & 0 & 0 \\ 0 & 0 & 1 \\ 0 & 1 & 0 \end{bmatrix} \quad (13) \to \begin{bmatrix} 0 & 0 & 1 \\ 0 & 1 & 0 \\ 1 & 0 & 0 \end{bmatrix}$$

$$(1.3)$$

The symbol (123) means that the first root, z_1, is replaced by z_2, z_2 is replaced by z_3, and z_3 is replaced by z_1

$$\begin{bmatrix} z_1 \\ z_2 \\ z_3 \end{bmatrix} \xrightarrow{(123)} \begin{bmatrix} z_2 \\ z_3 \\ z_1 \end{bmatrix} \tag{1.4}$$

The permutation matrix associated with this group operation carries out the same permutation

$$\begin{bmatrix} z_2 \\ z_3 \\ z_1 \end{bmatrix} = \begin{bmatrix} 0 & 1 & 0 \\ 0 & 0 & 1 \\ 1 & 0 & 0 \end{bmatrix} \begin{bmatrix} z_1 \\ z_2 \\ z_3 \end{bmatrix} \tag{1.5}$$

More generally, a **matrix representation** of a group is a mapping of each group operation into an $n \times n$ matrix that preserves the group multiplication operation

$$\begin{array}{ccccc} g_i & \cdot & g_j & = & g_i \cdot g_j \\ \downarrow & & \downarrow & & \downarrow \\ \Gamma(g_i) & \times & \Gamma(g_j) & = & \Gamma(g_i \cdot g_j) \end{array} \tag{1.6}$$

Here $\cdot$ represents the multiplication operation in the group (i.e., composition of substitutions in S_n) and $\times$ represents the multiplication operation among the matrices (i.e., matrix multiplication). The condition (1.6) that defines a matrix representation of a group, $G \to \Gamma(G)$, is that the product of matrices representing two group operations ($\Gamma(g_i) \times \Gamma(g_j)$) is equal to the matrix representing the product of these operations in the group ($\Gamma(g_i \cdot g_j)$) for all group operations $g_i, g_j \in G$.

This permutation representation of S_3 is 1:1, or a **faithful representation** of S_3, since knowledge of the 3×3 matrix uniquely identifies the original group operation in S_3.

A **subgroup** H of the group G is a subset of group operations in G that is closed under the group multiplication in G.

Example The subset of operations I, (123), (321) forms a subgroup of S_3. This particular subgroup is denoted A_3 (**alternating group**). It consists of those operations in S_3 whose determinants, in the permutation representation, are $+1$. The group S_3 has three two-element subgroups:

$$S_2(12) = \{I, (12)\}$$
$$S_2(23) = \{I, (23)\}$$
$$S_2(13) = \{I, (13)\}$$

as well as the subgroup consisting of the identity alone. The alternating subgroup $A_3 \subset S_3$ and the three two-element subgroups $S_2(ij)$ of S_3 are illustrated in Fig. 1.1.

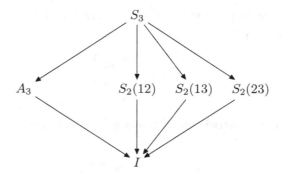

Figure 1.1. Subgroups of S_3.

The set of operations I, (123), (12) does not constitute a subgroup because products of operations in this subset do not lie in this subset: $(123) \cdot (123) = (321)$, $(123) \cdot (12) = (23)$, etc. In fact, the two operations (123), (12) **generate** S_3 by taking products of various lengths in various order.

A group G is **commutative**, or **abelian**, if

$$g_i \cdot g_j = g_j \cdot g_i \qquad (1.7)$$

for all group operations g_i, $g_j \in G$.

Example S_3 is not commutative, while A_3 is. For S_3 we have

$$(12)(23) = (321)$$
$$(123) \neq (321) \qquad (1.8)$$
$$(23)(12) = (123)$$

Two subgroups of G, $H_1 \subset G$ and $H_2 \subset G$ are **conjugate** if there is a group element $g \in G$ with the property

$$g H_1 g^{-1} = H_2 \qquad (1.9)$$

Example The subgroups $S_2(12)$ and $S_2(13)$ are conjugate in S_3 since

$$(23)S_2(12)(23)^{-1} = (23)\{I, (12)\}(23)^{-1} = \{I, (13)\} = S_2(13) \qquad (1.10)$$

On the other hand, the alternating group $A_3 \subset S_3$ is **self-conjugate**, since any operation in $G = S_3$ serves merely to permute the group operations in A_3 among themselves:

$$(23)A_3(23)^{-1} = (23)\{I, (123), (321)\}(23)^{-1} = \{I, (321), (123)\} = A_3 \qquad (1.11)$$

A subgroup $H \subset G$ which is self-conjugate under all operations in G is called an **invariant subgroup** of G, or **normal subgroup** of G.

1.3 Group theory background

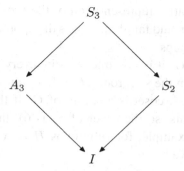

Figure 1.2. Subgroups of S_3, combining conjugate subgroups.

In constructing group-subgroup diagrams, it is customary to show only one of the mutually conjugate subgroups. This simplifies Fig. 1.1 to Fig. 1.2.

A mapping f from a group G with group operations $g_1, g_2, \ldots$ and group multiplication $\cdot$ to a group H with group operations $h_1, h_2, \ldots$ and group multiplication $\times$ is called a **homomorphism** if it preserves group multiplication:

$$
\begin{array}{ccccc}
g_i & \cdot & g_j & = & g_i \cdot g_j \\
\downarrow & \downarrow & \downarrow & & \downarrow \\
f(g_i) & \times & f(g_j) & = & f(g_i \cdot g_j)
\end{array}
\qquad (1.12)
$$

The group H is called a **homomorphic image** of G. Several different group elements in G may map to a single group element in H. Every element $h_i \in H$ has the same number of inverse images $g_j \in G$. If each group element $h \in H$ has a unique inverse image $g \in G$ ($h_1 = f(g_1)$ and $h_2 = f(g_2)$, $h_1 = h_2 \Rightarrow g_1 = g_2$) the mapping f is an **isomorphism**.

Example The 3:1 mapping f of S_3 onto S_2 given by

$$
\begin{array}{rcl}
S_3 & \xrightarrow{f} & S_2 \\
I, (123), (321) & \longrightarrow & I \\
(12), (23), (31) & \longrightarrow & (12)
\end{array}
\qquad (1.13)
$$

is a homomorphism.

Example The 1:1 mapping of S_3 onto the six 3×3 matrices given in (1.3) is an isomorphism.

Remark Homomorphisms of groups to matrix groups, such as that in (1.3), are called *matrix representations*. The representation in (1.3) is 1:1 or faithful, since the mapping is an isomorphism.

Remark Isomorphic groups are indistinguishable at the algebraic level. Thus, when an isomorphism exists between a group and a matrix group, it is often preferable to study the matrix representation of the group since the properties of matrices are so well known and familiar. This is the approach we pursue in Chapter 3 when discussing Lie groups.

If H is a subgroup of G, it is possible to write every group element in G as a product of an element h in the subgroup H with a group element in a "quotient," or *coset* (denoted G/H). A coset is a subset of G. If the *order* of G is $|G|$ (S_3 has $3! = 6$ group elements, so the order of S_3 is 6), then the order of G/H is $|G/H| = |G|/|H|$. For example, for subgroups $H = A_3 = \{I, (123), (321)\}$ and $H = S_2(23) = \{I, (23)\}$ we have

$$
\begin{array}{rcl}
G/H \cdot H & = & G \\
\{I, (12)\} \cdot \{I, (123), (321)\} & = & \{I, (123), (321), (12), (13), (23)\} \\
\{I, (12), (321)\} \cdot \{I, (23)\} & = & \{I, (23), (12), (123), (321), (13)\}
\end{array}
\tag{1.14}
$$

The choice of the $|G|/|H|$ group elements in the quotient space is not unique. For the subgroup A_3 we could equally well have chosen $G/H = S_3/A_3 = \{I, (13)\}$ or $\{I, (23)\}$; for $S_2(23)$ we could equally well have chosen $G/H = S_3/S_2(23) = \{I, (123), (321)\}$.

In general, it is not possible to choose the group elements in G/H so that they form a subgroup of G. However, if H is an invariant subgroup of G, it is always possible to choose the group elements in the quotient space G/H in such a way that they form a subgroup in G. This group is called the **factor group**, also denoted G/H. Since A_3 is an invariant subgroup of S_3, the coset S_3/A_3 is a group, and this group is isomorphic to S_2. More generally, if H is an invariant subgroup of G, then the group G is the **direct product** of the invariant subgroup H with the factor group G/H: $G = G/H \times H$.

1.4 Approach to solving polynomial equations

The general nth degree polynomial equation over the complex field can be expressed in terms of the kth order symmetric functions I_k of the roots z_i as follows:

$$(z - z_1)(z - z_2) \cdots (z - z_n) = z^n - I_1 z^{n-1} + I_2 z^{n-2} - \cdots + (-)^n I_n = 0$$

$$I_1 = \sum_{i=1}^{n} z_i = z_1 + z_2 + \cdots + z_n$$

$$I_2 = \sum_{i<j}^{n} z_i z_j = z_1 z_2 + z_1 z_3 + \cdots + z_1 z_n + z_2 z_3 + \cdots + z_{n-1} z_n$$

$$\vdots \quad \vdots \quad \vdots \tag{1.15}$$

$$I_n = \sum_{i<j<\cdots<k}^{n} z_i z_j \cdots z_k = z_1 z_2 \cdots z_n$$

The n functions I_k ($k = 1, 2, \ldots, n$) of the n roots ($z_1, z_2, \ldots, z_n$) are symmetric: this means that they are invariant under the Galois group S_n of this equation. Further, any function $f(z_1, z_2, \ldots, z_n)$ that is invariant under S_n can be written as a function of the invariants $I_1, I_2, \ldots, I_n$. The invariants are easily expressed in terms of the roots (see Eq. (1.15)). The inverse step, that of expressing the roots in terms of the invariants, or coefficients of the polynomial equation, is the problem of solving the polynomial equation.

Galois' theorem states that a polynomial equation over the complex field can be solved if and only if its Galois group G contains a chain of subgroups (Lang, 1984; Stewart, 1989)

$$G = G_0 \supset G_1 \supset \cdots \supset G_\omega = I \qquad (1.16)$$

with the properties

(i) G_{i+1} is an invariant subgroup of G_i,
(ii) G_i/G_{i+1} is commutative.

The procedure for solving polynomial equations is constructive. First, the last group-subgroup pair in this chain is isolated: $G_{\omega-1} \supset G_\omega = I$. The **character table** for the commutative group $G_{\omega-1}/G_\omega = G_{\omega-1}$ is constructed. This lists the $|G_{\omega-1}|/|G_\omega|$ inequivalent one-dimensional representations of $G_{\omega-1}$. Linear combinations of the roots z_i are identified that transform under (i.e., are basis functions for) the one-dimensional irreducible representations of $G_{\omega-1}$. These functions are

(i) symmetric under $G_\omega = I$,
(ii) not all symmetric under $G_{\omega-1}$.

Next, the next pair of groups $G_{\omega-2} \supset G_{\omega-1}$ is isolated. Starting from the set of functions in the previous step, one constructs from them functions that are

(i) symmetric under $G_{\omega-1}$,
(ii) not all symmetric under $G_{\omega-2}$.

This bootstrap procedure continues until the last group-subgroup pair $G = G_0 \supset G_1$ is treated. At this stage the last set of functions can be solved by radicals. These solutions are then fed down the group-subgroup chain until the last pair $G_{\omega-1} \supset G_\omega = I$ is reached. When this occurs, we obtain a *linear* relation between the roots $z_1, z_2, \ldots, z_n$ and functions of the invariants $I_1, I_2, \ldots, I_n$.

This brief description will now be illustrated by using Galois theory to solve quadratic, cubic, and quartic equations by radicals.

Figure 1.3. Group chain for the Galois group S_2 of the general quadratic equation.

1.5 Solution of the quadratic equation

The general quadratic equation has the form

$$(z - r_1)(z - r_2) = z^2 - I_1 z + I_2 = 0$$

$$I_1 = r_1 + r_2 \tag{1.17}$$
$$I_2 = r_1 r_2$$

The Galois group is S_2 with subgroup chain shown in Fig. 1.3.

The character table for the commutative group S_2 is

$$\begin{array}{c|cc|l}
 & I & (12) & \text{Basis functions} \\
\hline
\Gamma^1 & 1 & 1 & u_1 = r_1 + r_2 \\
\Gamma^2 & 1 & -1 & u_2 = r_1 - r_2
\end{array} \tag{1.18}$$

Linear combinations of the roots that transform under the one-dimensional irreducible representations Γ^1, Γ^2 are

$$\begin{bmatrix} u_1 \\ u_2 \end{bmatrix} = \begin{bmatrix} 1 & 1 \\ 1 & -1 \end{bmatrix} \begin{bmatrix} r_1 \\ r_2 \end{bmatrix} = \begin{bmatrix} r_1 + r_2 \\ r_1 - r_2 \end{bmatrix} \tag{1.19}$$

That is, the function $r_1 - r_2$ is mapped into itself by the identity, and into its negative by (12)

$$(r_1 - r_2) \left\} \begin{array}{l} \xrightarrow{I} +(r_1 - r_2) \\ \xrightarrow{(12)} (r_2 - r_1) = -(r_1 - r_2) \end{array} \right. \tag{1.20}$$

As a result, $(r_1 - r_2)$ is not symmetric under the action of the group S_2. It transforms under the irreducible representation Γ^2, not the identity representation Γ^1.

Since the square $(r_1 - r_2)^2$ is symmetric (transforms under the identity representation of S_2), it can be expressed in terms of the two invariants I_1, I_2 as follows

$$(r_1 - r_2)^2 = r_1^2 - 2r_1 r_2 + r_2^2$$
$$= r_1^2 + 2r_1 r_2 + r_2^2 - 4r_1 r_2 = I_1^2 - 4I_2 = D \tag{1.21}$$

where D is the **discriminant** of the quadratic equation. Since $(r_1 - r_2) = \pm\sqrt{D}$, we have the following linear relation between roots and symmetric functions:

$$\begin{bmatrix} 1 & 1 \\ 1 & -1 \end{bmatrix} \begin{bmatrix} r_1 \\ r_2 \end{bmatrix} = \begin{bmatrix} I_1 \\ \pm [I_1^2 - 4I_2]^{1/2} \end{bmatrix} \quad (1.22)$$

Inversion of a square matrix involves a sequence of linear operations. We find

$$\begin{bmatrix} r_1 \\ r_2 \end{bmatrix} = \frac{1}{2} \begin{bmatrix} 1 & 1 \\ 1 & -1 \end{bmatrix} \begin{bmatrix} I_1 \\ \pm\sqrt{D} \end{bmatrix} \quad (1.23)$$

The roots are

$$r_1, r_2 = \frac{1}{2}(I_1 \pm \sqrt{D}) \quad (1.24)$$

We solve the quadratic equation by another procedure, which we use in the following two sections to simplify the cubic and quartic equations. This method is to move the origin to the mean value of the roots by defining a new variable, x, in terms of z (see Eq. (1.15)) by a **Tschirnhaus transformation**

$$z = x + \frac{1}{2}I_1 \quad (1.25)$$

The quadratic equation for the new coordinate is

$$x^2 - I_1'x + I_2' = x^2 + I_2' = 0$$
$$I_1' = 0 \quad (1.26)$$
$$I_2' = I_2 - \left(\frac{1}{2}I_1\right)^2$$

The solutions for this **auxiliary** equation are constructed by radicals

$$x = \pm\sqrt{-I_2'} \quad (1.27)$$

from which we easily construct the roots of the original equation

$$r_{1,2} = \frac{1}{2}\left(I_1 \pm \sqrt{I_1^2 - 4I_2}\right) \quad (1.28)$$

1.6 Solution of the cubic equation

The general cubic equation has the form

$$(z - s_1)(z - s_2)(z - s_3) = z^3 - I_1 z^2 + I_2 z - I_3 = 0$$
$$I_1 = s_1 + s_2 + s_3$$
$$I_2 = s_1 s_2 + s_1 s_3 + s_2 s_3 \quad (1.29)$$
$$I_3 = s_1 s_2 s_3$$

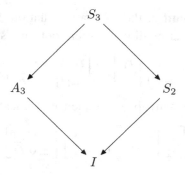

Figure 1.4. Group chain for the Galois group S_3 of the general cubic equation.

The Galois group is S_3 with subgroup chain shown in Fig. 1.4.

Since A_3 is an invariant subgroup of S_3 and I is an invariant subgroup of A_3, the first of the two conditions of the Galois theorem (there exists a chain of invariant subgroups) is satisfied. Since $S_3/A_3 = S_2$ is commutative and $A_3/I = A_3$ is commutative, the second condition is also satisfied. This means that the general cubic equation can be solved.

We begin the solution with the last group-subgroup pair in this chain: $A_3 \supset I$. The character table for the commutative group A_3 is

$$
\begin{array}{c|ccc l}
 & I & (123) & (321) & \text{Basis functions} \\
\hline
\Gamma^1 & 1 & 1 & 1 & v_1 = s_1 + s_2 + s_3 \\
\Gamma^2 & 1 & \omega & \omega^2 & v_2 = s_1 + \omega s_2 + \omega^2 s_3 \\
\Gamma^3 & 1 & \omega^2 & \omega & v_3 = s_1 + \omega^2 s_2 + \omega s_3
\end{array}
\quad (1.30)
$$

where

$$\omega^3 = +1 \qquad \omega = e^{2\pi i/3} = \frac{-1 + i\sqrt{3}}{2} \quad (1.31)$$

Linear combinations of the roots that transform under each of the three one-dimensional irreducible representations are easily constructed

$$\begin{bmatrix} v_1 \\ v_2 \\ v_3 \end{bmatrix} = \begin{bmatrix} 1 & 1 & 1 \\ 1 & \omega & \omega^2 \\ 1 & \omega^2 & \omega \end{bmatrix} \begin{bmatrix} s_1 \\ s_2 \\ s_3 \end{bmatrix} = \begin{bmatrix} s_1 + s_2 + s_3 \\ s_1 + \omega s_2 + \omega^2 s_3 \\ s_1 + \omega^2 s_2 + \omega s_3 \end{bmatrix} \quad (1.32)$$

For example, the action of $(123)^{-1}$ on v_2 is

$$(123)^{-1} v_2 = (321) v_2 = (321)(s_1 + \omega s_2 + \omega^2 s_3)$$
$$= s_3 + \omega s_1 + \omega^2 s_2 = \omega(s_1 + \omega s_2 + \omega^2 s_3) = \omega v_2 \quad (1.33)$$

1.6 Solution of the cubic equation

Since v_1 is symmetric under both A_3 and S_3, it can be expressed in terms of the invariants I_k:

$$v_1 = I_1 \tag{1.34}$$

The remaining functions, v_2 and v_3, are symmetric under I but not under A_3.

We now proceed to the next group-subgroup pair: $S_3 \supset A_3$. To construct functions symmetric under A_3 but not under S_3 we observe that the cubes of v_2 and v_3 are symmetric under A_3 but not under S_3:

$$\begin{aligned}
(12)(v_2)^3 &= (12)(s_1 + \omega s_2 + \omega^2 s_3)^3 = (s_2 + \omega s_1 + \omega^2 s_3)^3 \\
&= \omega^3(s_1 + \omega^2 s_2 + \omega s_3)^3 = (v_3)^3 \\
(12)(v_3)^3 &= (12)(s_1 + \omega^2 s_2 + \omega s_3)^3 = (s_2 + \omega^2 s_1 + \omega s_3)^3 \\
&= \omega^6(s_1 + \omega s_2 + \omega^2 s_3)^3 = (v_2)^3
\end{aligned} \tag{1.35}$$

Since $S_2 = S_3/A_3$ permutes the functions v_2^3 and v_3^3, it is the Galois group of the **resolvent** quadratic equation whose two roots are v_2^3 and v_3^3. This equation has the form

$$\begin{aligned}
(x - v_2^3)(x - v_3^3) &= x^2 - J_1 x + J_2 = 0 \\
J_1 &= v_2^3 + v_3^3 \\
J_2 &= v_2^3 v_3^3
\end{aligned} \tag{1.36}$$

Since J_1, J_2 are symmetric under S_3, they can be expressed in terms of the invariants I_1, I_2, I_3 of the original cubic. Since J_1 has order 3 and J_2 has order 6, we can write the invariants of the quadratic equation (1.36) in terms of the invariants I_1, I_2, I_3 (of orders 1, 2, 3) of the original cubic equation (1.29) as follows:

$$\begin{aligned}
J_1 &= \sum_{i+2j+3k=3} A_{ijk} I_1^i I_2^j I_3^k \\
J_2 &= \sum_{i+2j+3k=6} B_{ijk} I_1^i I_2^j I_3^k
\end{aligned} \tag{1.37}$$

These relations can be computed, but they simplify considerably if $I_1 = s_1 + s_2 + s_3 = 0$. This can be accomplished by shifting the origin using a Tschirnhaus transformation as before, with

$$z = y + \frac{1}{3} I_1 \tag{1.38}$$

The *auxiliary* cubic equation has the structure

$$y^3 - 0y^2 + I_2'y - I_3' = 0$$

$$\begin{aligned} I_1' &= s_1' + s_2' + s_3' &&= 0 \\ I_2' &= s_1's_2' + s_1's_3' + s_2's_3' &&= I_2 - (1/3)I_1^2 \\ I_3' &= s_1's_2's_3' &&= I_3 - (1/3)I_2I_1 + (2/27)I_1^3 \end{aligned} \quad (1.39)$$

The invariants $J_1 = v_2^3 + v_3^3$ and $J_2 = v_2^3 v_3^3$ can be expressed in terms of I_2', I_3' as follows

$$\begin{aligned} J_1 &= v_2^3 + v_3^3 = -27I_3' \\ J_2 &= v_2^3 v_3^3 = -27I_2'^3 \end{aligned} \quad (1.40)$$

The resolvent quadratic equation whose solution provides v_2^3, v_3^3 is

$$x^2 - (-27I_3')x + (-27I_2'^3) = 0 \quad (1.41)$$

The two solutions to this resolvent quadratic equation are

$$v_2^3, v_3^3 = -\frac{27}{2}I_3' \pm \frac{1}{2}\left[(27I_3')^2 + 4 \times 27I_2'^3\right]^{1/2} \quad (1.42)$$

The roots v_2 and v_3 are obtained by taking cube roots of v_2^3 and v_3^3.

$$\begin{matrix} v_2 \\ v_3 \end{matrix} = \left\{-\frac{27}{2}I_3' \pm \frac{1}{2}\left[(27I_3')^2 + 4 \times 27I_2'^3\right]^{1/2}\right\}^{1/3}$$

Finally, the roots s_1, s_2, s_3 are linearly related to v_1, v_2, v_3 by

$$\begin{bmatrix} 1 & 1 & 1 \\ 1 & \omega & \omega^2 \\ 1 & \omega^2 & \omega \end{bmatrix} \begin{bmatrix} s_1 \\ s_2 \\ s_3 \end{bmatrix} = \begin{bmatrix} v_1 \\ v_2 \\ v_3 \end{bmatrix} \quad (1.43)$$

Again, determination of the roots is accomplished by solving a set of simultaneous linear equations

$$\begin{bmatrix} s_1 \\ s_2 \\ s_3 \end{bmatrix} = \frac{1}{3}\begin{bmatrix} 1 & 1 & 1 \\ 1 & \omega^2 & \omega \\ 1 & \omega & \omega^2 \end{bmatrix} \begin{bmatrix} I_1 \\ v_2 \\ v_3 \end{bmatrix} = \frac{1}{3}\begin{bmatrix} v_1 + v_2 + v_3 \\ v_1 + \omega^2 v_2 + \omega v_3 \\ v_1 + \omega v_2 + \omega^2 v_3 \end{bmatrix} \quad (1.44)$$

1.7 Solution of the quartic equation

The general quartic equation has the form

$$(z - t_1)(z - t_2)(z - t_3)(z - t_4) = z^4 - I_1 z^3 + I_2 z^2 - I_3 z + I_4 = 0$$

$$I_1 = t_1 + t_2 + t_3 + t_4$$
$$I_2 = t_1 t_2 + t_1 t_3 + t_1 t_4 + t_2 t_3 + t_2 t_4 + t_3 t_4 \quad (1.45)$$
$$I_3 = t_1 t_2 t_3 + t_1 t_2 t_4 + t_1 t_3 t_4 + t_2 t_3 t_4$$
$$I_4 = t_1 t_2 t_3 t_4$$

For later convenience we will construct the auxiliary quartic by shifting the origin of coordinates through the Tschirnhaus transformation $z = z' + \frac{1}{4} I_1$

$$(z' - t_1)(z' - t_2)(z' - t_3)(z' - t_4) = z'^4 - I_1' z'^3 + I_2' z'^2 - I_3' z' + I_4' = 0$$

$$I_1' = 0$$
$$I_2' = I_2 - \tfrac{3}{8} I_1^2$$
$$I_3' = I_3 - \tfrac{1}{2} I_2 I_1 + \tfrac{1}{8} I_1^3 \quad (1.46)$$
$$I_4' = I_4 - \tfrac{1}{4} I_3 I_1 + \tfrac{1}{16} I_2 I_1^2 - \tfrac{3}{4^4} I_1^4$$

The Galois group is S_4. This has the subgroup chain shown in Fig. 1.5. The alternating group A_4 consists of the twelve group operations that have determinant $+1$ in the permutation matrix representation. The *four-group (vierergruppe, Klein group, Klein four-group)* V_4 is $\{I, (12)(34), (13)(24), (14)(23)\}$. The chain

$$S_4 \supset A_4 \supset V_4 \supset I$$

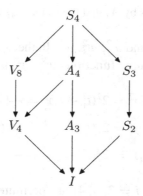

Figure 1.5. Group chain for the Galois group S_4 of the general quartic equation.

satisfies both conditions of Galois' theorem. In particular

(i) A_4 is invariant in S_4 and $S_4/A_4 = S_2$,
(ii) V_4 is invariant in A_4 and $A_4/V_4 = C_3 = \{I, (234), (432)\}$,
(iii) I is invariant in V_4 and $V_4/I = V_4 = \{I, (12)(34), (13)(24), (14)(23)\}$.

We again begin at the end of the chain with the commutative group V_4 whose character table is

	I	(12)(34)	(13)(24)	(14)(23)	Basis functions
Γ^1	1	1	1	1	$w_1 = t_1 + t_2 + t_3 + t_4$
Γ^2	1	1	-1	-1	$w_2 = t_1 + t_2 - t_3 - t_4$
Γ^3	1	-1	1	-1	$w_3 = t_1 - t_2 + t_3 - t_4$
Γ^4	1	-1	-1	1	$w_4 = t_1 - t_2 - t_3 + t_4$

(1.47)

The linear combinations of these roots that transform under each of the irreducible representations are

$$\begin{bmatrix} w_1 \\ w_2 \\ w_3 \\ w_4 \end{bmatrix} = \begin{bmatrix} 1 & 1 & 1 & 1 \\ 1 & 1 & -1 & -1 \\ 1 & -1 & 1 & -1 \\ 1 & -1 & -1 & 1 \end{bmatrix} \begin{bmatrix} t_1 \\ t_2 \\ t_3 \\ t_4 \end{bmatrix} = \begin{bmatrix} t_1 + t_2 + t_3 + t_4 \\ t_1 + t_2 - t_3 - t_4 \\ t_1 - t_2 + t_3 - t_4 \\ t_1 - t_2 - t_3 + t_4 \end{bmatrix}$$

(1.48)

These basis vectors are symmetric under I but the basis vectors w_2, w_3, w_4 are not symmetric under V_4.

We now advance to the next group-subgroup pair: $A_4 \supset V_4$. It is a simple matter to construct from these linear combinations functions that are

(i) symmetric under V_4,
(ii) permuted among themselves by A_4 and the group A_4/V_4.

These functions are $w_1 = I_1$ and w_2^2, w_3^2, w_4^2. In the coordinate system in which the sum of the roots is zero, the three functions w_2^2, w_3^2, w_4^2 are

$$w_2^2 = (t'_1 + t'_2 - t'_3 - t'_4)^2 = 2^2(t'_1 + t'_2)^2 = -4(t'_1 + t'_2)(t'_3 + t'_4)$$
$$w_3^2 = (t'_1 - t'_2 + t'_3 - t'_4)^2 = 2^2(t'_1 + t'_3)^2 = -4(t'_1 + t'_3)(t'_2 + t'_4) \quad (1.49)$$
$$w_4^2 = (t'_1 - t'_2 - t'_3 + t'_4)^2 = 2^2(t'_1 + t'_4)^2 = -4(t'_1 + t'_4)(t'_2 + t'_3)$$

It is clear that the three w_j^2 ($j = 2, 3, 4$) are permuted among themselves by the factor group $C_3 = A_4/V_4$, which is a subgroup of the Galois group of a resolvent cubic equation whose three roots are w_2^2, w_3^2, w_4^2:

$$(y - w_2^2)(y - w_3^2)(y - w_4^2) = y^3 - J_1 y^2 + J_2 y - J_3 = 0$$
$$J_1 = w_2^2 + w_3^2 + w_4^2$$
$$J_2 = w_2^2 w_3^2 + w_2^2 w_4^2 + w_3^2 w_4^2 \qquad (1.50)$$
$$J_3 = w_2^2 w_3^2 w_4^2$$

Since the three J_k are invariant under C_3, they can be expressed in terms of the symmetric functions (coefficients) of the original quartic equation (1.45) or (1.46). We find by direct calculation

$$J_1 = (-4)^1 (2I_2')$$
$$J_2 = (-4)^2 (I_2'^2 - 4I_4') \qquad (1.51)$$
$$J_3 = (-4)^3 (-I_3'^2)$$

This cubic equation is solved by proceeding to the first group-subgroup pair in the chain: $S_4 \supset A_4$, with $S_4/A_4 = S_2$. The cubic is solved by introducing the resolvent quadratic, as described in the previous section.

If the three solutions of the resolvent cubic equation are called y_2, y_3, y_4, then the functions w_2, w_3, w_4 are

$$w_2 = \pm\sqrt{y_2}$$
$$w_3 = \pm\sqrt{y_3} \qquad (1.52)$$
$$w_4 = \pm\sqrt{y_4}$$

A simple computation shows that $w_2 w_3 w_4 = 8 I_3'$. The signs $\pm\sqrt{y_j}$ are chosen so that their product is $8I_3'$. The simple linear relation between the roots t_i and the invariants I_1 and functions $w_j(I')$ is easily inverted:

$$\begin{bmatrix} t_1 \\ t_2 \\ t_3 \\ t_4 \end{bmatrix} = \frac{1}{4} \begin{bmatrix} 1 & 1 & 1 & 1 \\ 1 & 1 & -1 & -1 \\ 1 & -1 & 1 & -1 \\ 1 & -1 & -1 & 1 \end{bmatrix} \begin{bmatrix} I_1 \\ w_2 \\ w_3 \\ w_4 \end{bmatrix} \qquad (1.53)$$

where the w_j are square roots of the solutions of the resolvent cubic equation whose coefficients are functions (1.51) of the auxiliary quartic equation.

1.8 The quintic cannot be solved

To investigate whether the typical quintic equation is solvable (and if so, how), it is sufficient to study the structure of its Galois group S_5. The alternating subgroup

A_5 of order 60 is an invariant subgroup. S_5 has no invariant subgroups except A_5 and I. Further, A_5 has only I as an invariant subgroup. The only chain of invariant subgroups in S_5 is

$$S_5 \supset A_5 \supset I \tag{1.54}$$

Although $S_5/A_5 = S_2$ is commutative, $A_5/I = A_5$ is not. Therefore the quintic equation does not satisfy the conditions of Galois' theorem, so cannot be solved by radicals. General polynomial equations of degree greater than five also cannot be solved by radicals.

1.9 Example

To illustrate the solution of a polynomial equation by radicals using the machinery introduced above, we begin with a quartic equation whose roots are: $-2, -1, 2, 5$. We will carry out the algorithm on the corresponding quartic equation. As we proceed through the algorithm, we indicate the numerical values of the functions present. Those values that would not be available at each stage of the computation are indicated by arrows.

The fourth degree equation is

$$(z+2)(z+1)(z-2)(z-5) = z^4 - 4z^3 - 9z^2 + 16z + 20 = 0$$
$$I_1 = 4$$
$$I_2 = -9 \tag{1.55}$$
$$I_3 = -16$$
$$I_4 = 20$$

We now center the roots by making a Tschirnhaus transformation

$$z = z' + \frac{1}{4}I_1 = z' + 1$$

The new roots are $-3, -2, 1, 4$ and the auxiliary quartic equation is

$$(z'+1)^4 - 4(z'+1)^3 - 9(z'+1)^2 + 16(z'+1) + 20$$
$$= (z'+3)(z'+2)(z'-1)(z'+4) = z'^4 - 15z'^2 - 10z' + 24 = 0$$
$$I'_1 = 0$$
$$I'_2 = -15$$
$$I'_3 = 10 \tag{1.56}$$
$$I'_4 = 24$$

1.9 Example

Next, we introduce linear combinations of the four roots $t'_1 = -3$, $t'_2 = -2$, $t'_3 = 1$, $t'_4 = 4$

$$\begin{bmatrix} w_1 \\ w_2 \\ w_3 \\ w_4 \end{bmatrix} = \begin{bmatrix} 1 & 1 & 1 & 1 \\ 1 & 1 & -1 & -1 \\ 1 & -1 & 1 & -1 \\ 1 & -1 & -1 & 1 \end{bmatrix} \begin{bmatrix} t'_1 \\ t'_2 \\ t'_3 \\ t'_4 \end{bmatrix} \rightarrow \begin{bmatrix} 0 \\ -10 \\ -4 \\ 2 \end{bmatrix} \quad (1.57)$$

Observe at this stage that $w_2 w_3 w_4 = 8I'_3$.

Now we compute the squares of these numbers

$$\begin{aligned} w_2^2 &= y_2 \rightarrow (-10)^2 = 100 \\ w_3^2 &= y_3 \rightarrow (-4)^2 = 16 \\ w_4^2 &= y_4 \rightarrow (+2)^2 = 4 \end{aligned} \quad (1.58)$$

From the auxiliary quartic (1.56) the resolvent cubic equation can be constructed

$$y^3 - J_1 y^2 + J_2 y - J_3 = 0$$

$$\begin{aligned} J_1 &= (-4)^1 [2I'_2] & &= (-4)(-30) & &= 120 \\ J_2 &= (-4)^2 [I'^2_2 - 4I'_4] &&= 16(225 - 4 \times 24) &&= 2064 \\ J_3 &= (-4)^3 [-I'^2_3] & &= (-64)(-100) & &= 6400 \end{aligned} \quad (1.59)$$

Note that these are the coefficients of the equation

$$(y - 2^2)(y - 4^2)(y - 10^2) = y^3 - 120y^2 + 2064y - 6400 = 0 \quad (1.60)$$

Now we construct the cubic equation auxiliary to this cubic. This is done by defining $y = y' + \frac{1}{3}J_1 = y' + \frac{1}{3}(4 + 16 + 100) = y' + 40$. The roots are now

$$\begin{aligned} y'_1 &= y_1 - 40 \rightarrow & 4 - 40 &= -36 \\ y'_2 &= y_2 - 40 \rightarrow & 16 - 40 &= -24 \\ y'_3 &= y_3 - 40 \rightarrow & 100 - 40 &= 60 \end{aligned} \quad (1.61)$$

The auxiliary cubic is

$$y'^3 - J'_1 y'^2 + J'_2 y' - J'_3 = 0$$

$$\begin{aligned} J'_1 &= 0 \\ J'_2 &= -2736 \\ J'_3 &= 51840 \end{aligned} \quad (1.62)$$

We note that these are the coefficients of the equation

$$(y' + 36)(y' + 24)(y' - 60) = 0 \quad (1.63)$$

These coefficients are obtained directly from the coefficients of the resolvent cubic, in principle without knowledge of the values of the roots.

Next we construct the functions v_1, v_2, v_3

$$\begin{bmatrix} v_1 \\ v_2 \\ v_3 \end{bmatrix} = \begin{bmatrix} 1 & 1 & 1 \\ 1 & \omega & \omega^2 \\ 1 & \omega^2 & \omega \end{bmatrix} \begin{bmatrix} s_1 \\ s_2 \\ s_3 \end{bmatrix} \xrightarrow[s_2 = -36]{s_1 = -24} \begin{bmatrix} 0 \\ -36 - i48\sqrt{3} \\ -36 + i48\sqrt{3} \end{bmatrix} \quad (1.64)$$

We can express $v_2^3 + v_3^3$, $v_2^3 v_3^3$ in terms of J_2', J_3':

$$\begin{aligned} v_2^3 + v_3^3 &= 27 J_3' = 27 \times 518400 = 1399680 \\ v_2^3 v_3^3 &= -27 J_2'^3 = -27 \times (-2736)^3 = 552983334912 \end{aligned} \quad (1.65)$$

The quadratic resolvent for the auxiliary cubic is

$$x^2 - 1399680x + 552983334912 = 0$$

$$\begin{aligned} K_1 &= 1399680 \\ K_2 &= 552983334912 \end{aligned} \quad (1.66)$$

A Tschirnhaus transformation $x = x' + \frac{1}{2} K_1$ produces the auxiliary quadratic

$$x'^2 + 63207309312 = 0$$

$$\begin{aligned} K_1' &= 0 \\ K_2 &= 63207309312 \end{aligned} \quad (1.67)$$

The square of the difference between the two roots of this equation is easily determined:

$$\begin{aligned} x_1' - x_2' &= x_1 - x_2 &= \pm 2\sqrt{-K_2} = \pm 2i\sqrt{K_2} \\ &= \pm 2i \times 145152\sqrt{3} &= \pm i \times 290304\sqrt{3} \end{aligned} \quad (1.68)$$

Now we work backwards. The solutions of the resolvent quadratic are given by the linear equation

$$\begin{bmatrix} x_1 \\ x_2 \end{bmatrix} = \frac{1}{2} \begin{bmatrix} 1 & 1 \\ 1 & -1 \end{bmatrix} \begin{bmatrix} K_1 = 1399680 \\ 2\sqrt{-K_2} = i \times 290304\sqrt{3} \end{bmatrix}$$

$$= 699840 \pm i \times 145152\sqrt{3} \quad (1.69)$$

These solutions are the values of v_2^3 and v_3^3:

$$\begin{aligned} v_2^3 &= 699840 + i\, 145152\sqrt{3} \\ v_3^3 &= 699840 - i\, 145152\sqrt{3} \end{aligned} \quad (1.70)$$

Next, we take cube roots of these quantities. These are unique up to a factor of ω

$$v_2 = -36 + i48\sqrt{3}$$
$$v_3 = -36 - i48\sqrt{3} \qquad (1.71)$$

The values y_1, y_2, y_3 of the resolvent cubic are complex linear combinations of v_2, v_3

$$\begin{bmatrix} y_1 \\ y_2 \\ y_3 \end{bmatrix} = \frac{1}{3} \begin{bmatrix} 1 & 1 & 1 \\ 1 & \omega^2 & \omega \\ 1 & \omega & \omega^2 \end{bmatrix} \begin{bmatrix} J_1 = 120 \\ v_2 = -36 + i\,48\sqrt{3} \\ v_3 = -36 - i\,48\sqrt{3} \end{bmatrix} = \begin{bmatrix} 16 \\ 100 \\ 4 \end{bmatrix} \qquad (1.72)$$

$$\begin{array}{ll} w_2^2 = y_1 & w_2 = \pm 4 \\ w_3^2 = y_2 & w_3 = \pm 10 \\ w_4^2 = y_3 & w_4 = \pm 2 \end{array} \qquad (1.73)$$

Since $w_2 w_3 w_4 = 8I_3' = 80$, an even number of these signs must be negative. The simplest choice is to take all signs positive. This is different from the results shown in (1.57); this choice of signs serves only to permute the order of the roots. In the final step, the roots of the original quartic are linear combinations of w_2, w_3, w_4 and the linear symmetric function $w_1 = I_1$

$$\begin{bmatrix} x_1 \\ x_2 \\ x_3 \\ x_4 \end{bmatrix} = \frac{1}{4} \begin{bmatrix} 1 & 1 & 1 & 1 \\ 1 & 1 & -1 & -1 \\ 1 & -1 & 1 & -1 \\ 1 & -1 & -1 & 1 \end{bmatrix} \begin{bmatrix} I_1 = 4 \\ w_2 = 4 \\ w_3 = 10 \\ w_4 = 2 \end{bmatrix} = \begin{bmatrix} 20/4 = +5 \\ -4/4 = -1 \\ 8/4 = +2 \\ -8/4 = -2 \end{bmatrix} \qquad (1.74)$$

We have recovered the four roots of the original quartic equation using Galois' algorithm, based on the structure of the invariance group S_4 of the quartic equation.

1.10 Conclusion

One of the many consequences of Galois' study of algebraic equations and the symmetries that leave them invariant is the proof that an algebraic equation can be solved by radicals if and only if its invariance group has a certain structure. This proof motivated Lie to search for analogous results involving differential equations and their symmetry groups, now called Lie groups. We have described in this chapter how the structure of the discrete symmetry group (Galois group) of a polynomial equation determines whether or not that equation can be solved by radicals. If the answer is "yes," we have shown how the structure of the Galois group determines the structure of the algorithm for constructing solutions. This algorithm has been developed for the cubic and quartic equations, and illustrated by example for a quartic equation.

1.11 Problems

1. Compute S_4/A_4, A_4/V_4, V_4 and show that they are commutative.

2. Construct the group V_8 with the property $S_4 \supset V_8 \supset V_4$ (see Fig. 1.5). (Hint: include a cyclic permutation).

3. For the cubic equation $z^3 - 7z + 6 = 0$ $((z-1)(z-2)(z+3) = 0)$ show

$$I_1 = 0 \quad J_1 = 162$$
$$I_2 = -7 \quad J_2 = 9261$$
$$I_3 = -6$$

   Show that the resolvent equation for v_2^3, v_3^3 is $(x - v_2^3)(x - v_3^3) = x^2 - 162x + 9261 = 0$. Solve this quadratic to find $v_2^3, v_3^3 = 81 \pm i30\sqrt{3}$, so that $v_2, v_3 = \frac{1}{2}(3 \pm i5\sqrt{3})$. Invert Eq. (1.43) to determine the three roots of the original equation: $(1, 2, -3)$.

4. Ruler and compass can be used to construct an orthogonal pair of axes in the plane (Euclid). A compass is used to establish a unit of length 1. Then by ruler and compass it is possible to construct intervals of length x, where x is integer. From there it is possible to construct intervals of lengths $x + y$, $x - y$, $x y$ and x/y using ruler and compass. It is also possible to construct intervals of length $\sqrt{x}$ by these means. The set of all numbers that can be constructed from integers by addition, subtraction, multiplication, division, and extraction of square roots is called the set of constructable numbers. This forms a subset of the numbers $x + iy = (x, y)$ in the complex plane. If a number is (is not) constructable the point representing that number can (cannot) be constructed by ruler and compass alone. Since repeated square roots can be taken, a constructable number satisfies an algebraic equation of degree K with integer coefficients, where $K = 2^n$ must be some power of two.

   The three geometry problems of antiquity are as follows.

   a. **Square a circle?** For the circle of radius 1 the area is π. Squaring a circle means finding an interval of length x, where $x^2 - \pi = 0$. This is of degree 2 but π is not rational (not even algebraic). Argue that it is impossible to square the circle by ruler and compass alone.

   b. **Double the cube?** A cube with edge length 1 has volume $1^3 = 1$. A cube with twice the volume has edge length x, where x satisfies $x^3 - 2 = 0$. Although the coefficients are integers this equation is of degree $3 \neq 2^n$ for any integer n. Argue that it is impossible to double the volume of a cube by ruler and compass alone.

   c. **Trisect an angle?** If 3θ is some angle, the trigonometric functions of 3θ and $\frac{1}{3}(3\theta) = \theta$ are related by

$$e^{i3\theta} = (e^{i\theta})^3$$
$$\cos(3\theta) + i\sin(3\theta) = (\cos^3(\theta) - 3\cos(\theta)\sin^2(\theta))$$
$$+ i(3\cos^2(\theta)\sin(\theta) - \sin^3(\theta))$$

In particular

$$\cos(3\theta) = 4\cos^3(\theta) - 3\cos(\theta)$$

Whether $\cos(3\theta)$ is rational or irrational, the equation for $\cos(\theta)$:

$$4\cos^3(\theta) - 3\cos(\theta) - \cos(3\theta) = 0$$

is cubic. Argue that it is impossible to trisect an angle unless $\cos(3\theta)$ is such that the cubic factors into the form $(x^2 + ax + b)(x + c) = 0$, where a, b, c are rational. For example, if $\cos(3\theta) = 0$, $c = 0$ so that $a = 0$ and $b = -3/4$. Then $\cos(\theta) = 0$ or $\pm\sqrt{3}/2$ for $3\theta = \pi/2$ (+), $3\pi/2$ (0), or $5\pi/2$ (−).

2
Lie groups

Lie groups are beautiful, important, and useful because they have one foot in each of the two great divisions of mathematics – algebra and geometry. Their algebraic properties derive from the group axioms. Their geometric properties derive from the identification of group operations with points in a topological space. The rigidity of their structure comes from the continuity requirements of the group composition and inversion maps. In this chapter we present the axioms that define a Lie group.

2.1 Algebraic properties

The algebraic properties of a Lie group originate in the axioms for a group.

Definition A set $g_i, g_j, g_k, \ldots$ (called **group elements** or **group operations**) together with a combinatorial operation $\circ$ (called **group multiplication**) form a group G if the following axioms are satisfied.

(i) **Closure:** if $g_i \in G$, $g_j \in G$, then $g_i \circ g_j \in G$.
(ii) **Associativity:** $g_i \in G$, $g_j \in G$, $g_k \in G$, then

$$(g_i \circ g_j) \circ g_k = g_i \circ (g_j \circ g_k)$$

(iii) **Identity:** there is an operator e (the **identity operation**) with the property that for every group operation $g_i \in G$

$$g_i \circ e = g_i = e \circ g_i$$

(iv) **Inverse:** every group operation g_i has an inverse (called g_i^{-1}) with the property

$$g_i \circ g_i^{-1} = e = g_i^{-1} \circ g_i$$

2.2 Topological properties

Example We consider the set of real 2×2 matrices $SL(2; \mathbb{R})$:

$$A = \begin{bmatrix} \alpha & \beta \\ \gamma & \delta \end{bmatrix} \qquad \det(A) = \alpha\delta - \beta\gamma = +1 \tag{2.1}$$

where $\alpha, \beta, \gamma, \delta$ are real numbers. This set forms a group under matrix multiplication. This is verified by checking that the group axioms are satisfied.

(i) **Closure** if A and B are real 2×2 matrices, and $A \circ B = C$ (where $\circ$ now represents matrix multiplication), then C is a real 2×2 matrix. If $\det(A) = +1$ and $\det(B) = +1$, then $\det(C) = \det(A)\det(B) = +1$.

(ii) **Associativity:** $(A \circ B) \circ C$ and $A \circ (B \circ C)$ are given explicitly by

$$\sum_k \left(\sum_j A_{ij} B_{jk} \right) C_{kl} \stackrel{?}{=} \sum_j A_{ij} \left(\sum_k B_{jk} C_{kl} \right)$$

$$\sum_k \sum_j A_{ij} B_{jk} C_{kl} \stackrel{\text{ok}}{=} \sum_j \sum_k A_{ij} B_{jk} C_{kl} \tag{2.2}$$

(iii) **Identity:** the unit matrix is the identity

$$e \longrightarrow I_2 = \begin{bmatrix} 1 & 0 \\ 0 & 1 \end{bmatrix}$$

(iv) **Inverse:** the unique matrix inverse of A is

$$\begin{bmatrix} A_{11} & A_{12} \\ A_{21} & A_{22} \end{bmatrix} \rightarrow \begin{bmatrix} A_{11} & A_{12} \\ A_{21} & A_{22} \end{bmatrix}^{-1} = \frac{1}{A_{11}A_{22} - A_{12}A_{21}} \begin{bmatrix} A_{22} & -A_{12} \\ -A_{21} & A_{11} \end{bmatrix}$$

2.2 Topological properties

The geometric structure of a Lie group comes from the identification of each element in the group with a point in some topological space: $g_i \to g(x)$. In other words, the index i depends on one or more continuous real variables.

The topological space that parameterizes the elements in a Lie group is a manifold. A **manifold** is a space that looks Euclidean on a small scale everywhere. For example, every point on the surface of a unit sphere $S^2 \subset \mathbb{R}^3$: $x^2 + y^2 + z^2 = 1$, has a neighborhood that looks, over small distances, like a piece of the plane $\mathbb{R}^2$ (see Fig. 2.1). Locally, the two spaces S^2 and $\mathbb{R}^2$ are topologically equivalent but globally they are different (Columbus).

Definition An n-dimensional differentiable manifold M^n consists of the following.

(i) A topological space T. This includes a collection of open sets U_α (a topology) that cover T: $\cup_\alpha U_\alpha = T$.

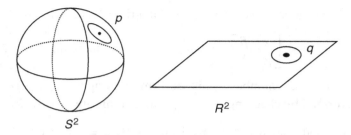

Figure 2.1. Every point p on a sphere S^2 is surrounded by an open neighborhood that is indistinguishable from an open neighborhood of any point in the plane R^2. Locally the two spaces are indistinguishable. Globally they are distinguishable.

(ii) A collection of charts ϕ_α, with $\phi_\alpha(U_\alpha) = V_\alpha \subset R^n$. Each ϕ_α is a homeomorphism of U_α to V_α.
(iii) Smoothness conditions. The homeomorphisms $\phi_\alpha \circ \phi_\beta^{-1}$: $\phi_\beta(U_\alpha \cap U_\beta) \to \phi_\alpha(U_\alpha \cap U_\beta)$ of open sets in R^n to open sets in R^n are 1:1, invertible, and differentiable.

Remarks The charts ϕ_α allow construction of coordinate systems on the open sets U_α. It is often not possible to find a single coordinate system on the entire manifold, as the example of the sphere in Fig. 2.1 shows. Since the "transition functions" $\phi_\alpha \circ \phi_\beta^{-1}$ map $R^n \to R^n$, all the definitions of elementary multivariable calculus are applicable to them. For example, the adjective "differentiable" can be replaced by other adjectives (C^k, smooth, analytic, ...) in the definition above.

Example Real 2×2 matrices are identified by four real variables. The unimodular condition $\det(A) = +1$ places one constraint on these four real variables. Therefore every group element in $SL(2; \mathbb{R})$ is determined by a point in some real three-dimensional space. One possible parameterization is

$$(x_1, x_2, x_3) \longrightarrow \begin{bmatrix} x_1 & x_2 \\ x_3 & \dfrac{1 + x_2 x_3}{x_1} \end{bmatrix} \quad x_1 \neq 0 \quad (2.3)$$

Parameterization of the operations in a group by real numbers is a nontrivial problem, as is clear when one asks: "what happens as $x_1 \to 0$?" We will consider this question in Chapter 5.

The manifold that parameterizes the group $SL(2; \mathbb{R})$ is the direct product manifold R^2 (plane) $\times S^1$ (circle) (see Fig. 2.2). This is not at all obvious, but will become clear when we discuss the infinitesimal properties of Lie groups in Chapter 4.

The dimension of the manifold that parameterizes a Lie group is the dimension of the Lie group. It is the number of continuous real parameters required to describe each operation in the group uniquely.

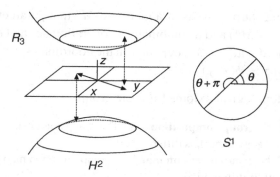

Figure 2.2. Every matrix in $SL(2;\mathbb{R})$ can be written as the product of a symmetric matrix and a rotation matrix, both unimodular. The symmetric matrix is parameterized by a two-dimensional manifold, the two-sheeted hyperboloid $z^2 - x^2 - y^2 = 1$. The rotation matrix is parameterized by a point on a circle. The parameterization manifold, $H^2 \times S^1$, is three dimensional.

It is useful at this point to introduce the ideas of compactness and noncompactness. Roughly speaking, a compact space is in some sense finite and a noncompact space is not finite.

Definition A topological space T is compact if every open cover (set of open sets U_α) has a finite subcover: $\cup_\alpha^{\text{finite}} T \subset U_\alpha$.

In spaces R^n with a Euclidean notion of distance ($|x - x'|^2 = |x_1 - x'_1|^2 + \cdots + |x_n - x'_n|^2$), this definition is equivalent to an older definition of compact spaces: a space is compact if every infinite sequence of points has a subsequence that converges to a point in the space.

Example In Fig. 2.1 the sphere S^2 is compact and the plane R^2 is not compact. In Fig. 2.2, the circle is compact and the hyperboloid is not compact.

Remark In R^n every bounded closed subset is compact. "Closed" means that the set contains all its limit points.

Remark Compactness is an important topological property because it means that the space is in some sense like a bounded, closed space. For Lie groups it is important because all irreducible representations of compact Lie groups are finite dimensional and can be constructed by rather simple means (tensor product constructions).

2.3 Unification of algebra and topology

The rigidity of Lie group structures comes from combining the algebraic and topological properties through smoothness (differentiability) requirements.

Definition A **Lie group** consists of a manifold M^n that parameterizes the group operations ($g(x)$, $x \in M^n$) and a combinatorial operation defined by $g(x) \circ g(y) = g(z)$, where the coordinate $z \in M^n$ depends on the coordinates $x \in M^n$ and $y \in M^n$ through a function $z = \phi(x, y)$.

There are two topological axioms for a Lie group.

(i) **Smoothness of the group composition map** The group composition map $z = \phi(x, y)$, defined by $g(x) \circ g(y) = g(z)$, is differentiable.
(ii) **Smoothness of the group inversion map** The group inversion map $y = \psi(x)$, defined by $g(x)^{-1} = g(y)$, is differentiable.

It is possible to combine these two axioms into a single axiom, but there is no advantage to this.

Example For $SL(2; \mathbb{R})$ with parameterization given by (2.3) the composition function $z = \phi(x, y)$ is constructed easily by matrix multiplication $g(x) \circ g(y) = g(\phi(x, y))$

$$g(x_1, x_2, x_3) \quad \circ \quad g(y_1, y_2, y_3) \quad = \quad g(z_1, z_2, z_3)$$

$$\begin{bmatrix} x_1 & x_2 \\ x_3 & \dfrac{1+x_2x_3}{x_1} \end{bmatrix} \times \begin{bmatrix} y_1 & y_2 \\ y_3 & \dfrac{1+y_2y_3}{y_1} \end{bmatrix} = \begin{bmatrix} z_1 & z_2 \\ z_3 & \dfrac{1+z_2z_3}{z_1} \end{bmatrix}$$

where

$$g(\phi(x_1, x_2, x_3; y_1, y_2, y_3)) \quad = \quad g(z_1, z_2, z_3)$$

$$\begin{bmatrix} x_1 y_1 + x_2 y_3 & x_1 y_2 + x_2 \dfrac{1+y_2y_3}{y_1} \\ x_3 y_1 + \dfrac{1+x_2x_3}{x_1} y_3 & * \end{bmatrix} = \begin{bmatrix} z_1 & z_2 \\ z_3 & \dfrac{1+z_2z_3}{z_1} \end{bmatrix} \quad (2.4)$$

The result is easily read off, matrix element by matrix element:

$$\begin{aligned} z_1 &= \phi_1(x_1, x_2, x_3; y_1, y_2, y_3) = x_1 y_1 + x_2 y_3 \\ z_2 &= \phi_2(x_1, x_2, x_3; y_1, y_2, y_3) = x_1 y_2 + x_2 \frac{1+y_2 y_3}{y_1} \\ z_3 &= \phi_3(x_1, x_2, x_3; y_1, y_2, y_3) = x_3 y_1 + \frac{1+x_2 x_3}{x_1} y_3 \end{aligned} \quad (2.5)$$

The function ϕ is analytic in its two pairs of arguments provided x_1 and y_1 are bounded away from the x_2–x_3 plane $x_1 = 0$ and the y_2–y_3 plane $y_1 = 0$. In the neighborhood of these values an alternative parameterization of the group is needed.

It is also useful to determine the mapping that takes a group operation into its inverse. We can determine the coordinates (y_1, y_2, y_3) of $[g(x_1, x_2, x_3)]^{-1}$ by setting $(z_1, z_2, z_3) = (1, 0, 0)$ and solving for (y_1, y_2, y_3) in terms of (x_1, x_2, x_3). Or more simply we can compute the inverse of the matrix (2.3)

$$\begin{bmatrix} x_1 & x_2 \\ x_3 & (1+x_2 x_3)/x_1 \end{bmatrix}^{-1} = \begin{bmatrix} (1+x_2 x_3)/x_1 & -x_2 \\ -x_3 & x_1 \end{bmatrix} \quad (2.6)$$

The inverse mapping $[g(x)]^{-1} = g(y) = g(\psi(x))$ is

$$\begin{aligned} \psi_1(x_1, x_2, x_3) = y_1 &= (1+x_2 x_3)/x_1 \\ \psi_2(x_1, x_2, x_3) = y_2 &= -x_2 \\ \psi_3(x_1, x_2, x_3) = y_3 &= -x_3 \end{aligned} \quad (2.7)$$

This mapping is analytic except at $x_1 = 0$, where an alternative parameterization is required. The parameterization shown in Fig. 2.2 handles this problem quite well. Every matrix in $SL(2; \mathbb{R})$ can be written as the product of a symmetric matrix and a rotation matrix, both 2×2 and unimodular. The symmetric matrix is parameterized by a two-dimensional manifold, the two-sheeted hyperboloid $z^2 - x^2 - y^2 = 1$. The rotation matrix is parameterized by a point on a circle. Two points $(x, y, |z|, \theta)$ and $(-x, -y, -|z|, \theta + \pi)$ map to the same matrix in $SL(2; \mathbb{R})$. The manifold that parameterizes $SL(2; \mathbb{R})$ is three dimensional. It is $H^{2+} \times S^1$, where H^{2+} is the upper sheet of the two-sheeted hyperboloid.

2.4 Unexpected simplification

Almost every Lie group that we will encounter is either a matrix group or else equivalent to a matrix group. This simplifies the description of the algebraic, topological, and continuity properties of these groups. Algebraically, the only group operations that we need to consider are matrix multiplication and matrix inversion. Geometrically, the only manifolds we encounter are those manifolds that can be constructed from matrices by imposing algebraic constraints (*algebraic manifolds*) on the matrix elements. The continuity properties on the matrix elements are simple consequences of matrix multiplication and inversion.

2.5 Conclusion

Lie groups lie at the intersection of the two great divisions of mathematics: algebra and topology. The group elements are points in a manifold, and as such are parameterized by continuous real variables. These points can be combined by an operation that obeys the group axioms. The combinatorial operation $\phi(x, y)$ defined by $g(x) \circ g(y) = g(z) = g(\phi(x, y))$ is differentiable in both sets of variables.

In addition, the mapping $y = \psi(x)$ of a group operation to its inverse $[g(x)]^{-1} = g(y) = g(\psi(x))$ is also differentiable.

Unexpectedly, almost all of the Lie groups encountered in applications are matrix groups. This effects an enormous simplification in our study of Lie groups. Almost all of what we would like to learn about Lie groups can be determined by studying matrix groups.

2.6 Problems

1. Construct the analytic mapping $\phi(x, y)$ for the parameterization of $SL(2; \mathbb{R})$ illustrated in Fig. 2.2.

2. Construct the inversion mapping for the parameterization of $SL(2; \mathbb{R})$ given in Fig. 2.2. Show that

$$\begin{bmatrix} x' \\ y' \\ \theta' \end{bmatrix} = - \begin{bmatrix} \cos(2\theta) & -\sin(2\theta) & 0 \\ \sin(2\theta) & \cos(2\theta) & 0 \\ 0 & 0 & 1 \end{bmatrix} \begin{bmatrix} x \\ y \\ \theta \end{bmatrix}$$

3. Convince yourself that every matrix M in the group $SL(n; \mathbb{R})$ can be written as the product of an $n \times n$ real symmetric unimodular matrix S and an orthogonal matrix O in $SO(n)$: $M = SO$. Devise an algorithm for constructing these matrices. Show $S = (MM^t)^{1/2}$ and $O = S^{-1}M$. How do you compute the square root of a matrix? Show that O is compact while S and M are not compact.

4. Construct the most general linear transformation $(x, y, z) \to (x', y', z')$ that leaves invariant (unchanged) the quadratic form $z^2 - x^2 - y^2 = 1$. Show that this linear transformation can be expressed in the form

$$\begin{bmatrix} x' \\ y' \\ z' \end{bmatrix} = \begin{bmatrix} M_1 & \begin{matrix} a \\ b \end{matrix} \\ \begin{matrix} a & b \end{matrix} & M_2 \end{bmatrix} \begin{bmatrix} SO(2) & \begin{matrix} 0 \\ 0 \end{matrix} \\ \begin{matrix} 0 & 0 \end{matrix} & 1 \end{bmatrix} \begin{bmatrix} x \\ y \\ z \end{bmatrix}$$

where the real symmetric matrices M_1 and M_2 satisfy

$$M_1^2 = I_2 + \begin{bmatrix} a \\ b \end{bmatrix} [a \ b] = \begin{bmatrix} 1 + a^2 & ab \\ ba & 1 + b^2 \end{bmatrix} \text{ and }$$

$$M_2^2 = I_1 + [a \ b] \begin{bmatrix} a \\ b \end{bmatrix} = [1 + a^2 + b^2]$$

5. Construct the group of linear transformations $[SO(1, 1)]$ that leaves invariant the quantity $(ct)^2 - x^2$. Compare this with the group of linear transformations $[SO(2)]$ that leaves invariant the radius of the circle $x^2 + y^2$. (This comparison involves mapping trigonometric functions to hyperbolic functions by analytic continuation.)

6. Construct the group of linear transformations that leaves invariant the quantity $(ct)^2 - x^2 - y^2 - z^2$. This is the Lorentz group $O(3, 1)$. Four disconnected manifolds parameterize this group. These contain the four different group operations

$$\begin{bmatrix} \pm 1 & 0 & 0 & 0 \\ 0 & \pm 1 & 0 & 0 \\ 0 & 0 & 1 & 0 \\ 0 & 0 & 0 & 1 \end{bmatrix}$$

where the $\pm$ signs are incoherent.

7. The group of 2×2 complex matrices with determinant $+1$ is named $SL(2; \mathbb{C})$. Matrices in this group have the structure $[\begin{smallmatrix} \alpha & \beta \\ \gamma & \delta \end{smallmatrix}]$, where $\alpha, \beta, \gamma, \delta$ are complex numbers and $\alpha\delta - \beta\gamma = 1$. Define the matrix X by

$$X = H(x, y, z, ct) = \begin{bmatrix} ct + z & x - iy \\ x + iy & ct - z \end{bmatrix} = ct I_2 + \sigma \cdot \mathbf{x}$$

where $\mathbf{x}$ is the three vector $\mathbf{x} = (x, y, z)$ and $\sigma = (\sigma_1, \sigma_2, \sigma_3) = (\sigma_x, \sigma_y, \sigma_z)$ are the Pauli spin matrices.

a. Show that X is hermitian: $X^\dagger \equiv (X^t)^* = X$.
b. Show that the most general 2×2 hermitian matrix can be written in the form used to construct X.
c. If $g \in SL(2; \mathbb{C})$, show that $g^\dagger X g = X' = H(x', y', z', ct')$.
d. How are the new space-time coordinates (x', y', z', ct') related to the original coordinates (x, y, z, ct)? (They are linearly related by coefficients that are bilinear in the matrix elements $\alpha, \beta, \gamma, \delta$ of g and $\alpha^*, \beta^*, \gamma^*, \delta^*$ of its adjoint matrix $g^\dagger$.)
e. Find the subgroup of $SL(2; \mathbb{C})$ that leaves $t' = t$. (It is $SU(2) \subset SL(2; \mathbb{C})$).
f. For any $g \in SL(2; \mathbb{C})$ write $g = kh$, where $h \in SU(2)$, $h^\dagger = h^{-1}$, h has the form $h = \text{EXP}(\frac{i}{2}\sigma \cdot \theta)$ and $k \in SL(2; \mathbb{C})/SU(2)$, $k^\dagger = k^{+1}$, k has the form $k = \text{EXP}(\frac{1}{2}\sigma \cdot \mathbf{b})$. The three vector $\mathbf{b}$ is called a boost vector. The three vectors θ and $\mathbf{b}$ are real. Construct $k^\dagger H(x, y, z, ct)k = H(x', y', z', ct')$. If this is too difficult, choose $\mathbf{b}$ along the z-axis, $\mathbf{b} = (0, 0, b)$.
g. Show that the usual Lorentz transformation law results.
h. Applying $k(b')$ after applying $k(b)$ results in (a) $k(b' + b)$, (b) two successive Lorentz transformations. Show that the velocity addition law for colinear boosts results.
i. If $\mathbf{b}$ and $\mathbf{b}'$ are not colinear, $k(\mathbf{b}')k(\mathbf{b}) = k(\mathbf{b}'')h(\theta)$. Compute $\mathbf{b}''$, θ. The angle θ is related to the Thomas precession (Gilmore 1974b).

8. The circumference of the unit circle is mapped into itself under the transformation $\theta \rightarrow \theta' = \theta + k + f(\theta)$, where k is a real number, $0 \leq k < 2\pi$, and $f(\theta)$ is periodic, $f(\theta + 2\pi) = f(\theta)$. The mapping must be 1:1, so an additional condition is imposed on $f(\theta)$: $df(\theta)/d\theta > -1$ everywhere. Does this set of transformations form a group? What are the properties of this group?

9. Rational fractional transformations (a, b, c, d) map points on the real line (real projective line RP^1) to the real line as follows:

$$x \to x' = (a, b, c, d)x = \frac{ax+b}{cx+d}$$

The transformations (a, b, c, d) and $(\lambda a, \lambda b, \lambda c, \lambda d) = \lambda(a, b, c, d)$ ($\lambda \neq 0$) generate identical mappings.

a. Compose two successive rational fractional transformations

$$(A, B, C, D) = (a', b', c', d') \circ (a, b, c, d)$$

and show that the composition is a rational fractional transformation. Compute the values of A, B, C, D.

b. Show that the transformations $(\lambda, 0, 0, \lambda)$ map x to itself.

c. Construct the inverse transformation $x' \to x$, and show that it is $\lambda(d, -b, -c, a)$ provided $\lambda \neq 0$. Such transformations exist if $D = ad - bc \neq 0$.

d. Show that the transformation degeneracy $x' = (a, b, c, d)x = \lambda(a, b, c, d)x$ can be lifted by requiring that the four parameters a, b, c, d describing these transformations satisfy the constraint $D = ad - bc = 1$.

e. It is useful to introduce homogeneous coordinates (y, z) and define the real projective coordinate x as the ratio of these homogeneous coordinates: $x = y/z$. If the homogeneous coordinates transform linearly under $SL(2; \mathbb{R})$ then the real projective coordinates x transform under rational fractional transformations:

$$\begin{bmatrix} y' \\ z' \end{bmatrix} = \begin{bmatrix} a & b \\ c & d \end{bmatrix} \begin{bmatrix} y \\ z \end{bmatrix} \Rightarrow x' = \frac{y'}{z'} = \frac{a(y/z) + b}{c(y/z) + d} = \frac{ax+b}{cx+d}$$

f. Show that a rational fractional transformation can be constructed that maps three distinct points x_1, x_2, x_3 on the real line to the three standard positions $(0, 1, \infty)$, and that this mapping is

$$x \to x' = \frac{(x-x_1)(x_2-x_3)}{(x-x_3)(x_2-x_1)}$$

What matrix in $SL(2; \mathbb{R})$ describes this mapping? (Careful of the condition $D = 1$.)

g. Use this construction to show that there is a unique mapping of any triple of distinct points (x_1, x_2, x_3) to any other triple of distinct points (x'_1, x'_2, x'_3).

10. The real projective space RP^n is the space of all straight lines through the origin in R^{n+1}. The group $SL(n+1; \mathbb{R})$ maps $x = (x_1, x_2, \ldots, x_{n+1}) \in R^{n+1}$ to $x' \in R^{n+1}$, with $x' \neq 0 \leftrightarrow x \neq 0$ and $x' = 0 \leftrightarrow x = 0$. A straight line through the origin contains $x \neq 0$ and $y \neq 0$ if (and only if) $y = \lambda x$ for some real scale factor $\lambda \neq 0$. The scale factor can always be chosen so that y is in the unit sphere in R^{n+1}: $y \in S^n \subset R^{n+1}$. In fact, two values of λ can be chosen: $\lambda = \pm 1/(\sum_{i=1}^{n+1} x_i^2)^{1/2}$. In R^3 the straight line containing (x, y, z) can be represented by homogeneous coordinates $(X, Y) = (x/z, y/z)$ if $z \neq 0$. Straight lines through the origin of R^3 are mapped to straight lines in R^3 by $x \to x' = Mx$, $M \in SL(3; \mathbb{R})$. Show that the homogeneous coordinates

representing the two lines containing x and x' are related by the linear fractional transformation

$$\begin{bmatrix} X \\ Y \end{bmatrix} \to \begin{bmatrix} X' \\ Y' \end{bmatrix}$$

$$= \left(\begin{bmatrix} m_{11} & m_{12} \\ m_{21} & m_{22} \end{bmatrix} \begin{bmatrix} X \\ Y \end{bmatrix} + \begin{bmatrix} m_{13} \\ m_{23} \end{bmatrix} \right) \Big/ \left(\begin{bmatrix} m_{31} & m_{32} \end{bmatrix} \begin{bmatrix} X \\ Y \end{bmatrix} + m_{33} \right)$$

Generalize for linear fractional transformations $RP^n \to RP^n$.

11. The hyperbolic two-space $SL(2; \mathbb{R})/SO(2) \simeq [\begin{smallmatrix} z+x & y \\ y & z-x \end{smallmatrix}]$ consists of the algebraic submanifold in the Minkowski $2+1$ dimensional space-time with metric $(+1, -1, -1)$

$$z^2 - (x^2 + y^2) = 1$$

This submanifold inherits the metric

$$ds^2 = dz^2 - (dx^2 + dy^2)$$

a. Show that

$$-ds^2 = dx^2 + dy^2 - (d\sqrt{1 + x^2 + y^2})^2$$

$$= \frac{1}{1 + x^2 + y^2} (dx \quad dy) \begin{bmatrix} 1 + y^2 & -xy \\ -yx & 1 + x^2 \end{bmatrix} \begin{pmatrix} dx \\ dy \end{pmatrix}$$

b. Introduce polar coordinates $x = r\cos\phi$, $y = r\sin\phi$, and show

$$-ds^2 = \frac{dr^2}{1 + r^2} + (r\,d\phi)^2$$

c. Show that the volume element on this surface is

$$dV = \frac{r\,dr\,d\phi}{\sqrt{1 + r^2}}$$

d. Repeat this calculation for $SO(3)/SO(2)$. This space is a sphere $S^2 \subset R^3$: the algebraic manifold in R^3 that satisfies $z^2 + (x^2 + y^2) = 1$ and inherits the metric $ds^2 = dz^2 + (dx^2 + dy^2)$ from this Euclidean space. Show that the metric and measure on S^2 are obtained from the results above for H^2 by the substitutions $1 + r^2 \to 1 - r^2$. Show that the disk $0 \le r \le 1$, $0 \le \phi \le 2\pi$ maps onto the upper hemisphere of the sphere, with $r = 0$ mapping to the north pole and $r = 1$ mapping to the equator. Show that the geodesic length from the north pole to the equator along the longitude $\phi = 0$ is $s = \int_0^1 dr/\sqrt{1 - r^2} = \pi/2$ and the volume of the hemisphere surface is $V = \int_{r=0}^{r=1} \int_{\phi=0}^{\phi=2\pi} dV(r, \phi) = \int_0^1 r\,dr/\sqrt{1 - r^2} \int_0^{2\pi} d\phi = 2\pi$.

3
Matrix groups

Almost all Lie groups encountered in the physical sciences are matrix groups. In this chapter we describe most of the matrix groups that are typically encountered. These include the general linear groups $GL(n; \mathbb{F})$ of nonsingular $n \times n$ matrices over the fields $\mathbb{F}$ of real numbers, complex numbers, and quaternions, and various of their subgroups obtained by imposing linear, bilinear and quadratic, and n-linear constraints on these matrix groups.

3.1 Preliminaries

It is first useful to state a simple theorem.

Definition A subgroup H of G (also $H \subset G$) is a subset of G that is also a group under the group multiplication of G.

Example The set of matrices

$$\begin{bmatrix} a & b \\ 0 & \frac{1}{a} \end{bmatrix} \quad (3.1)$$

is a subgroup of $SL(2; \mathbb{R})$.

Theorem If $H_1 \subset G$ and $H_2 \subset G$ are subgroups of G then their intersection $H_{12} = H_1 \cap H_2$ is a subgroup of G.

Proof Verify that the four group axioms are satisfied for all operations in $H_1 \cap H_2$.

Example If H_1 is the two-dimensional subgroup of $SL(2; \mathbb{R})$ described in (3.1) above and H_2 is the one-dimensional subgroup of 2×2 orthogonal matrices

$$H_2 = SO(2) = \begin{bmatrix} \cos\theta & \sin\theta \\ -\sin\theta & \cos\theta \end{bmatrix} \quad \theta \in [0, 2\pi) \quad (3.2)$$

3.2 No constraints

then the intersection $H_1 \cap H_2$ is the zero-dimensional subgroup containing the two discrete group operations $\pm I_2$.

The matrix groups that we consider are defined over the fields of real numbers ($\mathbb{F} = \mathbb{R}$), complex numbers ($\mathbb{F} = \mathbb{C}$), and quaternions ($\mathbb{F} = \mathbb{Q}$). The complex numbers can be constructed from pairs of real numbers by adjoining a square root of -1. Their multiplication properties can be analyzed by mapping the pair of real numbers into 2×2 matrices

$$c = (a, b) = a + ib \qquad a \in \mathbb{R}, \ b \in \mathbb{R}, \ i^2 = -1$$

$$(a, b) \longrightarrow \begin{bmatrix} a & b \\ -b & a \end{bmatrix} \qquad i = (0, 1) \longrightarrow \begin{bmatrix} 0 & 1 \\ -1 & 0 \end{bmatrix} \qquad (3.3)$$

In an analogous way, the quaternions can be constructed from pairs of complex numbers by adjoining another square root of -1, and their multiplication properties analyzed by mapping the pair of complex numbers into 2×2 matrices

$$q = (c_1, c_2) = c_1 + jc_2 \qquad \begin{array}{l} c_1 = a_1 + ib_1 \in \mathbb{C} \\ c_2 = a_2 + ib_2 \in \mathbb{C} \\ i^2 = -1, \ j^2 = -1, \ ij + ji = 0 \end{array}$$

$$(c_1, c_2) \longrightarrow \begin{bmatrix} c_1 & c_2 \\ -c_2^* & c_1^* \end{bmatrix} \qquad (3.4)$$

The mapping of two complex numbers into a 2×2 matrix representing a quaternion can also be expressed as a mapping of four real numbers into a 2×2 matrix representing a quaternion:

$$q_0 + q_1 \mathcal{I} + q_2 \mathcal{J} + q_3 \mathcal{K} \rightarrow \begin{bmatrix} q_0 + iq_3 & iq_1 + q_2 \\ iq_1 - q_2 & q_0 - iq_3 \end{bmatrix}$$

The four basis vectors $1, \mathcal{I}, J, K$ for this map are related to the four Pauli spin matrices, and i is the usual square root of -1 introduced above in Eq. (3.3). The details are presented in Problem 1 at the end of this chapter.

We list, in order, matrix groups on which no constraints are imposed (1), on which only linear constraints are imposed ((2)–(7)), on which bilinear and quadratic constraints are imposed ((8)–(11)), and on which n-linear or multilinear constraints $[\det(M) = +1]$ are imposed (12).

3.2 No constraints

1. $GL(n; \mathbb{F})$. General linear groups consist of nonsingular $n \times n$ matrices over the real, complex, or quaternion fields. The group $GL(1; \mathbb{Q})$ consists of 1×1

quaternion, or 2×2 complex matrices that satisfy

$$\det \begin{bmatrix} a_1 + ib_1 & a_2 + ib_2 \\ -a_2 + ib_2 & a_1 - ib_1 \end{bmatrix} = a_1^2 + b_1^2 + a_2^2 + b_2^2 \neq 0 \quad (3.5)$$

The determinant of an $n \times n$ matrix A with matrix elements A_i^j is defined by

$$\det(A) = \sum_I \sum_J \frac{1}{n!} \epsilon^{i_1 i_2 \cdots i_n} A_{i_1}^{j_1} A_{i_2}^{j_2} \cdots A_{i_n}^{j_n} \epsilon_{j_1 j_2 \cdots j_n}$$

Here $\epsilon^{i_1 i_2 \cdots i_n}$ and its covariant version are the Levi–Civita symbols: $+1$ for an even permutation of the integers $1, 2, \ldots, n$; -1 for an odd permutation; and 0 if two or more values of the indices i_* are equal. With this definition there is no difficulty computing the determinant of a matrix containing matrix elements that do not commute (quaternions).

All remaining matrix groups in this list are subgroups of $GL(n; \mathbb{F})$.

3.3 Linear constraints

These matrix groups all have a block structure or an echelon block structure. The linear constraints simply require specific blocks of matrix elements to vanish, or require some diagonal matrix elements to be $+1$. The structures of all these matrix groups are exhibited in Fig. 3.1.

2. $UT(p, q)$. **Upper triangular** groups. The $n \times n$ ($n = p + q$) matrix is partitioned into block form and an off-diagonal block is constrained to be zero

$$m_{i\alpha} = 0 \qquad \begin{array}{l} p + 1 \leq i \leq p + q \\ 1 \leq \alpha \leq p \end{array} \quad (3.6)$$

Example The action of transformations in $UT(1, 1)$ on the plane R^2 is as follows:

$$\begin{bmatrix} x' \\ y' \end{bmatrix} = \begin{bmatrix} a & b \\ 0 & d \end{bmatrix} \begin{bmatrix} x \\ y \end{bmatrix} = \begin{bmatrix} ax + by \\ dy \end{bmatrix} \quad (3.7)$$

The x-axis $y = 0$ remains invariant. It is an invariant subspace ($y = 0 \to y' = 0$), mapped into itself by all group operations in $UT(1, 1)$. The y-axis $x = 0$ is not invariant. More generally, if $UT(p, q)$ acts on the direct sum vector space $V_p \oplus V_q$, the subspace V_q is invariant while V_p is not. For lower triangular matrices reverse p and q.

3.3 Linear constraints

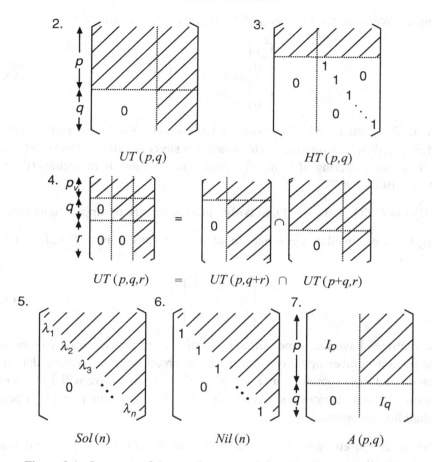

Figure 3.1. Structure of the matrix groups defined by linear constraints.

3. $HT(p, q)$. This is a subgroup of $UT(p, q)$ obtained by imposing the additional linear constraints on the matrix elements of a diagonal block

$$m_{ij} - \delta_{ij} = 0 \qquad \begin{array}{l} p+1 \le i \le p+q \\ p+1 \le j \le p+q \end{array} \qquad (3.8)$$

Example Affine transformations in $HT(1, 1)$ ($m_{22} = 1$) act on the x-axis by $x \to x' = ax + b$:

$$\begin{bmatrix} x' \\ 1 \end{bmatrix} = \begin{bmatrix} a & b \\ 0 & 1 \end{bmatrix} \begin{bmatrix} x \\ 1 \end{bmatrix} = \begin{bmatrix} ax+b \\ 1 \end{bmatrix} \qquad (3.9)$$

4. $UT(p, q, r)$. This matrix group consists of upper triangular matrices that are the intersection of the matrix groups $UT(p, q+r) \cap UT(p+q, r)$.

Example We consider 4×4 complex matrices with the structure

$$\begin{bmatrix} 1 & * & * & * \\ 0 & & & * \\ 0 & \multicolumn{2}{c}{SU(1,1)} & * \\ 0 & & & * \\ 0 & 0 & 0 & 1 \end{bmatrix} \tag{3.10}$$

where the 2×2 matrix $SU(1, 1)$ is defined below in (3.30). Matrix groups with the structure (3.10) are encountered in treatments of squeezed states of the electromagnetic field and scattering of projectiles from simple diatomic molecules (Gilmore and Yuan, 1987, 1989).

5. $Sol(n) = UT(1, 1, 1, \ldots, 1)$. **Solvable** groups are strictly upper triangular.

Example We consider the subgroup of 3×3 matrices in $UT(1, 1, 1)$ of the form

$$\begin{bmatrix} 1 & l & d \\ 0 & \eta & r \\ 0 & 0 & 1 \end{bmatrix} \tag{3.11}$$

These matrices have the same structure as the group generated by exponentials of the photon number operator ($\hat{n} = a^\dagger a$), the creation ($a^\dagger$) and annihilation (a) operators, and their commutator ($I = aa^\dagger - a^\dagger a = [a, a^\dagger]$). We will use this identification between operator and matrix groups to develop some powerful operator disentangling theorems.

6. $Nil(n)$. **Nilpotent** groups are subgroups of $Sol(n)$ whose diagonal matrix elements are all $+1$.

Example Matrices in $Nil(3)$ of the form

$$\begin{bmatrix} 1 & l & d \\ 0 & 1 & r \\ 0 & 0 & 1 \end{bmatrix} \tag{3.12}$$

are closely related to the photon creation and annihilation operators ($a^\dagger, a, I$) and the group generated by the exponentials of the position and momentum operators (p and q) and their commutator $[p, q] = \hbar/i$. This 3×3 matrix group is called the Heisenberg group. (It is technically the covering group of the Heisenberg group.) The set of change of basis transformations $\langle p|q \rangle = \frac{1}{\sqrt{2}} e^{2\pi i pq/h}$ encountered in quantum mechanics is a unitary representation of this group of 3×3 matrices.

7. $A(p, q)$. This group consists of matrices that are the sum of an identity matrix and the upper right-hand off-diagonal block of a (p, q) blocked matrix. Its matrix elements satisfy

$$A_{i,j} = \delta_{i,j} \quad 1 \le i, j \le p$$
$$A_{\alpha,\beta} = \delta_{\alpha,\beta} \quad p+1 \le \alpha, \beta \le p+q$$
$$A_{\alpha,j} = 0$$
$$A_{i,\beta} = \text{arbitrary}$$

This group is abelian or commutative: $AB = BA$ for all elements (matrices) in this group.

Example We consider the translation subgroup $A(1, 1)$ of the affine group of transformations of the x-axis (3.9): $x \to x' = x + a$. Successive transformations of this type commute

$$\begin{bmatrix} 1 & a \\ 0 & 1 \end{bmatrix} \begin{bmatrix} 1 & b \\ 0 & 1 \end{bmatrix} = \begin{bmatrix} 1 & a+b \\ 0 & 1 \end{bmatrix} = \begin{bmatrix} 1 & b \\ 0 & 1 \end{bmatrix} \begin{bmatrix} 1 & a \\ 0 & 1 \end{bmatrix} \quad (3.13)$$

3.4 Bilinear and quadratic constraints

In (8)–(11) we treat groups that preserve a metric, represented by a matrix G. They all satisfy the bilinear or quadratic constraint condition $M^\dagger G M = G$. If G is symmetric positive-definite we can set $G = I_n$ (8). If G is nonsingular and symmetric but indefinite we can set $G = I_{p,q}$ (9). If G is nonsingular and antisymmetric, we can take (10)

$$G = \begin{bmatrix} 0 & I_n \\ -I_n & 0 \end{bmatrix}$$

These are the groups that leave Hamilton's equations of motion invariant in form. A large spectrum of interesting groups occurs if G is singular (11). The matrix elements in these cases are defined by both bilinear *and* linear conditions.

8. Compact metric-preserving groups Matrices M in these groups satisfy the quadratic condition $M^\dagger G M = G$, where G is symmetric positive-definite, and which we can take as I_n

$$G = I_n \quad \begin{array}{ll} \mathbb{R} & O(n) \text{ orthogonal group} \\ \mathbb{C} & U(n) \text{ unitary group} \\ \mathbb{Q} & Sp(n) \text{ symplectic group} \end{array} \quad (3.14)$$

These are groups of rotations that leave invariant a positive-definite metric in a real, complex, or quaternion valued n-dimensional linear vector space. The manifolds that parameterize these groups are compact because the condition $M^\dagger G M = G$

defines matrices that form closed bounded subsets of the manifolds that parameterize the matrix groups $GL(n; \mathbb{F})$, $\mathbb{F} = \mathbb{R}, \mathbb{C}, \mathbb{Q}$.

Example As examples we introduce real 3×3 matrices of rigid rotations (and inversions) in R^3, complex 2×2 matrices that preserve inner products in a complex two-dimensional linear vector space C^2 (of spin states, for example), and quaternion valued 1×1 matrices that preserve length in a one-dimensional linear vector space over $\mathbb{Q}$

$$\begin{aligned} M^\dagger I_3 M &= I_3 & M &\in O(3) & \mathbb{F} &= \mathbb{R} \\ M^\dagger I_2 M &= I_2 & M &\in U(2) & \mathbb{F} &= \mathbb{C} \\ M^\dagger I_1 M &= I_1 & M &\in Sp(1) & \mathbb{F} &= \mathbb{Q} \end{aligned} \qquad (3.15)$$

The group $SU(1; \mathbb{Q})$ is the subgroup of $GL(1; \mathbb{Q})$ (3.5) subject to the condition

$$a_1^2 + b_1^2 + a_2^2 + b_2^2 = 1 \qquad (3.16)$$

This group is geometrically equivalent to the three-dimensional sphere embedded in R^4

$$SU(1; \mathbb{Q}) \sim S^3 \subset R^4 \qquad (3.17)$$

We will see many other relations between groups and geometry.

9. Noncompact metric-preserving groups Matrices in these groups leave invariant a nonsingular symmetric but indefinite metric G, which we take as $G = I_{p,q}$, $p + q = n$. This is a diagonal matrix with p elements $+1$ and q elements -1 along the diagonal. Matrices M in these groups satisfy the quadratic condition $M^\dagger G M = G$, where

$$G = I_{p,q} \quad \begin{array}{ll} \mathbb{R} & O(p,q) \text{ orthogonal group} \\ \mathbb{C} & U(p,q) \text{ unitary group} \\ \mathbb{Q} & Sp(p,q) \text{ symplectic group} \end{array} \qquad (3.18)$$

The manifolds that parameterize these groups are noncompact when $p \neq 0, q \neq 0$. These noncompact groups are related by analytic continuation to corresponding compact metric-preserving groups.

Example The Lorentz group preserves the invariant $x^2 + y^2 + z^2 - (ct)^2$ and is thus defined by the condition

$$M^t I_{3,1} M = I_{3,1}$$

$$\begin{bmatrix} A^t & C^t \\ B^t & D^t \end{bmatrix} \begin{bmatrix} I_3 & 0 \\ 0 & -1 \end{bmatrix} \begin{bmatrix} A & B \\ C & D \end{bmatrix}$$

$$= \begin{bmatrix} A^t A - C^t C & A^t B - C^t D \\ B^t A - D^t C & B^t B - D^t D \end{bmatrix} = \begin{bmatrix} I_3 & 0 \\ 0 & -1 \end{bmatrix} \qquad (3.19)$$

3.4 Bilinear and quadratic constraints

There are much better ways to parameterize this group. These involve exponentiating its Lie algebra.

10. Antisymmetric metric-preserving groups The metric G is an $N \times N$ nonsingular antisymmetric matrix

$$M^t GM = G \qquad \mathbb{F} = \begin{cases} \mathbb{R} & Sp(N, \mathbb{R}) \\ \mathbb{C} & Sp(N, \mathbb{C}) \end{cases} \qquad (3.20)$$

Since $\det(G) = \det(G^t) = \det(-G) = (-)^N \det(G)$, N must be even: $N = 2n$. The metric matrix can be chosen to have the canonical forms

$$G = \begin{bmatrix} 0 & I_n \\ -I_n & 0 \end{bmatrix}$$

or

$$G = \sum_{\alpha=1}^{n} \oplus [i\sigma_y]_\alpha$$

This consists of n copies of the matrix $i\sigma_y = \begin{bmatrix} 0 & 1 \\ -1 & 0 \end{bmatrix}$ along the diagonal. Symplectic transformations in $Sp(2n; \mathbb{R})$ leave invariant the form of the classical hamiltonian equations of motion.

Example The symplectic group $Sp(2; \mathbb{R}) \subset GL(2; \mathbb{R})$ satisfies the constraint

$$\begin{bmatrix} a & c \\ b & d \end{bmatrix} \begin{bmatrix} 0 & 1 \\ -1 & 0 \end{bmatrix} \begin{bmatrix} a & b \\ c & d \end{bmatrix} = \begin{bmatrix} 0 & ad-bc \\ bc-ad & 0 \end{bmatrix} = \begin{bmatrix} 0 & 1 \\ -1 & 0 \end{bmatrix} \qquad (3.21)$$

The constraint is $ad - bc = +1$. Thus, $Sp(2; \mathbb{R}) = SL(2; \mathbb{R})$.

11. General metric-preserving groups Matrices in these groups leave invariant a singular metric G.

$$\begin{array}{ll} \mathbb{R} & O(n; G) \\ \mathbb{C} & U(n; G) \\ \mathbb{Q} & Sp(n; G) \end{array} \qquad (3.22)$$

Example We consider 4×4 real matrices and choose

$$G = \begin{bmatrix} I_3 & 0 \\ 0 & 0 \end{bmatrix} \qquad (3.23)$$

Partitioning M into blocks and imposing the condition $MGM^t = G$, we find

$$\begin{bmatrix} A & B \\ C & D \end{bmatrix} \begin{bmatrix} I_3 & 0 \\ 0 & 0 \end{bmatrix} \begin{bmatrix} A^t & C^t \\ B^t & D^t \end{bmatrix} = \begin{bmatrix} AA^t & AC^t \\ CA^t & CC^t \end{bmatrix} = \begin{bmatrix} I_3 & 0 \\ 0 & 0 \end{bmatrix} \qquad (3.24)$$

This results in the conditions

$$AA^t = I_3 \quad \text{quadratic constraints, } A \in O(3)$$
$$C = 0 \quad \text{linear constraints}$$
$$B, D \quad \text{arbitrary} \quad \text{no constraints}$$

The subgroup obtained by setting the 1×1 submatrix D equal to $+1$ is the Euclidean group $E(3)$ whose action on the coordinates (x, y, z) of a point in R^3 is

$$\begin{bmatrix} x' \\ y' \\ z' \\ \hline 1 \end{bmatrix} = \begin{bmatrix} A & \begin{array}{|c} t_1 \\ t_2 \\ t_3 \end{array} \\ \hline 0\ 0\ 0 & 1 \end{bmatrix} \begin{bmatrix} x \\ y \\ z \\ \hline 1 \end{bmatrix} = \begin{bmatrix} A \begin{bmatrix} x \\ y \\ z \end{bmatrix} + \begin{array}{c} t_1 \\ t_2 \\ t_3 \end{array} \\ \hline 1 \end{bmatrix} \quad (3.25)$$

That is, the coordinates are rotated by the matrix A and translated by the vector $\mathbf{t}$. By closely similar arguments the Poincaré group, consisting of Lorentz transformations ($A \in SO(3, 1)$, $AI_{3,1}A^t = I_{3,1}$ (3.17)) and space-time displacements is isomorphic to the real 5×5 matrix group

$$\text{Poincaré group} \quad \begin{bmatrix} O(3, 1) & \mathbf{t} \\ \hline 0 & 1 \end{bmatrix} \quad (3.26)$$

The Galilei group consists of rotations in R^3, transformations to a coordinate system moving with velocity $\mathbf{v}$, and displacements of space ($\mathbf{t}$) and time (t_4) coordinates. It is isomorphic to the group of 5×5 matrices with the structure

$$\text{Galilei group} \quad \begin{bmatrix} O(3) & \mathbf{v} & \mathbf{t} \\ \hline 0 & 1 & t_4 \\ 0 & 0 & 1 \end{bmatrix} \quad (3.27)$$

3.5 Multilinear constraints

It is possible to impose trilinear, four-linear, ..., constraints on $n \times n$ matrices. This requires a great deal of effort, and leads to few results, principal among which are the five exceptional Lie groups that we will meet in Chapter 10. The only multilinear constraint that leads systematically to a large class of Lie groups is the n-linear constraint, defined by the determinant.

12. Special linear groups or unimodular groups These are defined by the condition

$$\det M = +1 \quad \mathbb{F} = \begin{cases} \mathbb{R} & SL(n, \mathbb{R}) \\ \mathbb{C} & SL(n, \mathbb{C}) \\ \mathbb{Q} & SL(n, \mathbb{Q}) \end{cases} \quad (3.28)$$

Example The group $SL(2; \mathbb{R})$ has previously been encountered. The subset of matrices $\begin{bmatrix} a & b \\ c & d \end{bmatrix} \in SL(2; \mathbb{R}) \subset GL(2; \mathbb{R})$ satisfies the constraint $ad - bc = +1$, which is bilinear.

3.6 Intersections of groups

Some important groups are intersections of those listed above

$$\begin{aligned} SO(n) &= O(n) \cap SL(n; \mathbb{R}) \\ SO(p, q) &= O(p, q) \cap SL(p + q; \mathbb{R}) \\ SU(n) &= U(n) \cap SL(n; \mathbb{C}) \\ SU(p, q) &= U(p, q) \cap SL(p + q; \mathbb{C}) \end{aligned} \quad (3.29)$$

Example We construct the three-dimensional noncompact group $SU(1, 1)$ by taking the intersection of $U(1, 1)$ with $SL(2; \mathbb{C})$:

$$SU(1, 1) = U(1, 1) \cap SL(2; \mathbb{C}) \to \begin{bmatrix} a & b \\ b^* & a^* \end{bmatrix} \quad (3.30)$$

where $a^*a - b^*b = +1$.

3.7 Embedded groups

The unitary group $U(n)$ consists of $n \times n$ complex matrices that obey the constraint $U^\dagger U = I_n$. For some purposes it is useful to represent this group as a group of real matrices. This is done by replacing each of the complex entries in $U(n)$ by a real 2×2 matrix according to the prescription given in Eq. (3.3). The resulting matrix is a real $2n \times 2n$ matrix M. This matrix inherits the constraint that comes with the unitary group, $U^\dagger U = I_n$. This constraint now appears in the form $M^t M = I_{2n}$. We have been able to replace $\dagger$ by t since the matrices are real, and must replace I_n by I_{2n} since the matrices are $2n \times 2n$. In other words, the matrices M obey the condition that determines orthogonal groups. This group of $2n \times 2n$ matrices forms an *orthogonal* representation of the *unitary* group. It is a subgroup of $SO(2n)$. This matrix group is called $OU(2n)$. Symbolically,

$$U(n) \xrightarrow{\mathbb{C} \to 2 \times 2 \; \mathbb{R}} OU(2n) \subset SO(2n) \quad (3.31)$$

There is an even more compelling reason to carry out the same type of replacement of quaternions by 2×2 complex matrices. Quaternions do not commute, as do real and complex numbers. Rather than worry about the order in which quaternions are written down in carrying out computations (such as constructing the determinant of a matrix), it is usually safer and more convenient to replace each quaternion in an $n \times n$ matrix by a 2×2 complex matrix using the embedding shown in Eq. (3.4). For the metric-preserving quaternion group $U(n; \mathbb{Q}) = Sp(n)$ whose matrices obey $U^\dagger U = I_n$, this process generates $2n \times 2n$ complex matrices M that inherit the constraint in the form $M^\dagger M = I_{2n}$. In other words, the matrices M obey the condition that determines unitary groups (over $\mathbb{C}$). This group of $2n \times 2n$ matrices forms a *unitary* representation of the *symplectic* group. It is a subgroup of $SU(2n)$. This matrix group is called $USp(2n)$. Symbolically,

$$Sp(n) \xrightarrow{\mathbb{Q} \to 2 \times 2 \; \mathbb{C}} USp(2n) \subset SU(2n) \tag{3.32}$$

The groups $OU(2n)$ and $USp(2n)$ will appear in Chapter 11 (see, Table 11.1) in the classification of the real forms of the simple Lie groups.

3.8 Modular groups

We close with a useful aside. We have not considered matrices over the integers because they lack the geometric structure contributed by the continuous fields $\mathbb{R}$, $\mathbb{C}$, and $\mathbb{Q}$. However, matrices over the integers play an important role in some areas of Lie group theory (representation theory of noncompact unimodular groups).

There are in fact three distinct groups over the integers that are sometimes confused

(i) $GL(n; \mathbb{Z})$: if $m \in GL(n; \mathbb{Z})$, $\det(m) = \pm 1$.
(ii) $SL(n; \mathbb{Z})$: if $m \in SL(n; \mathbb{Z})$, $\det(m) = +1$.
(iii) $PSL(n; \mathbb{Z})$, n even: $PSL(n; \mathbb{Z}) = SL(n; \mathbb{Z})/\{I_n, -I_n\}$.

For $n = 2$ these groups of matrices have the form $\begin{bmatrix} a & b \\ c & d \end{bmatrix}$, with a, b, c, d all integers. If $\det(m) = n$, with n an integer, then $\det(m^{-1}) = 1/n$. Since the determinant of any matrix composed of integers must be an integer, the condition is that $\det(m) = \pm 1$. The subset of $GL(2; \mathbb{Z})$ with determinant $+1$ forms the subgroup $SL(2; \mathbb{Z}) \subset GL(2; \mathbb{Z})$. The modular group $PSL(2; \mathbb{Z})$ is obtained by identifying each pair of matrices in $SL(2; \mathbb{Z})$ of the form $\begin{bmatrix} -a & -b \\ -c & -d \end{bmatrix} \simeq \begin{bmatrix} a & b \\ c & d \end{bmatrix}$.

As a hint of the useful properties of these groups, we consider the matrix

$$\begin{bmatrix} 1 & 1 \\ 1 & 0 \end{bmatrix} \in GL(n; \mathbb{Z}) \tag{3.33}$$

3.8 Modular groups

Then

$$\begin{bmatrix} 1 & 1 \\ 1 & 0 \end{bmatrix}^n = \begin{bmatrix} F(n+1) & F(n) \\ F(n) & F(n-1) \end{bmatrix} \quad (3.34)$$

where $F(n)$ is the nth Fibonacci number, defined recursively by

$$F(n) = F(n-1) + F(n-2)$$

n	0	1	2	3	4	5	6	7	$\cdots$
$F(n)$	0	1	1	2	3	5	8	13	$\cdots$

The proof by induction is simple. It proceeds by computation

$$\begin{aligned} \begin{bmatrix} 1 & 1 \\ 1 & 0 \end{bmatrix}^{n+1} &= \begin{bmatrix} 1 & 1 \\ 1 & 0 \end{bmatrix} \begin{bmatrix} F(n+1) & F(n) \\ F(n) & F(n-1) \end{bmatrix} \\ &= \begin{bmatrix} F(n+1)+F(n) & F(n)+F(n-1) \\ F(n+1) & F(n) \end{bmatrix} \\ &= \begin{bmatrix} F(n+2) & F(n+1) \\ F(n+1) & F(n) \end{bmatrix} \end{aligned} \quad (3.35)$$

and by comparison of initial conditions for $n = 1$ ($F(0) = 0, F(1) = 1$). Many other recursive relations among the integers are possible using different matrices in the groups $GL(2; \mathbb{Z})$, $GL(3; \mathbb{Z})$, etc.

The group $GL(n; \mathbb{Z})$ has important subgroups defined by imposing linear, quadratic, and multilinear constraints on the matrix elements, in exact analogy with $GL(n; \mathbb{R})$.

Imposing linear constraints generates subgroups with the structures given in Examples (2) through (7) above. The only remark necessary is that for the analogs of Example (5) (solvable groups) the diagonal matrix elements can only be ± 1.

Imposing quadratic constraints, for example $M^t I_n M = I_n$, generates a subgroup for which the sum of the squares of the matrix elements in each row or column is $+1$. Since the matrix elements themselves can only be $\pm 1, 0$, this group, $O(n; \mathbb{Z})$, consists of $n \times n$ matrices in which all but one matrix element in every row or column is zero, and the nonzero matrix element is ± 1. An important subgroup of $O(n; \mathbb{Z})$ is S_n, in which the nonzero matrix elements are all $+1$. This is the $n \times n$ faithful permutation representation P_n of the symmetric group S_n.

Finally, the multilinear condition $\det(m) = +1$ defines the unimodular subgroup $SL(n; \mathbb{Z})$ of $GL(n; \mathbb{Z})$.

Additional important groups are intersections of those just described. For example, the alternating group A_n consists of unimodular matrices in P_n:

$$A_n = P_n \cap SL(n; \mathbb{Z}) \qquad (3.36)$$

Example The group $O(2; \mathbb{Z})$ consists of the $8 = 2^2 \times 2!$ matrices

$$\pm \begin{bmatrix} 1 & 0 \\ 0 & 1 \end{bmatrix} \quad \pm \begin{bmatrix} 1 & 0 \\ 0 & -1 \end{bmatrix} \quad \pm \begin{bmatrix} 0 & 1 \\ 1 & 0 \end{bmatrix} \quad \pm \begin{bmatrix} 0 & 1 \\ -1 & 0 \end{bmatrix} \qquad (3.37)$$

The group $O(3; \mathbb{Z})$ has order $2^3 \times 3! = 48$. Its subgroup S_3 of order $6 = 3!$ consists of the six matrices

$$\begin{bmatrix} 1 & 0 & 0 \\ 0 & 1 & 0 \\ 0 & 0 & 1 \end{bmatrix} \begin{bmatrix} 0 & 1 & 0 \\ 0 & 0 & 1 \\ 1 & 0 & 0 \end{bmatrix} \begin{bmatrix} 0 & 0 & 1 \\ 1 & 0 & 0 \\ 0 & 1 & 0 \end{bmatrix}$$

$$\begin{bmatrix} 0 & 1 & 0 \\ 1 & 0 & 0 \\ 0 & 0 & 1 \end{bmatrix} \begin{bmatrix} 0 & 0 & 1 \\ 0 & 1 & 0 \\ 1 & 0 & 0 \end{bmatrix} \begin{bmatrix} 1 & 0 & 0 \\ 0 & 0 & 1 \\ 0 & 1 & 0 \end{bmatrix} \qquad (3.38)$$

Its alternating subgroup $A_3 \subset S_3 \subset O(3; \mathbb{Z})$ consists of the three matrices with positive determinant, contained in the first row.

3.9 Conclusion

In this chapter we have taken advantage of a surprising observation: most of the Lie groups encountered in applied (as well as pure) mathematics, the physical sciences, and the engineering disciplines are matrix groups. Most of the matrix groups typically encountered have been listed here. They consist of the general linear groups of $n \times n$ nonsingular matrices over the fields of real numbers, complex numbers, and quaternions, as well as subgroups obtained by imposing linear conditions, bilinear and quadratic conditions, and multilinear conditions on the matrix elements of the $n \times n$ matrices. Lie groups not encountered in the simple construction presented here consist primarily of some real forms (analytic continuations, encountered in Chapter 11) of those encountered here, the exceptional Lie groups G_2, F_4, E_6, E_7, E_8 and their real forms (encountered in Chapters 10 and 11), and covering groups of noncompact simple Lie groups such as $SL(2; \mathbb{R})$ (encountered in Chapter 7). We have in addition opened a door to analogs of Lie groups over the integers, $GL(n; \mathbb{Z})$, $SL(n; \mathbb{Z})$, and $PSL(n; \mathbb{Z})$. Matrix groups over finite fields are also of great interest, but fall outside the scope of our discussions.

3.10 Problems

1. Use the mapping (3.4) to construct a 2 × 2 matrix representation of the quaternions over the field of complex numbers. In particular, make the following associations, where $\mathcal{I}\mathcal{J} = -\mathcal{K}$:

$$1 = \begin{bmatrix} 1 & 0 \\ 0 & 1 \end{bmatrix} \quad \mathcal{I} = i\begin{bmatrix} 0 & 1 \\ 1 & 0 \end{bmatrix} \quad \mathcal{J} = i\begin{bmatrix} 0 & -i \\ i & 0 \end{bmatrix} \quad \mathcal{K} = i\begin{bmatrix} 1 & 0 \\ 0 & -1 \end{bmatrix} \quad (3.39)$$

$$I_2 \qquad\qquad i\sigma_x \qquad\qquad i\sigma_y \qquad\qquad i\sigma_z$$

Here $\sigma_x, \sigma_y, \sigma_z$ are the Pauli spin matrices, and i is the usual square root of -1. Show that any pair of the unit quaternions anticommute: i.e., $\{\mathcal{I}, \mathcal{J}\} = \mathcal{I}\mathcal{J} + \mathcal{J}\mathcal{I} = 0$.

2. Show that the unit quaternions $\mathcal{I}, \mathcal{J}, \mathcal{K}$ generate a group of order 8 under multiplication. Show that this group is isomorphic to $O(2; \mathbb{Z})$. Exhibit this isomorphism explicitly.

3. Show that $SU(1; \mathbb{Q}) \sim SU(2; \mathbb{C})$.

4. Show that the dimensionalities (over the real field) of the general linear groups and their special linear subgroups are

$$\begin{aligned} GL(n; \mathbb{R}) &= n^2 & SL(n; \mathbb{R}) &= n^2 - 1 \\ GL(n; \mathbb{C}) &= 2n^2 & SL(n; \mathbb{C}) &= 2n^2 - 2 \\ GL(n; \mathbb{Q}) &= 4n^2 & & \end{aligned}$$

5. Show that if the $n \times n$ metric matrix G is symmetric, nonsingular, and positive definite, then we can set $G = I_n$ in the definitions in Example (8). If the $n \times n$ metric matrix G is symmetric, nonsingular, and indefinite, then we can set $G = I_{p,q}$ in the definitions in Example (9), for suitable positive integers p and q, with $p + q = n$.

6. Show that it is possible to define subgroups $SL_i(n; \mathbb{C})$ of $GL(n; \mathbb{C})$ by the conditions

$$\begin{aligned} SL_1(n; \mathbb{C}) \quad \det(M) &= e^{i\phi} & 2n^2 - 1 \\ SL_2(n; \mathbb{C}) \quad \det(M) &= e^\lambda & 2n^2 - 1 \\ SL_3(n; \mathbb{C}) \quad \det(M) &= r & 2n^2 - 1 \\ SL(n; \mathbb{C}) \quad \det(M) &= +1 & 2n^2 - 2 \end{aligned}$$

where ϕ, λ, r are real and $r \neq 0$. Show that the dimensions of these three subgroups are $2n^2 - 1$ and that $SL_3(n; \mathbb{C})$ is disconnected. It consists of two topologically identical copies of a $2n^2 - 1$ dimensional manifold, one of which contains the identity. Show that $SL(n; \mathbb{C}) = SL_1(n; \mathbb{C}) \cap SL_2(n; \mathbb{C})$. Do these results extend under the field restriction $\mathbb{C} \to \mathbb{R}$? and the field extension $\mathbb{C} \to \mathbb{Q}$?

7. A subgroup of $UT(1, 1)$ includes matrices of the form $\begin{bmatrix} -1 & a \\ 0 & 1 \end{bmatrix}$, $a \in R$. Show that the underlying group manifold consists of two copies of the real line R^1. If matrices of the form $\begin{bmatrix} 1 & a \\ 0 & -1 \end{bmatrix}$ are also included, then the parameterizing manifold consists of how many copies of R^1?

8. Compute the dimensions of the real matrix groups in Examples (2)–(7) over the real field and show:

Group	Dimension
$UT(p,q)$	$p^2 + q^2 + pq$
$HT(p,q)$	$p(p+q)$
$UT(p,q,r)$	$p^2 + q^2 + r^2 + pq + pr + qr$
$Sol(n)$	$n(n+1)/2$
$Nil(n)$	$n(n-1)/2$
$A(p,q)$	pq

What happens to these dimensions if the matrix groups are over the field of complex numbers? Quaternions?

9. Newton's equations of motion are $\mathbf{F} = d\mathbf{p}/dt$. In the Lorentz gauge Maxwell's equations can be written in the form

$$\left(\nabla^2 - \frac{1}{c^2}\frac{\partial^2}{\partial t^2}\right) A_\mu(x,t) = -\frac{4\pi}{c} j_\mu$$

These equations can be expressed in a different coordinate system usisng either Galilean or Poincaré transformations. Verify that the equations do or do not remain invariant in form under these transformations, as follows:

Transformation	$\mathbf{F} = d\mathbf{p}/dt$	$(\nabla^2 - \frac{1}{c^2}\frac{\partial^2}{\partial t^2}) A_\mu = -\frac{4\pi}{c} j_\mu$
Galilean	invariant	not invariant
Poincaré	not invariant	invariant

How do you reconcile these results?

10. Show that the group of 2×2 matrices $SU(2)$ is parameterized by two complex numbers $c_1 = a_1 + ib_1$ and $c_2 = a_2 + ib_2$, so that

$$SU(2) = \begin{bmatrix} c_1 & c_2 \\ -c_2^* & c_1^* \end{bmatrix}$$

subject to the condition $a_1^2 + b_1^2 + a_2^2 + b_2^2 = 1$. Convince yourself (a) that topologically this group (i.e., its parameterizing manifold) is equivalent to a three-sphere $S^3 \subset R^4$; and (b) algebraically it is equivalent to $SU(1; \mathbb{Q})$ (cf. (3.16)).

11. The group of 2×2 matrices $SU(1,1)$ is parameterized by two complex numbers $c_1 = a_1 + ib_1$ and $c_2 = a_2 + ib_2$, so that

$$SU(1,1) = \begin{bmatrix} c_1 & c_2 \\ c_2^* & c_1^* \end{bmatrix}$$

subject to the condition $a_1^2 + b_1^2 - a_2^2 - b_2^2 = 1$. Identify the parameterizing manifold.

12. The group $SO(2)$ is one dimensional. Show that every matrix in $SO(2)$ can be written in the form $\begin{bmatrix} m_{11} & x \\ m_{21} & m_{22} \end{bmatrix}$, where $m_{11}^2 + x^2 = 1$, so that $m_{11} = \pm\sqrt{1-x^2}$. The second row

is orthogonal to the first, so that $m_{21}m_{11} + m_{22}x = 0$. As a result, we find

$$SO(2) \longrightarrow \begin{bmatrix} \pm\sqrt{1-x^2} & x \\ -x & \pm\sqrt{1-x^2} \end{bmatrix}$$

The $\pm$ signs are coherent. Each choice of sign ($\pm$) covers half the group.

13. The group $SO(3)$ is three dimensional. Show that every matrix in $SO(3)$ can be written in the form

$$SO(3) \longrightarrow \begin{bmatrix} m_{11} & x & y \\ m_{21} & m_{22} & z \\ m_{31} & m_{32} & m_{33} \end{bmatrix}$$

Use arguments similar to those used in Problem 12 to express the matrix elements m_{ij} $i \geq j$ in terms of the three parameters (x, y, z).

14. An alternative parameterization of $SO(3)$ is given by

$$SO(3) \longrightarrow \begin{bmatrix} Z_2 & \begin{matrix} x \\ y \end{matrix} \\ -x \quad -y & Z_1 \end{bmatrix} \times \begin{bmatrix} \pm\sqrt{1-z^2} & z & 0 \\ -z & \pm\sqrt{1-z^2} & 0 \\ 0 & 0 & 1 \end{bmatrix}$$

Express the 2×2 and 1×1 submatrices Z_2 and Z_1 in terms of the coordinates (x, y). Determine the range of the parameters (x, y, z). How many square roots ("sheets") are necessary to cover $SO(3)$ completely?

15. If $M \in GL(n; \mathbb{Z})$, show that $\det(M)$ must be ± 1.

16. Show that the orders of $O(n; \mathbb{Z}) \supset S_n \supset A_n$ are $2^n \times n!$, $n!$, $\frac{1}{2}n!$.

17. Estimate the Fibonacci number $F(n)$ from the eigenvalues $\lambda_{\pm} = \frac{1}{2}(1 \pm \sqrt{5})$ of the generating matrix (3.33). What happens to this sequence if different initial conditions (other than $F(0) = 0$, $F(1) = 1$) are introduced?

18. Derive other Fibonacci-type series using other symmetric generating matrices in $GL(2; \mathbb{Z})$ (for example, $\begin{bmatrix} 2 & -1 \\ -1 & 1 \end{bmatrix}$) and other initial conditions.

19. The energy levels $|nlm\rangle$ of the nonrelativistic hydrogen atom exhibit an n^2-fold degeneracy under the Lie group $SO(4)$. All bound states with the same principal quantum number n have the same energy $E(nlm) = -E_0/n^2$ ($E_0 = 13.6$ eV). If the Coulomb symmetry is broken by placing one or more electrons in the Coulomb potential, the overall symmetry reduces to that of the rotation group: there is a symmetry reduction $SO(4) \downarrow SO(3)$. The representations of $SO(4)$ that enter into the description of the hydrogen atom bound states are indexed by the principal quantum number n ($n = 1, 2, 3, \ldots$). The $SO(4)$ representation with quantum number n splits into angular momentum representations that are indexed with quantum number l, $l = 0, 1, 2 \ldots, n-1$, with $\sum_{l=0}^{l=n-1}(2l+1) = n^2$. The $SO(3)$ multiplet with quantum number l is $2l+1$-fold degenerate. An empirical hamiltonian with $SO(4) \downarrow SO(3)$ broken symmetry that describes the filling order when electrons are introduced into a Coulomb potential established by a central charge $+Ze$ can be

chosen to have the form:

$$E = -E_0 Z^2 \{1 + \delta * (n - l - 1)\}/n^2$$

This hamiltonian depends only on the quantum numbers of the representations of $SO(4)$ and its subgroup $SO(3)$. Show that this phenomenological energy spectrum with $\delta = 0.28$ provides the filling ordering that accounts for Mendeleev's periodic table of the chemical elements: $(n, l) \to$ $1s$; $2s, 2p$; $3s, 3p$; $4s, 3d, 4p$; $5s, 4d, 5p$; $6s, 4f, 5d, 6p$; $7s, 5f, 6d, 7p$; $8s, 6f, 7d, 8p$; ....

20. **Symmetries** Show the following equivalences:

$$UT(p, q) = UT(q, p) \qquad SO(p, q) = SO(q, p)$$
$$A(p, q) = A(q, p) \qquad U(p, q) = U(q, p)$$
$$Sp(p, q) = Sp(q, p)$$

21. G_1 and G_2 are two metrics on a real $2n$-dimensional linear vector space that are defined by

$$G_1 = \begin{bmatrix} I_n & 0 \\ 0 & I_n \end{bmatrix} \qquad G_2 = \begin{bmatrix} 0 & I_n \\ -I_n & 0 \end{bmatrix}$$

Show that the $2n \times 2n$ matrices M that satisfy the bilinear constraints $M^t G_i M = G_i$ are:

G_1	G_2	G_1 and G_2
$O(2n; \mathbb{R})$	$Sp(2n; \mathbb{R})$	$OU(2n; \mathbb{R})$

22. In an n-dimensional linear vector space two coordinate systems x and y are related by a linear transformation: $y^j = x^i M_i{}^j$. Show that the derivatives are related by the same transformation (covariance–contravariance)

$$\frac{\partial}{\partial x^i} = \frac{\partial y^j}{\partial x^i} \frac{\partial}{\partial y^j} = M_i{}^j \frac{\partial}{\partial y^j}$$

As a result, a transformation that preserves a metric when acting on the coordinates preserves the same metric when acting on the derivatives.

23. The Poisson brackets between two functions $f(q, p)$ and $g(q, p)$ on a classical phase space of dimension $2n$ are defined by

$$\{f, g\} = \sum_k \frac{\partial f}{\partial q_k} \frac{\partial g}{\partial p_k} - \frac{\partial g}{\partial q_k} \frac{\partial f}{\partial p_k}$$

a. Show that these relations can be written in simple matrix form as

$$\{f, g\} = (Df)^t G(Dg) \quad \text{where} \quad G = \begin{bmatrix} 0 & I_n \\ -I_n & 0 \end{bmatrix} \quad \text{and} \quad (Dg) = \begin{bmatrix} \partial g/\partial q \\ \partial g/\partial p \end{bmatrix}$$

3.10 Problems

b. Introduce a new coordinate system (Q, P), related to the original by a linear transformation of the form

$$\begin{bmatrix} \partial g/\partial Q \\ \partial g/\partial P \end{bmatrix} = \begin{bmatrix} A & B \\ C & D \end{bmatrix} \begin{bmatrix} \partial g/\partial q \\ \partial g/\partial p \end{bmatrix}$$

Find the conditions on this $2n \times 2n$ matrix that preserves the structure of the Poisson brackets. Show $A^t C$ and $B^t D$ must be symmetric and $A^t D - B^t C = I_n$.

c. Show that the same conditions hold for linear transformations and the quantumn mechanical commutator bracket: $[q_j, q_k] = [p_j, p_k] = 0$ and $[q_j, p_k] = i\hbar \delta_{jk}$.
Note: The transformation from classical mechanics to quantum mechanics is made by identifying the classical Poisson bracket $\{,\}$ with the quantum commutator bracket $[,]$ according to

$$\{u(q, p), v(q, p)\} \leftrightarrow \frac{[u(\hat{q}, \hat{p}), v(\hat{q}, \hat{p})]}{i\hbar}$$

The hat ˆ indicates an operator.

24. **Transfer matrices** Figure 3.2 shows a potential in one dimension. The wavefunction to the left of the interaction region has the form

$$\psi_L(x) = A_L e^{+ikx} + B_L e^{-ikx} = \begin{bmatrix} e^{+ikx} & e^{-ikx} \end{bmatrix} \begin{bmatrix} A_L \\ B_L \end{bmatrix}$$

with a similar expression for the wavefunction on the right. The exponential e^{+ikx} describes a particle of mass m moving to the right (+) with momentum $\hbar k$ and energy $E = (\hbar k)^2/2m$. The complex number A_L is the probability amplitude for finding a particle moving to the right with this momentum. The expected value of the momentum in the left-hand region is $\langle \hat{p} \rangle = (|A_L|^2 - |B_L|^2)\hbar k$, where the operator $\hat{p} = \frac{\hbar}{i}\frac{d}{dx}$.

a. Show that conservation of momentum leads to the equation

$$|A_L|^2 - |B_L|^2 = |A_R|^2 - |B_R|^2$$

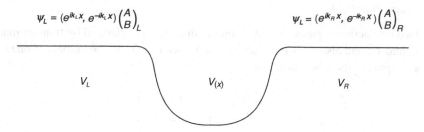

Figure 3.2. The potentials to the left and right of the interaction region are constant, with $V_L = V_R$. The wavefunctions to the left and right of this region are represented in the form $\psi_\sigma(x) = A_\sigma e^{+ikx} + B_\sigma e^{-ikx}$, where $\sigma = L, R$.

when the asymptotic value of the potential to the left of the interaction region is the same as the value on the right.

b. Since the Schrödinger equation is second order the four amplitudes A_L, A_R, B_L, B_R are not independent. Only two are independent. Two linear relations exist among them. Show that they can be expressed in terms of a matrix relation of the form

$$\begin{bmatrix} A_L \\ B_L \end{bmatrix} = \begin{bmatrix} t_{11} & t_{12} \\ t_{21} & t_{22} \end{bmatrix} \begin{bmatrix} A_R \\ B_R \end{bmatrix}$$

The 2×2 matrix T is called a **transfer matrix**. The transfer matrix is a function of energy E. Show that $T(E) \in U(1, 1)$.

c. Show that $T \in SU(1, 1)$ by appropriate choice of phase.

25. **Crossing symmetry:** A transfer matrix T for a one-dimensional potential relates amplitudes for the wavefunction on the left of the interaction region with the amplitudes on the right. A scattering matrix (S-matrix) S relates the incoming amplitudes with the outgoing amplitudes:

$$\begin{bmatrix} A_L \\ B_L \end{bmatrix} = T \begin{bmatrix} A_R \\ B_R \end{bmatrix} \qquad \begin{bmatrix} A_R \\ B_L \end{bmatrix} = S \begin{bmatrix} A_L \\ B_R \end{bmatrix}$$

a. Invoke conservation of momentum arguments to conclude $S \in U(2)$.

b. Show that the matrix elements of S and T are related by

$$\begin{bmatrix} s_{11} & s_{12} \\ s_{21} & s_{22} \end{bmatrix} = \begin{bmatrix} \dfrac{1}{t_{11}} & -\dfrac{t_{12}}{t_{11}} \\ \dfrac{t_{21}}{t_{11}} & \dfrac{t_{11}t_{22} - t_{12}t_{21}}{t_{11}} \end{bmatrix}$$

c. Show that the poles of $S(E)$ are the zeroes of $T(E)$, specifically of $t_{11}(E)$. Poles along the real energy axis describe bound states. Poles off the real axis of the form $r_j/[(E - E_j) + i(\Gamma_j/2)]$ describe resonances at energy E_j with characteristic decay time $\Gamma_j/\hbar$.

26. Two interaction regions V_1 and V_2 on the line are characterized by transfer matrices T_1 and T_2, and also by S-matrices S_1 and S_2 (see Fig. 3.3). The outputs of one region are inputs to the other, as follows:

$$\begin{bmatrix} i_2 \\ i_3 \end{bmatrix} = \begin{bmatrix} 0 & 1 \\ 1 & 0 \end{bmatrix} \begin{bmatrix} o_1 \\ o_4 \end{bmatrix}$$

a. The transfer matrices for the two regions are defined by

$$\begin{bmatrix} i_1 \\ o_2 \end{bmatrix} = T_1 \begin{bmatrix} o_1 \\ i_2 \end{bmatrix} \qquad \begin{bmatrix} i_3 \\ o_4 \end{bmatrix} = T_2 \begin{bmatrix} o_3 \\ i_4 \end{bmatrix}$$

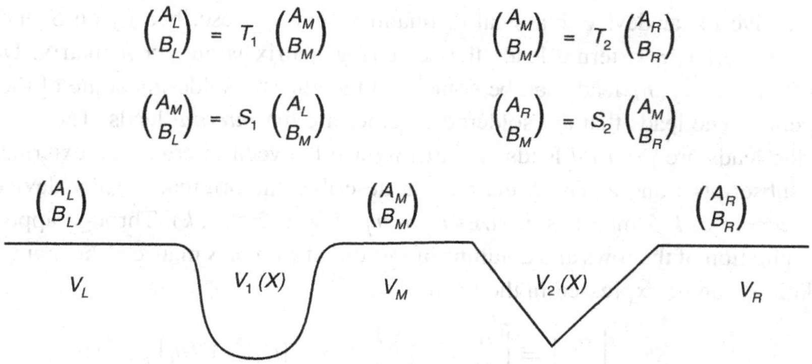

3.10 Problems

$$\begin{pmatrix} A_L \\ B_L \end{pmatrix} = T_1 \begin{pmatrix} A_M \\ B_M \end{pmatrix} \qquad \begin{pmatrix} A_M \\ B_M \end{pmatrix} = T_2 \begin{pmatrix} A_R \\ B_R \end{pmatrix}$$

$$\begin{pmatrix} A_M \\ B_L \end{pmatrix} = S_1 \begin{pmatrix} A_L \\ B_M \end{pmatrix} \qquad \begin{pmatrix} A_R \\ B_M \end{pmatrix} = S_2 \begin{pmatrix} A_M \\ B_R \end{pmatrix}$$

$$\begin{pmatrix} A_L \\ B_L \end{pmatrix} \qquad\qquad \begin{pmatrix} A_M \\ B_M \end{pmatrix} \qquad\qquad \begin{pmatrix} A_R \\ B_R \end{pmatrix}$$

$V_L \qquad\quad V_1(X) \qquad\quad V_M \qquad\quad V_2(X) \qquad\quad V_R$

Figure 3.3. Two potentials on the line are characterized by their T and S matrices.

Show that the transfer matrix for the entire interaction region is

$$\begin{bmatrix} i_1 \\ o_2 \end{bmatrix} = T_{\text{Tot}} \begin{bmatrix} o_3 \\ i_4 \end{bmatrix} \qquad T_{\text{Tot}} = T_1 T_2$$

b. The S-matrices for the two regions relate inputs to outputs as follows

$$\begin{bmatrix} o_1 \\ o_2 \\ o_3 \\ o_4 \end{bmatrix} = \begin{bmatrix} s_{11} & s_{12} & 0 & 0 \\ s_{21} & s_{22} & 0 & 0 \\ 0 & 0 & s_{33} & s_{34} \\ 0 & 0 & s_{43} & s_{44} \end{bmatrix} \begin{bmatrix} i_1 \\ i_2 \\ i_3 \\ i_4 \end{bmatrix}$$

Show that the scattering matrix for the entire region is

$$S_{\text{Tot}} = \begin{bmatrix} 0 & s_{34} \\ s_{21} & 0 \end{bmatrix} + \frac{1}{1 - s_{12}s_{43}} \begin{bmatrix} s_{33}s_{22} & s_{33}s_{11}s_{44} \\ s_{22}s_{43}s_{11} & s_{22}s_{44} \end{bmatrix}$$

c. Show that S_{Tot} is unitary.

d. Interpret the matrix S_{Tot} in terms of a Feynman-like sum over all paths. Do this by expanding the fraction $1/(1 - s_{12}s_{43})$ as a geometric sum and interpreting each term in this expansion as a path through the two scattering potentials.

27. If the potentials V_1 and V_2 are modified to V_1' and V_2' their transfer matrices and their scattering matrices will also be modified $T_i(E) \to T_i'(E)$ and $S_i(E) \to S_i'(E)$, $i = 1, 2$. It is possible that for some energy E, $S_{\text{Tot}}'(E) = S_{\text{Tot}}(E)$. Find the set of all modified scattering matrices $S_1'(E)$ and $S_2'(E)$ with the property that the modified pair maps into the original total S-matrix $S_{\text{Tot}}(E)$. In fancy terms, find the fiber in $U(4) \supset U(2) \otimes U(2) \downarrow U(2)$. (Hint: if this seems daunting, note that to satisfy $T_1(E)T_2(E) = T_{\text{Tot}}(E) = T_1'(E)T_2'(E)$ we can take $T_1'(E) = T_1(E)R$ and $T_2'(E) = R^{-1}T_2(E)$ for any $R \in U(1,1)$. The fiber in $U(2,2) \supset U(1,1) \otimes U(1,1) \downarrow U(1,1)$ over $T_{\text{Tot}}(E)$ consists of the matrices $(T_1(E)R, R^{-1}T_2(E))$.) Now map this into the fiber $(S_1'(E), S_2'(E))$ over $S_{\text{Tot}}(E)$.

28. A passive linear device, classical or quantum, can be described by an S matrix. If the device has n external leads the scattering matrix is an $n \times n$ matrix. Devices with $n_1, n_2, \ldots, n_k$ leads can be connected together by soldering some of the leads together. The leads that are soldered together are the *internal* leads. The remainder of the leads are *external* leads. We distinguish between internal and external leads by subscripts i and e. The S matrix that describes the original set of k devices is a direct sum of k S matrices of sizes $n_j \times n_j$ ($j = 1, 2, \ldots, k$). Through appropriate permutation of the rows and columns of this direct sum of S matrices the input-output relations can be expressed in the form

$$\begin{bmatrix} o_i \\ o_e \end{bmatrix} = \begin{bmatrix} A & B \\ C & D \end{bmatrix} \begin{bmatrix} i_i \\ i_e \end{bmatrix} \qquad [i_i] = \Gamma [o_i]$$

The matrix Γ that relates internal outputs to internal inputs describes the topology, or connectivity, of the network.

   a. Show that the S matrix that describes the network, defined by $[o_e] = S_{\text{Network}} [i_e]$, is given by (cf., Problem 3.26c)

$$S_{\text{Network}} = D + C\Gamma(I - A\Gamma)^{-1} B$$

   b. Show that S_{Newtork} is unitary: $S_{\text{Newtork}}^\dagger = S_{\text{Newtork}}$, $S_{\text{Newtork}} \subset U(d)$.
   c. Expand S_{Newtork} to show that

$$S_{\text{Network}} = D + C\Gamma B + C\Gamma A\Gamma B + C\Gamma A\Gamma A\Gamma B + C\Gamma A\Gamma A\Gamma A\Gamma B + \cdots$$

   Interpret this expansion in terms of a Feynman sum over all possible scattering paths through the network.

29. A mathematical description of the preceeding problem involves a subgroup restriction $U(\sum_{j=1}^{k} n_j) \supset \Pi_{j=1}^{k} \otimes U(n_j)$ and a projection to the total network scattering matrix in $U(d)$, where d is the number of the network's external leads. The connectivity is determined by the permutation matrix Γ. Determine the fiber in $\Pi_{j=1}^{k} \otimes U(n_j)$ over each group operation in $U(d)$.

30. All the matrices in this problem are square $n \times n$, with: H hermitian; U unitary; A antihermitian. Show the right-hand column follows from the definition in the left-hand column.

$$H_2 = \frac{H_1 + I_n}{H_1 - I_n} \qquad [H_1, H_2] = 0 \qquad H_1 = \frac{H_2 + I_n}{H_2 - I_n}$$

$$U = \frac{I_n + iH}{I_n - iH} \qquad [H, U] = 0 \qquad H = i\frac{I_n - U}{I_n + U}$$

$$A = \frac{I_n + iU}{I_n - iU} \qquad [U, A] = 0 \qquad U = i\frac{I_n - A}{I_n + A}$$

$$H = \frac{I_n - iA}{I_n + iA} \qquad [A, H] = 0 \qquad A = i\frac{H - I_n}{H + I_n}$$

4
Lie algebras

The study of Lie groups can be greatly facilitated by linearizing the group in the neighborhood of its identity. This results in a structure called a Lie algebra. The Lie algebra retains most, but not quite all, of the properties of the original Lie group. Moreover, most of the Lie group properties can be recovered by the inverse of the linearization operation, carried out by the EXPonential mapping. Since the Lie algebra is a linear vector space, it can be studied using all the standard tools available for linear vector spaces. In particular, we can define convenient inner products and make standard choices of basis vectors. The properties of a Lie algebra in the neighborhood of the origin are identified with the properties of the original Lie group in the neighborhood of the identity. These structures, such as inner product and volume element, are extended over the entire group manifold using the group multiplication operation.

4.1 Why bother?

Two Lie groups are isomorphic if:

(i) their underlying manifolds are topologically equivalent;
(ii) the functions defining the group composition laws are equivalent.

Two manifolds are topologically equivalent if they can be smoothly deformed into each other. This requires that all their topological indices, such as dimension, Betti numbers, connectivity properties, etc., are equal.

Two group composition laws are equivalent if there is a smooth change of variables that deforms one function into the other.

Showing the topological equivalence of two manifolds is not necessarily an easy job. Showing the equivalence of two composition laws is typically a much more difficult task. It is difficult because the group composition law is generally nonlinear, and working with nonlinear functions is notoriously difficult.

The study of Lie groups would simplify greatly if the group composition law could somehow be linearized, and if this linearization retained a substantial part of the information inherent in the original group composition law. This in fact can be done.

Lie algebras are constructed by linearizing Lie groups.

A Lie group can be linearized in the neighborhood of any of its points, or group operations. Linearization amounts to Taylor series expansion about the coordinates that define the group operation. What is being Taylor expanded is the group composition function. This function can be expanded in the neighborhood of any group operation.

A Lie group is homogeneous – every point looks locally like every other point. This can be seen as follows. The neighborhood of group element a can be mapped into the neighborhood of group element b by multiplying a, and every element in its neighborhood, on the left by group element ba^{-1} (or on the right by $a^{-1}b$). This maps a into b and points near a into points near b.

It is therefore necessary to study the neighborhood of only one group operation in detail. Although geometrically all points are equivalent, algebraically one point is special – the identity. It is very useful and convenient to study the neighborhood of this special group element.

Linearization of a Lie group about the identity generates a new set of operators. These operators form a **Lie algebra**. A Lie algebra is a linear vector space, by virtue of the linearization process.

The composition of two group operations in the neighborhood of the identity reduces to vector addition. The construction of more complicated group products, such as the commutator, and the linearization of these products introduces additional structure in this linear vector space. This additional structure, the commutation relations, carries information about the original group composition law.

In short, the linearization of a Lie group in the neighborhood of the identity to form a Lie algebra brings about an enormous simplification in the study of Lie groups.

4.2 How to linearize a Lie group

We illustrate how to construct a Lie algebra for a Lie group in this section. The construction is relatively straightforward once an explicit parameterization of the underlying manifold and an expression for the group composition law is available. In particular, for the matrix groups the group composition law is matrix multiplication, and one can construct the Lie algebra immediately for the matrix Lie groups.

We carry this construction out for $SL(2;\mathbb{R})$. It is both customary and convenient to parameterize a Lie group so that the origin of the coordinate system maps to the identity of the group. Accordingly, we parameterize $SL(2; \mathbb{R})$ as follows

$$(a, b, c) \longrightarrow M(a, b, c) = \begin{bmatrix} 1+a & b \\ c & (1+bc)/(1+a) \end{bmatrix} \quad (4.1)$$

The group is linearized by investigating the neighborhood of the identity. This is done by allowing the parameters (a, b, c) to become infinitesimals and expanding the group operation in terms of these infinitesimals to first order

$$(a, b, c) \rightarrow (\delta a, \delta b, \delta c) \rightarrow M(\delta a, \delta b, \delta c)$$

$$= \begin{bmatrix} 1+\delta a & \delta b \\ \delta c & (1+\delta b \delta c)/(1+\delta a) \end{bmatrix} \quad (4.2)$$

The basis vectors in the Lie algebra are the coefficients of the first order infinitesimals. In the present case the basis vectors are 2×2 matrices

$$(\delta a, \delta b, \delta c) \rightarrow I_2 + \delta a X_a + \delta b X_b + \delta c X_c = \begin{bmatrix} 1+\delta a & \delta b \\ \delta c & 1-\delta a \end{bmatrix} \quad (4.3)$$

$$X_a = \begin{bmatrix} 1 & 0 \\ 0 & -1 \end{bmatrix} = \left.\frac{\partial M(a, b, c)}{\partial a}\right|_{(a,b,c)=(0,0,0)}$$

$$X_b = \begin{bmatrix} 0 & 1 \\ 0 & 0 \end{bmatrix} = \left.\frac{\partial M(a, b, c)}{\partial b}\right|_{(a,b,c)=(0,0,0)} \quad (4.4)$$

$$X_c = \begin{bmatrix} 0 & 0 \\ 1 & 0 \end{bmatrix} = \left.\frac{\partial M(a, b, c)}{\partial c}\right|_{(a,b,c)=(0,0,0)}$$

Lie groups that are isomorphic have Lie algebras that are isomorphic.

Remark The group composition function $\phi(x, y)$ is usually linearized in one of its arguments, say $\phi(x, y) \rightarrow \phi(x, 0 + \delta y)$. This generates a left-invariant vector field. The commutators of two left-invariant vector fields at a point x are independent of x, so that x can be taken in the neighborhood of the identity. It is for this reason that the linearization of the group in the neighborhood of the identity is so powerful.

4.3 Inversion of the linearization map: EXP

Linearization of a Lie group in the neighborhood of the identity to form a Lie algebra preserves the local group properties but destroys the global properties – that is, what happens far from the identity. It is important to know whether the linearization process can be reversed. Can one recover the Lie group from its Lie algebra?

To answer this question, assume X is some operator in a Lie algebra – such as a linear combination of the three matrices spanning the Lie algebra of $SL(2;\mathbb{R})$ given in (4.4). Then if ϵ is a small real number, $I + \epsilon X$ represents an element in the Lie group close to the identity. We can attempt to move far from the identity by iterating this group operation many times

$$\lim_{k\to\infty}\left(I + \frac{1}{k}X\right)^k = \sum_{n=0}^{\infty}\frac{X^n}{n!} = \text{EXP}(X) \tag{4.5}$$

The limiting and rearrangement procedures leading to this result are valid not only for real and complex numbers, but for $n \times n$ matrices and bounded operators as well.

Example We take an arbitrary vector X in the three-dimensional linear vector space of traceless 2×2 matrices spanned by the generators X_a, X_b, X_c of $SL(2;\mathbb{R})$ given in (4.4)

$$X = aX_a + bX_b + cX_c = \begin{bmatrix} a & b \\ c & -a \end{bmatrix} \tag{4.6}$$

The exponential of this matrix is

$$\text{EXP}(X) = \text{EXP}(aX_a + bX_b + cX_c) = \sum_{n=0}^{\infty}\frac{1}{n!}\begin{bmatrix} a & b \\ c & -a \end{bmatrix}^n = I_2 \cosh\theta + X\frac{\sinh\theta}{\theta}$$

$$= \begin{bmatrix} \cosh\theta + a\sinh(\theta)/\theta & b\sinh(\theta)/\theta \\ c\sinh(\theta)/\theta & \cosh\theta - a\sinh(\theta)/\theta \end{bmatrix}$$

$$\theta^2 = a^2 + bc \tag{4.7}$$

The actual computation can be carried out using either brute force or finesse.

With brute force, each of the matrices X^n is computed explicitly, a pattern is recognized, and the sum is carried out. The first few powers are $X^0 = I_2$, $X^1 = X$ (given in (4.6)), and $X^2 = \theta^2 I_2$. Since X^2 is a multiple of the identity, $X^3 = X^2X^1$ must be proportional to X ($= \theta^2 X$), X^4 is proportional to the identity, and so on.

Finesse involves use of the Cayley–Hamilton theorem, that every matrix satisfies its secular equation. This means that a 2×2 matrix must satisfy a polynomial equation of degree 2. Thus we can replace X^2 by a function of $X^0 = I_2$ and $X^1 = X$. Similarly, X^3 can be replaced by a linear combination of X^2 and X, and then X^2 replaced by I_2 and X. By induction, any function of the 2×2 matrix X can be written in the form

$$F(X) = f_0(a, b, c)X^0 + f_1(a, b, c)X^1 \tag{4.8}$$

Furthermore, the functions f_0, f_1 are not arbitrary functions of the three parameters (a, b, c), but rather functions of the invariants of the matrix X. These invariants are the coefficients of the secular equation. The only such invariant for the 2×2 matrix X is $\theta^2 = a^2 + bc$. As a result, we know from general and simple considerations that

$$\text{EXP}(X) = f_0(\theta^2) I_2 + f_1(\theta^2) X \qquad (4.9)$$

The two functions are $f_0(\theta^2) = 1 + \theta^2/2! + \theta^4/4! + \theta^6/6! + \cdots = \cosh \theta$ and $f_1(\theta^2) = 1 + \theta^2/3! + \theta^4/5! + \theta^6/7! + \cdots = \sinh(\theta)/\theta$. These arguments are applicable to the exponential of any matrix Lie algebra.

The EXPonential operation provides a natural parameterization of the Lie group in terms of linear quantities. This function maps the linear vector space – the Lie algebra – to the geometric manifold that parameterizes the Lie group. We can expect to find a lot of geometry in the EXPonential map.

Three important questions arise about the reversibility of the process represented by

$$\text{Lie group} \underset{\text{EXP}}{\overset{\ln}{\rightleftharpoons}} \text{Lie algebra} \qquad (4.10)$$

(i) Does the EXPonential function map the Lie algebra back onto the entire Lie group?
(ii) Are Lie groups with isomorphic Lie algebras themselves isomorphic?
(iii) Is the mapping from the Lie algebra to the Lie group unique, or are there other ways to parameterize a Lie group?

These are very important questions. In brief, the answer to each of these questions is "No." However, as is very often the case, exploring the reasons for the negative result produces more insight than a simple "Yes" response would have. They will be treated in more detail in Chapter 7.

4.4 Properties of a Lie algebra

We now turn to the properties of a Lie algebra. These are derived from the properties of a Lie group. A Lie algebra has three properties:

(i) the operators in a Lie algebra form a linear vector space;
(ii) the operators close under commutation: the commutator of two operators is in the Lie algebra;
(iii) the operators satisfy the Jacobi identity.

If X and Y are elements in the Lie algebra, then $g_1 = I + \epsilon X$ is an element in the Lie group near the identity for ϵ sufficiently small. In fact, so also is $I + \epsilon \alpha X$

for any real number α. We can form the product

$$(I + \epsilon\alpha X)(I + \epsilon\beta X) = I + \epsilon(\alpha X + \beta Y) + \text{higher order terms} \quad (4.11)$$

If X and Y are in the Lie algebra, then so is any linear combination of X and Y. The Lie algebra is therefore a linear vector space.

The commutator of two group elements is a group element:

$$\text{commutator of } g_1 \text{ and } g_2 \text{ is } g_1 g_2 g_1^{-1} g_2^{-1} \quad (4.12)$$

If X and Y are in the Lie algebra, then for any ϵ, δ sufficiently small, $g_1(\epsilon) = \text{EXP}(\epsilon X)$ and $g_1(\epsilon)^{-1} = \text{EXP}(-\epsilon X)$ are group elements near the identity, as are $g_2(\delta)^{\pm 1} = \text{EXP}(\pm \delta Y)$. Expanding the commutator to lowest order nonvanishing terms, we find

$$\text{EXP}(\epsilon X)\text{EXP}(\delta Y)\text{EXP}(-\epsilon X)\text{EXP}(-\delta Y)$$
$$= I + \epsilon\delta(XY - YX) = I + \epsilon\delta[X, Y] \quad (4.13)$$

Therefore, the commutator of two group elements, $g_1(\epsilon) = \text{EXP}(\epsilon X)$ and $g_2(\delta) = \text{EXP}(\delta Y)$, which is in the group G, requires the commutator of the operators X and Y, $[X, Y] = (XY - YX)$, to be in its Lie algebra $\mathfrak{g}$

$$g_1 g_2 g_1^{-1} g_2^{-1} \in G \Leftrightarrow [X, Y] \in \mathfrak{g} \quad (4.14)$$

The commutator (4.12) provides information about the structure of a group. If the group is commutative then the commutator in the group (4.12) is equal to the identity. The commutator in the algebra vanishes

$$g_1 g_2 g_1^{-1} g_2^{-1} = I \Rightarrow [X, Y] = 0 \quad (4.15)$$

If H is an invariant subgroup of G, then $g_1 H g_1^{-1} \subset H$. This means that if X is in the Lie algebra of G and Y is in the Lie algebra of H

$$g_1 H g_1^{-1} \in H \Rightarrow [X, Y] \in \text{Lie algebra of } H \quad (4.16)$$

If X, Y, Z are in the Lie algebra, then the **Jacobi identity** is satisfied

$$[X, [Y, Z]] + [Y, [Z, X]] + [Z, [X, Y]] = 0 \quad (4.17)$$

This identity involves the cyclic permutation of the operators in a double commutator. For matrices this identity can be proved by opening up the commutators ($[X, Y] = XY - YX$) and showing that the 12 terms so obtained cancel pairwise. This proof remains true when the operators X, Y, Z are not matrices but operators for which composition (e.g., XY is well defined, as are all other pairwise products) is defined. When operator products (as opposed to commutators) are not defined, this method of proof fails but the theorem (it is *not* an identity) remains true. This theorem represents an integrability condition on the functions that define the group multiplication operation on the underlying manifold.

To summarize, a Lie algebra $\mathfrak{g}$ has the following structure.

(i) It is a linear vector space under vector addition and scalar multiplication. If $X \in \mathfrak{g}$ and $Y \in \mathfrak{g}$ then every linear combination of X and Y is in $\mathfrak{g}$:

$$X \in \mathfrak{g}, \quad Y \in \mathfrak{g}, \quad \alpha X + \beta Y \in \mathfrak{g}$$

(ii) It is an algebra under commutation. If $X \in \mathfrak{g}$ and $Y \in \mathfrak{g}$ then their commutator is in $\mathfrak{g}$:

$$X \in \mathfrak{g}, \quad Y \in \mathfrak{g}, \quad [X, Y] \in \mathfrak{g}$$

This property is called "closure under commutation."

(iii) The Jacobi identity is satisfied. If $X \in \mathfrak{g}$, $Y \in \mathfrak{g}$, and $Z \in \mathfrak{g}$, then

$$[X, [Y, Z]] + [Y, [Z, X]] + [Z, [X, Y]] = 0$$

Example The three generators (4.4) of the Lie group $SL(2; \mathbb{R})$ obey the commutation relations

$$\begin{aligned}
[X_a, X_b] &= 2X_b \\
[X_a, X_c] &= -2X_c \\
[X_b, X_c] &= X_a
\end{aligned} \qquad (4.18)$$

It is an easy matter to verify that the Jacobi identity is satisfied for this Lie algebra.

4.5 Structure constants

Since a Lie algebra is a linear vector space we can introduce all the usual concepts of a linear vector space, such as dimension, basis, inner product. The dimension of the Lie algebra $\mathfrak{g}$ is equal to the dimension of the manifold that parameterizes the Lie group G. If the dimension is n, it is possible to choose n linearly independent vectors in the Lie algebra (a basis for the linear vector space) in terms of which any operator in $\mathfrak{g}$ can be expanded. If we call these basis vectors, or basis operators $X_1, X_2, \ldots, X_n$, then we can ask several additional questions such as: Is there a natural choice of basis vectors? Is there a reasonable definition of inner product (X_i, X_j)? We return to these questions shortly.

Since the linear vector space is closed under commutation, the commutator of any two basis vectors can be expressed as a linear superposition of basis vectors

$$[X_i, X_j] = C_{ij}{}^k X_k \qquad (4.19)$$

62 Lie algebras

The coefficients $C_{ij}{}^k$ in this expansion are called **structure constants**. The structure of the Lie algebra is completely determined by its structure constants. The antisymmetry of the commutator induces a corresponding antisymmetry in the structure constants

$$[X_i, X_j] + [X_j, X_i] = 0 \qquad C_{ij}{}^k + C_{ji}{}^k = 0 \tag{4.20}$$

Under a change of basis transformation

$$X_i = A_i{}^r Y_r \tag{4.21}$$

the structure constants change in a systematic way

$$C'_{rs}{}^t = (A^{-1})_r{}^i (A^{-1})_s{}^j C_{ij}{}^k A_k{}^t \tag{4.22}$$

(second order covariant, first order contravariant tensor). This piece of information is surprisingly useless.

Example The only nonzero structure constants for the three basis vectors X_a, X_b, X_c (4.4) in the Lie algebra $\mathfrak{sl}(2; R)$ for the Lie group $SL(2; \mathbb{R})$ are, from (4.18)

$$C_{ab}{}^b = -C_{ba}{}^b = +2 \qquad C_{ac}{}^c = -C_{ca}{}^c = -2 \qquad C_{bc}{}^a = -C_{cb}{}^a = +1 \tag{4.23}$$

4.6 Regular representation

A better way to look at a change of basis transformation is to determine how the change of basis affects the commutator of an arbitrary element Z in the algebra

$$[Z, X_i] = R(Z)_i{}^j X_j \tag{4.24}$$

Under the change of basis (4.21) we find

$$[Z, Y_r] = S(Z)_r{}^s Y_s \tag{4.25}$$

where

$$S_r{}^s(Z) = (A^{-1})_r{}^i R(Z)_i{}^j A_j{}^s \tag{4.26}$$

In this manner the effect of a change of basis on the structure constants is reduced to a study of similarity transformations.

The association of a matrix $R(Z)$ with each element of a Lie algebra is called the **regular representation**

$$Z \xrightarrow[\text{representation}]{\text{regular}} R(Z) \tag{4.27}$$

The regular representation of an n-dimensional Lie algebra is a set of $n \times n$ matrices. This representation contains exactly as much information as the structure constants, for the regular representation of a basis vector is

$$[X_i, X_j] = R(X_i)_j{}^k X_k = C_{ij}{}^k X_k \tag{4.28}$$

so that

$$R(X_i)_j{}^k = C_{ij}{}^k \tag{4.29}$$

The regular representation is an extremely useful tool for resolving a number of problems.

Example The regular representation of the Lie algebra $\mathfrak{sl}(2;\mathbb{R})$ is easily constructed, since the structure constants have been given in (4.23)

$$R(X) = R(aX_a + bX_b + cX_c) = aR(X_a) + bR(X_b) + cR(X_c)$$
$$= \begin{bmatrix} 0 & -2b & 2c \\ -c & 2a & 0 \\ b & 0 & -2a \end{bmatrix} \tag{4.30}$$

The rows and columns of this 3×3 matrix are labeled by the indices a, b and c, respectively.

4.7 Structure of a Lie algebra

The first step in the classification problem is to investigate the regular representation of the Lie algebra under a change of basis. We look for a choice of basis that brings the matrix representative of every element in the Lie algebra simultaneously to one of the three forms shown in Fig. 4.1. The first term (nonsemisimple, ...) is applied typically to Lie groups and algebras while the second term (reducible, ...) is typically applied to representations.

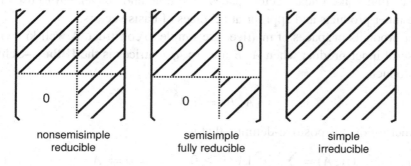

Figure 4.1. Standard forms into which a representation can be reduced.

Example It is not possible to reduce simultaneously the regular representatives of the three generators X_a, X_b, and X_c of $\mathfrak{sl}(2; \mathbb{R})$ to either the nonsemisimple or the semisimple form. This algebra is therefore simple. However, the Euclidean group $E(2)$ with structure

$$E(2) = \begin{bmatrix} \cos\theta & \sin\theta & t_1 \\ -\sin\theta & \cos\theta & t_2 \\ 0 & 0 & 1 \end{bmatrix} \tag{4.31}$$

has a Lie algebra with three infinitesimal generators

$$L_z = \begin{bmatrix} 0 & 1 & 0 \\ -1 & 0 & 0 \\ 0 & 0 & 0 \end{bmatrix} \quad P_x = \begin{bmatrix} 0 & 0 & 1 \\ 0 & 0 & 0 \\ 0 & 0 & 0 \end{bmatrix} \quad P_y = \begin{bmatrix} 0 & 0 & 0 \\ 0 & 0 & 1 \\ 0 & 0 & 0 \end{bmatrix} \tag{4.32}$$

and regular representation

$$R(\theta L_z + t_1 P_x + t_2 P_y) = \begin{bmatrix} 0 & -\theta & 0 \\ \theta & 0 & 0 \\ -t_2 & t_1 & 0 \end{bmatrix} \tag{4.33}$$

where the rows and columns are labeled successively by the basis vectors P_x, P_y, and L_z. This regular representation has the block diagonal structure of a nonsemisimple Lie algebra. The algebra, and the original group, are therefore nonsemisimple.

There is a beautiful structure theory for simple and semisimple Lie algebras. This will be discussed in Chapter 9. A structure theory exists for nonsemisimple Lie algebras. It is neither as beautiful nor as complete as the structure theory for simple Lie algebras.

4.8 Inner product

Since a Lie algebra is a linear vector space, we are at liberty to impose on it all the structures that make linear vector spaces so simple and convenient to use. These include inner products and appropriate choices of basis vectors.

Inner products in spaces of matrices are simple to construct. A well-known and very useful inner product when A, B are $p \times q$ matrices is the **Hilbert–Schmidt inner product**

$$(A, B) = \operatorname{tr} A^\dagger B \tag{4.34}$$

This inner product is positive-definite, that is

$$(A, A) = \sum_i \sum_j |A_i{}^j|^2 \geq 0 \qquad = 0 \Rightarrow A = 0 \tag{4.35}$$

4.8 Inner product

If we were to adopt the Hilbert–Schmidt inner product on the regular representation of $\mathfrak{g}$, then

$$(X_i, X_j) = \operatorname{tr} R(X_i)^\dagger R(X_j) = \sum_r \sum_s R(X_i)_r^{s*} R(X_j)_r^s = \sum_r \sum_s C_{ir}^{s*} C_{jr}^s \qquad (4.36)$$

This inner product is positive-semidefinite on $\mathfrak{g}$: it vanishes identically on those generators that commute with all operators in the Lie algebra (X_i, where $C_{i\star}^{\ \ *} = 0$) and also on all generators that are not representable as the commutator of two generators (X_i, where $C_{\star\star}^i = 0$).

The Hilbert–Schmidt inner product is a reasonable choice of inner product from an algebraic point of view. However, there is an even more useful choice of inner product that provides both algebraic and geometric information. This is defined by

$$(X_i, X_j) = \operatorname{Tr} R(X_i) R(X_j) = \sum_r \sum_s R(X_i)_r^s R(X_j)_s^r = \sum_r \sum_s C_{ir}^{\ s} C_{js}^{\ r} \qquad (4.37)$$

This inner product is called the **Cartan–Killing inner product**, or **Cartan–Killing form**. It is in general an indefinite inner product. It is used extensively in the classification theory of Lie algebras.

The Cartan–Killing metric can be used to advantage to make further refinements on the structure theory of a Lie algebra. The vector space of the Lie algebra can be divided into three subspaces under the Cartan–Killing inner product. The inner product is positive-definite, negative-definite, and identically zero on these three subspaces:

$$\mathfrak{g} = V_+ + V_- + V_0 \qquad (4.38)$$

The subspace V_0 is a subalgebra of $\mathfrak{g}$. It is the largest nilpotent invariant subalgebra of $\mathfrak{g}$. Under exponentiation, this subspace maps onto the maximal nilpotent invariant subgroup in the original Lie group.

The subspace V_- is also a subalgebra of $\mathfrak{g}$. It consists of compact (a topological property) operators. That is to say, the exponential of this subspace is a subset of the original Lie group that is parameterized by a compact manifold. It also forms a subalgebra in $\mathfrak{g}$ (not invariant).

Finally, the subspace V_+ is not a subalgebra of $\mathfrak{g}$. It consists of noncompact operators. The exponential of this subspace is parameterized by a noncompact submanifold in the original Lie group.

In short, a Lie algebra has the following decomposition under the Cartan–Killing inner product

$$\mathfrak{g} \xrightarrow[\text{inner product}]{\text{Cartan–Killing}} \begin{cases} V_0 & \text{nilpotent invariant subalgebra} \\ V_- & \text{compact subalgebra} \\ V_+ & \text{noncompact operators} \end{cases} \quad (4.39)$$

We return to the structure of Lie algebras in Chapter 8 and the classification of simple Lie algebras in Chapter 10.

Example The Cartan–Killing inner product on the regular representation (4.30) of $\mathfrak{sl}(2; R)$ is

$$(X, X) = \operatorname{tr} R(X)R(X) = \operatorname{tr} \begin{bmatrix} 0 & -2b & 2c \\ -c & 2a & 0 \\ b & 0 & -2a \end{bmatrix}^2 = 8(a^2 + bc) \quad (4.40)$$

From this we easily drive the form of the metric for the Cartan–Killing inner product:

$$8(a^2 + bc) = \begin{pmatrix} a & b & c \end{pmatrix} \begin{bmatrix} 8 & 0 & 0 \\ 0 & 0 & 4 \\ 0 & 4 & 0 \end{bmatrix} \begin{pmatrix} a \\ b \\ c \end{pmatrix} \quad (4.41)$$

A convenient choice of basis vectors is one that diagonalizes this metric matrix: X_a and $X_\pm = X_b \pm X_c$. In this basis the metric matrix is

$$\begin{bmatrix} 8 & 0 & 0 \\ 0 & 8 & 0 \\ 0 & 0 & -8 \end{bmatrix} \begin{matrix} X_a \\ X_+ \\ X_- \end{matrix} \quad (4.42)$$

In this representation it is clear that the operator X_- spans a one-dimensional compact subalgebra in $\mathfrak{sl}(2; \mathbb{R})$ and the generators X_a, X_+ are noncompact.

We should point out here that the inner product can also be computed even more simply in the defining 2×2 matrix representation of $\mathfrak{sl}(2; \mathbb{R})$

$$(X, X) = \operatorname{tr} \begin{bmatrix} a & b \\ c & -a \end{bmatrix}^2 = 2(a^2 + bc) \quad (4.43)$$

This gives an inner product that is proportional to the inner product derived from the regular representation. This is not an accident, and this observation can be used to compute the Cartan–Killing inner products very rapidly for all matrix Lie algebras.

4.9 Invariant metric and measure on a Lie group

The properties of a Lie algebra can be identified with the properties of the corresponding Lie group at the identity.

4.9 Invariant metric and measure on a Lie group

Once the properties of a Lie group have been determined in the neighborhood of the identity, these properties can be translated to the neighborhood of any other group operation. This is done by multiplying the identity and its neighborhood on the left (or right) by that group operation.

Two properties that are useful to define over the entire manifold are the **metric** and **measure**. We assume the coordinates of the identity are $(\alpha^1, \alpha^2, \ldots, \alpha^n)$ and the coordinates of a point near the identity are $(\alpha^1 + d\alpha^1, \alpha^2 + d\alpha^2, \ldots, \alpha^n + d\alpha^n)$. If $(x^1, x^2, \ldots, x^n)$ represents some other group operation, then the point $(\alpha^1 + d\alpha^1, \alpha^2 + d\alpha^2, \ldots, \alpha^n + d\alpha^n)$ is mapped to the point $(x^1 + dx^1, x^2 + dx^2, \ldots, x^n + dx^n)$ under left (right) multiplication by the group operation associated with $(x^1, x^2, \ldots, x^n)$. The displacements dx and $d\alpha$ are related by a position-dependent *linear* transformation

$$dx^r = M(x)^r{}_i d\alpha^i \qquad (4.44)$$

Suppose now that the distance ds between the identity and a point with coordinates $\alpha^i + d\alpha^i$ infinitesimally close to the identity is given by

$$ds^2 = g_{ij}(\text{Id}) d\alpha^i d\alpha^j \qquad (4.45)$$

Any metric can be chosen at the identity, but the most usual choice is the Cartan–Killing inner product. Can we define a metric at x, $g_{rs}(x)$, with the property that the arc length is an invariant?

$$g_{rs}(x) dx^r dx^s = g_{ij}(\text{Id}) d\alpha^i d\alpha^j \qquad (4.46)$$

In order to enforce the invariance condition, the metric at x, $g(x)$, must be related to the metric at the identity by

$$g(x) = M^{-1}(x)^t g(\text{Id}) M^{-1}(x) \qquad (4.47)$$

The volume elements at the identity and x are

$$dV(\text{Id}) = d\alpha^1 \wedge d\alpha^2 \wedge \cdots \wedge d\alpha^n$$
$$dV(x) = dx^1 \wedge dx^2 \wedge \cdots \wedge dx^n = \| M \| d\alpha^1 \wedge d\alpha^2 \wedge \cdots \wedge d\alpha^n \qquad (4.48)$$

The two volume elements can be made equal by introducing a measure over the manifold and defining an invariant volume

$$d\mu(x) = \rho(x) dV(x) = \rho(x) \| M(x) \| dV(\text{Id}) \Rightarrow \rho(x) = \| M(x) \|^{-1} \qquad (4.49)$$

Example Under the simple parameterization (4.1) of the group $SL(2; \mathbb{R})$ the neighborhood of the identity is parameterized by (4.3). We move a neighborhood of the identity to the neighborhood of the group operation parameterized by (x, y, z)

using left multiplication as follows

$$\begin{bmatrix} 1+x & y \\ z & 1+yz \\ & \overline{1+x} \end{bmatrix} \times \begin{bmatrix} 1+d\alpha^1 & d\alpha^2 \\ d\alpha^3 & 1-d\alpha^1 \end{bmatrix}$$

$$= \begin{bmatrix} 1+(x+dx) & y+dy \\ z+dz & \dfrac{1+(y+dy)(z+dz)}{1+(x+dx)} \end{bmatrix}$$

$$= \begin{bmatrix} (1+x)(1+d\alpha^1)+y\,d\alpha^3 & (1+x)d\alpha^2+y(1-d\alpha^1) \\ z(1+d\alpha^1)+\dfrac{(1+yz)d\alpha^3}{(1+x)} & zd\alpha^2+\dfrac{(1+yz)(1-d\alpha^1)}{(1+x)} \end{bmatrix} \quad (4.50)$$

The linear relation between the infinitesimals $(d\alpha^1, d\alpha^2, d\alpha^3)$ in the neighborhood of the identity and the infinitesimals (dx, dy, dz) in the neighborhood of the group operation (x, y, z) can now be read off, matrix element by matrix element

$$\begin{bmatrix} dx \\ dy \\ dz \end{bmatrix} = \begin{bmatrix} 1+x & 0 & y \\ -y & 1+x & 0 \\ z & 0 & \dfrac{1+yz}{1+x} \end{bmatrix} \begin{bmatrix} d\alpha^1 \\ d\alpha^2 \\ d\alpha^3 \end{bmatrix} \quad (4.51)$$

From this linear transformation we immediately compute the invariant measure by taking the inverse of the determinant

$$d\mu(x) = \rho(x, y, z)dx \wedge dy \wedge dz = \frac{dx \wedge dy \wedge dz}{1+x} \quad (4.52)$$

The invariant metric is somewhat more difficult, as it involves computing the inverse of the linear transformation (4.51). The result is

$$\begin{bmatrix} 2 & 0 & 0 \\ 0 & 0 & 1 \\ 0 & 1 & 0 \end{bmatrix} \xrightarrow[\text{by }(x,y,z)]{\text{left translation}} \begin{bmatrix} \dfrac{2(1+yz)}{(1+x)^2} & -\dfrac{z}{(1+x)} & -\dfrac{y}{(1+x)} \\ -\dfrac{z}{(1+x)} & 0 & 1 \\ -\dfrac{y}{(1+x)} & 1 & 0 \end{bmatrix} \quad (4.53)$$

The invariant measure (4.52) can be derived from the invariant metric (4.53) in the usual way (see Problem 4.11).

4.10 Conclusion

The structure that results from the linearization of a Lie group is called a Lie algebra. Lie algebras are linear vector spaces. They are endowed with an additional combinatorial operation, the commutator $[X, Y] = (XY - YX)$, and obey the Jacobi identity. Since they are linear vector spaces, many powerful tools are available for their study. It is possible to define an inner product that reflects not only the algebraic properties of the original Lie group, but its topological properties as well. The properties of a Lie algebra can be identified with the properties of the parent Lie group in the neighborhood of the identity. These structures can be moved to neighborhoods of other points in the group manifold by a suitable group multiplication.

The linearization procedure is more or less invertible (a little less than more). The inversion is carried out by the EXPonential mapping.

4.11 Problems

1. Carry out the commutator calculation for $g_1 = (I + \epsilon X)$, $g_1^{-1} = (I + \epsilon X)^{-1} = I - \epsilon X + \epsilon^2 X^2 - \cdots$, with similar expressions for g_2, to obtain the same result as in (4.13). In other words, this local result is independent of the parameterization in the neighborhood of the identity.

2. The inner product of two vectors X and X' in a linear vector space can be computed if the inner product of a vector with itself is known. This is done by the method of **polarization**. For a real linear vector space the argument is as follows:

$$(X', X) = \frac{1}{2}\left[(X' + X, X' + X) - (X', X') - (X, X)\right]$$

   a. Verify this.
   b. Extend to complex linear vector spaces.
   c. Use the result from Eq. (4.43) that $(X, X) = 2(a^2 + bc)$ to show

$$(X', X) = 2a'a + b'c + c'b$$

3. Suppose that the $n \times n$ matrix Y is defined as the exponential of an $n \times n$ matrix X in a Lie algebra: $Y = e^X$. Show that "for Y sufficiently close to the identity" the matrix X can be expressed as

$$X = -\sum_{n=1}^{\infty} \frac{(I - Y)^n}{n} \qquad (4.54)$$

   Show that this expansion converges when X and Y are symmetric if the real eigenvalues λ_i of Y all satisfy $0 < \lambda_i < +2$. Show that if $Y \in SL(2; \mathbb{R})$ and tr $Y < -2$ this expansion does not converge. That is, there is no 2×2 matrix $X \in \mathfrak{sl}(2; \mathbb{R})$ with the property tr $e^X < -2$.

4. The Lie algebra of $SO(3)$ is spanned by three 3×3 antisymmetric matrices $\mathbf{L} = (L_1, L_2, L_3) = (X_{23}, X_{31}, X_{12})$, with

$$\theta \cdot \mathbf{L} = \begin{bmatrix} 0 & \theta_3 & -\theta_2 \\ -\theta_3 & 0 & \theta_1 \\ \theta_2 & -\theta_1 & 0 \end{bmatrix} = \begin{bmatrix} 0 & \theta_{12} & \theta_{13} \\ \theta_{21} & 0 & \theta_{23} \\ \theta_{31} & \theta_{32} & 0 \end{bmatrix} = \mathbf{X} \quad (4.55)$$

Use the Cayley–Hamilton theorem to show

$$e^{\theta \cdot \mathbf{L}} = I_3 f_0(\theta) + \mathbf{X} f_1(\theta) + \mathbf{X}^2 f_2(\theta) \quad (4.56)$$

where θ is the single invariant that can be constructed from the matrix $\mathbf{X} = \theta \cdot \mathbf{L}$: $\theta^2 = \theta_1^2 + \theta_2^2 + \theta_3^2$. Show

$$\begin{aligned} f_0(\theta) &= \cos\theta \\ f_1(\theta) &= \sin(\theta)/\theta \\ f_2(\theta) &= (1 - \cos(\theta))/\theta^2 \end{aligned} \quad \text{or} \quad \begin{aligned} f_0(\theta) &= \cos\theta \\ \theta f_1(\theta) &= \sin(\theta) \\ \theta^2 f_2(\theta) &= 1 - \cos(\theta) \end{aligned}$$

5. The Lie algebra for the matrix group $SO(n)$ consists of antisymmetric $n \times n$ matrices. Show that a useful set of basis vectors (matrices) consists of the $n(n-1)/2$ matrices $X_{ij} = -X_{ji}$ ($1 \leq i \neq j \leq n$) with matrix elements $(X_{ij})_{\alpha\beta} = \delta_{i\alpha}\delta_{j\beta} - \delta_{i\beta}\delta_{j\alpha}$.
   a. Show that these matrices satisfy the commutation relations

   $$[X_{ij}, X_{rs}] = X_{is}\delta_{jr} + X_{jr}\delta_{is} - X_{ir}\delta_{js} - X_{js}\delta_{ir} \quad (4.57)$$

   b. Show that the operators $\mathcal{X}_{ij} = x^i \partial_j - x^j \partial_i$ satisfy isomorphic commutation relations.
   c. Show that bilinear products of boson creation and annihilation operators $\mathcal{B}_{ij} = b_i^\dagger b_j - b_j^\dagger b_i$ ($1 \leq i \neq j \leq n$) satisfy isomorphic commutation relations.
   d. Show that bilinear products of fermion creation and annihilation operators $\mathcal{F}_{ij} = f_i^\dagger f_j - f_j^\dagger f_i$ ($1 \leq i \neq j \leq n$) satisfy isomorphic commutation relations.

6. The Jacobi identity for operators D, Y, Z (replace X by D in Eq. (4.17)) can be rewritten in the form

$$[D, [Y, Z]] = [[D, Y], Z] + [Y, [D, Z]] \quad (4.58)$$

Show this. Compare with the expression for the differential operator

$$d(f \wedge g) = (df) \wedge g + f \wedge (dg)$$

It is for this reason that the Jacobi identity is sometimes called a differential identity.

7. For the matrix Lie algebra $\mathfrak{so}(4)$ the defining matrix representation consists of 4×4 antisymmetric matrices while the regular representation consists of 6×6 antisymmetric matrices. Construct the defining and regular matrix representations for the

element $a_{ij}X_{ij}$ in the Lie algebra:

$$X = \sum_{ij} a_{ij} X_{ij} \rightarrow \begin{cases} \mathfrak{def}(X) = \sum a_{ij} \mathfrak{def}(X_{ij}) \\ \mathfrak{reg}(X) = \sum a_{ij} \mathfrak{reg}(X_{ij}) \end{cases} \quad (4.59)$$

Construct the Cartan–Killing inner product using these two different matrix representations:

$$\operatorname{tr} \mathfrak{def}(X)\mathfrak{def}(X) \leftarrow (X, X) \rightarrow \operatorname{tr} \mathfrak{reg}(X)\mathfrak{reg}(X) \quad (4.60)$$

Show that the two inner products are proportional. What is the proportionality constant? How does this result extend to $SO(n)$? to $SO(p, q)$? Is there a simple relation between the proportionality constant and the dimensions of the defining and regular representations?

8. Assume a Lie algebra of $n \times n$ matrices is noncompact and its Cartan–Killing form splits this Lie algebra into three subspaces:

$$\mathfrak{g} \rightarrow V_0 + V_- + V_+$$

Show that the subspace V_- exponentiates onto a compact manifold. Do this by showing that the basis matrices in V_- have eigenvalues that are imaginary or zero, so that $\operatorname{EXP}(V_-)$ is multiply periodic. Apply this construction to the noncompact groups $SO(3, 1)$ and $SO(2, 2)$. Show $\operatorname{EXP}(V_-)$ is a two-sphere S^2 for $SO(3, 1)$ and a two-torus T^2 for $SO(2, 2)$.

9. Construct the infinitesimal generators for the group $SO(3)$ using the parameterizations proposed in Problems 13 and 14 in Chapter 3.

10. Use the exponential parameterization of $SO(3)$ to construct the linear transformation M (Eq. 4.44) describing displacements from the identity to displacements at the group operator $e^{\theta \cdot L} \in SO(3)$. From this construct the invariant density $\rho(\theta)$ and the metric tensor $g_{\mu\nu}(\theta)$. Give a reason for the strange behavior (singularities) that these invariant quantities exhibit.

11. Compute the determinant of the metric tensor (4.53) on the group $SL(2; \mathbb{R})$. Show that the square root of the determinant is equal to the measure, in accordance with the standard result of Riemannian geometry that $dV(x) = \| g(x) \|^{1/2} d^n x$. Discuss the additional factors of 2 and -1 that appear in this calculation.

12. An inner product is $(\mathbf{x}, \mathbf{x})$ is imposed on a real n-dimensional linear vector space. It is represented by a real symmetric nonsingular $n \times n$ matrix $g_{rs} = (\mathbf{e}_r, \mathbf{e}_s)$, where $\mathbf{x} = \mathbf{e}_i x^i$. The inverse matrix, g^{rs}, is well defined.
  a. Lie group G preserves inner products. If $\mathbf{y} = G\mathbf{x}$, $(\mathbf{y}, \mathbf{y}) = (\mathbf{x}, \mathbf{x})$. Show $G^t g G = g$.
  b. Show the Lie algebra H of G satisfies $H^t g + g H = 0$.
  c. Show that the infinitesimal generators of G are $X_{rs} = g_{rt} x^t \partial_s - g_{st} x^t \partial_r$.

72 Lie algebras

d. Show that the operators X_{rs} satisfy commutation relations

$$[X_{ij}, X_{rs}] = +X_{is}g_{jr} + X_{jr}g_{is} - X_{ir}g_{js} + X_{js}g_{ir}$$

13. Every real unimodular 2×2 matrix M can be written in the form $M = SO$, where S is a real symmetric unimodular matrix and O is a real orthogonal matrix.

In group		Relation	In algebra	
$S^t = S^{+1}$	$\det(S) = +1$	$S = e^\Sigma$	$\operatorname{Tr} \Sigma = 0$	$\Sigma^t = +\Sigma$
$O^t = O^{-1}$	$\det(O) = +1$	$O = e^A$	$\operatorname{Tr} A = 0$	$A^t = -A$

a. Show that $MM^t = S^2 = e^{2\Sigma}$.
b. Show $O = S^{-1}M = e^{-\Sigma}M$.
c. Write S as a power series expansion in Σ.
d. Write Σ as a power series expansion in $S - I_2$.
Under what conditions are these expansions valid?

14. Extend the result of the previous problem to complex $n \times n$ matrices $M = HU$, with M arbitrary but nonsingular, $H^\dagger = H^{+1}$ hermitian and $U^\dagger = U^{-1}$ unitary.

15. Transfer matrices have been described in Chapter 3, Problem 24. In one dimension the transfer matrix for a scattering potential, with free particles incident from the left or right with momentum $\hbar k_L$ or $-\hbar k_R$, has the form (Gilmore, 2004)

$$\begin{bmatrix} \alpha_R + i\alpha_I & \beta_R + i\beta_I \\ \beta_R - i\beta_I & \alpha_R - i\alpha_I \end{bmatrix} \tag{4.61a}$$

The matrix elements are given explicitly by

$$2\alpha_R = +m_{11} + \frac{k_R}{k_L}m_{22} \qquad 2\alpha_I = +m_{12}k_R - k_L^{-1}m_{21}$$

$$2\beta_R = +m_{11} - \frac{k_R}{k_L}m_{22} \qquad 2\beta_I = -m_{12}k_R - k_L^{-1}m_{21} \tag{4.61b}$$

The real quantities m_{ij} are the four matrix elements of a group operation in $SL(2; \mathbb{R})$. They are energy dependent. By appropriate choice of $\hbar k_L = \hbar k_R$ and the matrix elements m_{ij}, construct three infinitesimal generators for the group of the transfer matrix for scattering states. Show that they are

$$\begin{bmatrix} i & 0 \\ 0 & -i \end{bmatrix} \quad \begin{bmatrix} 0 & i \\ -i & 0 \end{bmatrix} \quad \begin{bmatrix} 0 & -1 \\ -1 & 0 \end{bmatrix} \tag{4.61c}$$

Show that these three matrices span the Lie algebra of the group $SU(1, 1)$.

16. The transfer matrix for a potential that possesses bound states has the form

$$\begin{bmatrix} \alpha_1 + \alpha_2 & \beta_1 + \beta_2 \\ \beta_1 - \beta_2 & \alpha_1 - \alpha_2 \end{bmatrix} \tag{4.62a}$$

4.11 Problems

The matrix elements are given explicitly by

$$2\alpha_1 = +m_{11} + \frac{\kappa_R}{\kappa_L}m_{22} \qquad 2\alpha_2 = -m_{12}\kappa_R - \kappa_L^{-1}m_{21}$$

$$2\beta_1 = +m_{11} - \frac{\kappa_R}{\kappa_L}m_{22} \qquad 2\beta_2 = +m_{12}\kappa_R - \kappa_L^{-1}m_{21}$$

(4.62b)

The parameters κ_R and κ_L describe the decay length of the exponentially decaying wavefunction in the asymptotic left- and right-hand regions of the potential. The real quantities m_{ij} are the four matrix elements of a group operation in $SL(2;\mathbb{R})$. They are energy dependent. By appropriate choice of $\kappa_L = \kappa_R$ and the matrix elements m_{ij}, construct the infinitesimal generators for the group of the transfer matrix for bound states. Show that they are

$$\begin{bmatrix} 1 & 0 \\ 0 & -1 \end{bmatrix} \quad \begin{bmatrix} 0 & 1 \\ -1 & 0 \end{bmatrix} \quad \begin{bmatrix} 0 & 1 \\ 1 & 0 \end{bmatrix} \qquad (4.62\text{c})$$

Show that these three matrices span the Lie algebra of the group $SL(2;\mathbb{R})$. Argue that there ought to be interesting relations (e.g., analytic continuations) between the scattering states (e.g., resonances) and bound states through the relation between the groups $SL(2;\mathbb{R})$ and $SU(1,1)$, which are isomorphic. How are the matrices (4.61a) and (4.62a), the matrix elements (4.61b) and (4.62b), and the infinitesimal generators (4.61c) and (4.62c) related to each other by analytic continuation? (Hint: $k_* = \sqrt{2m(E - V_*)/\hbar^2}$ for $E > V_*$ and $\kappa_* = \sqrt{2m(V_* - E)/\hbar^2}$ for $E < V_*$, $* = L, R$.)

5
Matrix algebras

The Lie algebras of the matrix Lie groups described in Chapter 3 are constructed. This is done by linearizing the constraints defining these matrix groups in the neighborhood of the identity operation.

5.1 Preliminaries

Lie algebras for the matrix groups treated in Chapter 3 are computed by linearizing the defining conditions in the neighborhood of the identity. The general linear groups $GL(n; \mathbb{F})$ have no defining condition (the only condition is $\det(M) \neq 0$), while Examples (2)–(7) are already defined by linear constraints. Examples (8)–(11) are defined by bilinear and quadratic constraints that are linearized by applying the constraint to matrices infinitesimally close to the identity: $I + \epsilon A$. The matrices in the Lie algebra are subject to easily derived linear constraints:

$$(I + \epsilon A)^\dagger G (I + \epsilon A) = G$$
$$G + \epsilon(A^\dagger G + GA) + \mathcal{O}(\epsilon^2) = G \qquad (5.1)$$
$$A^\dagger G + GA = 0$$

The special linear groups are defined by the n-linear constraint

$$\det(I + \epsilon A) = 1 + \epsilon \, \text{tr}(A) + \mathcal{O}(\epsilon^2) = 1$$
$$\text{tr}(A) = 0 \qquad (5.2)$$

The matrix Lie algebras of the matrix Lie groups given in Chapter 3 are summarized below.

5.2 No constraints

1. $\mathfrak{gl}(n; \mathbb{F})$. This algebra consists of arbitrary $n \times n$ matrices over the field $\mathbb{F}$. All remaining matrix algebras in this list are subalgebras of $\mathfrak{gl}(n; \mathbb{F})$.

5.3 Linear constraints

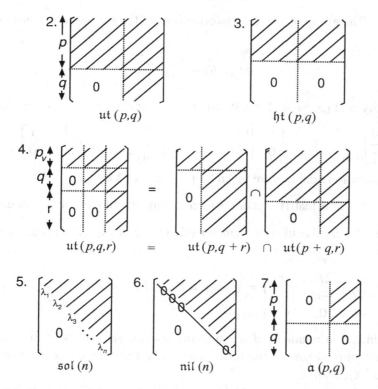

Figure 5.1. Structure of the matrix algebras for groups defined by linear constraints.

5.3 Linear constraints

The Lie algebras of the matrix groups have the same structures as the matrix groups. The only difference is that matrix elements that are constrained to be $+1$ in the groups are replaced by 0 in the algebra. All matrix algebras of matrix groups defined by linear constraints are summarized in Fig. 5.1.

2. $\mathfrak{ut}(p, q)$. Upper triangular algebras. The matrix algebra has the same structure as the group $UT(p, q)$:

$$m_{i\alpha} = 0 \qquad \begin{matrix} p+1 & \leq i & \leq p+q \\ 1 & \leq \alpha & \leq p \end{matrix} \qquad (5.3)$$

3. $\mathfrak{ht}(p, q)$. The algebra for this class of groups is defined by the condition

$$m_{ij} = 0 \qquad \begin{matrix} p+1 & \leq i & \leq p+q \\ 1 & \leq j & \leq p+q \end{matrix} \qquad (5.4)$$

Example The group of affine transformations of the straight line consists of matrices

$$M(a, b) = \begin{bmatrix} a & b \\ 0 & 1 \end{bmatrix} \quad (5.5)$$

The identity is at $(a, b) = (1, 0)$. Its algebra is spanned by the two operators

$$\left.\frac{\partial M}{\partial a}\right|_{(1,0)} = X_a = \begin{bmatrix} 1 & 0 \\ 0 & 0 \end{bmatrix} \qquad \left.\frac{\partial M}{\partial b}\right|_{(1,0)} = X_b = \begin{bmatrix} 0 & 1 \\ 0 & 0 \end{bmatrix} \quad (5.6)$$

The commutation relations are given by $[X_a, X_b] = X_b$.

4. $\mathfrak{ut}(p, q, r)$. This matrix algebra is identical in structure to the parent group.

Example A very useful six-parameter subalgebra of $\mathfrak{ut}(1, 2, 1)$ is given by

$$\begin{bmatrix} 0 & l & r & -2\delta \\ 0 & \eta & 2R & -r \\ 0 & -2L & -\eta & l \\ 0 & 0 & 0 & 0 \end{bmatrix} = \eta X_\eta + R X_R + L X_L + r X_r + l X_l + \delta X_\delta \quad (5.7)$$

The commutation relations of the six infinitesimal generators of this matrix Lie algebra are summarized in the table below. The operator in the ith row and jth column is $[X_i, X_j]$.

	X_η	X_R	X_L	X_r	X_l	X_δ
X_η	0	$2X_R$	$-2X_L$	X_r	$-X_l$	0
X_R		0	$-4X_\eta$	0	$-2X_r$	0
X_L			0	$2X_l$	0	0
X_r				0	$-X_\delta$	0
X_l					0	0
X_δ						0

	$\hat{n}+\tfrac{1}{2}I$	$a^\dagger a^\dagger$	aa	$a^\dagger$	a	I
$\hat{n}+\tfrac{1}{2}I$	0	$2a^\dagger a^\dagger$	$-2aa$	$a^\dagger$	$-a$	0
$a^\dagger a^\dagger$		0	$-4(\hat{n}+\tfrac{1}{2}I)$	0	$-2a^\dagger$	0
aa			0	$2a$	0	0
$a^\dagger$				0	$-I$	0
a					0	0
I						0

(5.8)

The table inherits the antisymmetry of the commutator, so only one half has been presented. It is clear that there is an isomorphism between this matrix algebra and the algebra of the photon energy operator $\hat{n} = a^\dagger a + \tfrac{1}{2}$, two-photon creation

5.3 Linear constraints

and annihilation operators $a^\dagger a^\dagger$ and aa, single-photon creation and annihilation operators $a^\dagger$ and a, and the identity operator $I = [a, a^\dagger]$. We observe that the 4×4 matrix X_δ representing the operator $I = [a, a^\dagger]$ is not diagonal. It need not be, as long as it obeys the appropriate commutation relations.

5. $\mathfrak{sol}(n) = \mathfrak{ut}(1, 1, 1, \ldots, 1)$. This matrix algebra is also identical in structure to its parent group. A very useful four-parameter subalgebra of $\mathfrak{ut}(1, 1, 1)$ is given by matrices of the following form

$$\begin{bmatrix} 0 & l & \delta \\ 0 & \eta & r \\ 0 & 0 & 0 \end{bmatrix} = \eta X_\eta + l X_l + r X_r + \delta X_\delta \qquad (5.9)$$

The following commutation properties are easily verified

$$\begin{aligned}
{[X_\eta, X_r]} &= +X_r & [a^\dagger a, a^\dagger] &= +a^\dagger \\
[X_\eta, X_l] &= -X_l & [a^\dagger a, a] &= -a \\
[X_l, X_r] &= X_\delta & [a, a^\dagger] &= I \\
[X_\eta, X_\delta] &= 0 & [a^\dagger a, I] &= 0
\end{aligned} \qquad (5.10)$$

6. $\mathfrak{nil}(n)$. Nilpotent matrices have an upper triangular structure, with $+1$ along the diagonal in the group and zeroes along the diagonal in the algebra. The three generators of the algebra of nilpotent 3×3 matrices have structure and commutation relations

$$\begin{bmatrix} 0 & l & \delta \\ 0 & 0 & r \\ 0 & 0 & 0 \end{bmatrix} = l X_l + r X_r + \delta X_\delta \qquad (5.11)$$

$$\begin{aligned}
{[X_l, X_r]} &= X_\delta & [a, a^\dagger] &= I \\
[X_l, X_\delta] &= 0 & [a, I] &= 0 \\
[X_r, X_\delta] &= 0 & [a^\dagger, I] &= 0
\end{aligned} \qquad (5.12)$$

These commutation relations are isomorphic to Heisenberg commutation relations. This is easily seen by setting $\eta = 0$ in (5.9). As a result, a number of difficult computations involving this algebra can be replaced by much simpler computations involving only 3×3 matrices.

7. $\mathfrak{a}(p, q)$. The matrix algebra for the commutative group of Example (7) in Chapter 3 (see (3.13)) consists of matrices having the form shown in Fig. 5.1.

5.4 Bilinear and quadratic constraints

The nonlinear constraints that define the metric-preserving matrix Lie groups are easily converted to linear constraints that define their Lie algebras following the procedure described in (5.1) above.

8. Compact metric-preserving groups Matrices M for the algebras of these groups satisfy

$$M^\dagger + M = 0 \quad \begin{array}{ll} \mathbb{R} & \mathfrak{o}(n) \quad \text{orthogonal} \\ \mathbb{C} & \mathfrak{u}(n) \quad \text{unitary} \\ \mathbb{Q} & \mathfrak{sp}(n) \quad \text{symplectic} \end{array} \quad (5.13)$$

The algebras of the orthogonal, unitary, and symplectic groups consist of $n \times n$ antihermitian matrices. The Lie algebras for the groups $O(3)$ and $U(2)$ are

$$\mathfrak{o}(3) = \begin{bmatrix} 0 & \theta_3 & -\theta_2 \\ -\theta_3 & 0 & \theta_1 \\ \theta_2 & -\theta_1 & 0 \end{bmatrix} = \sum_i \theta_i L_i$$

$$\mathfrak{u}(2) = \frac{1}{2} \begin{bmatrix} ix_0 + ix_3 & ix_1 + x_2 \\ ix_1 - x_2 & ix_0 - ix_3 \end{bmatrix} = \frac{i}{2} \sum_\mu x_\mu \sigma_\mu \quad (5.14)$$

The four 2×2 matrices σ_μ are called **Pauli spin matrices**.

9. Noncompact metric-preserving groups Matrices M for the algebras of these groups satisfy

$$M^\dagger I_{p,q} + I_{p,q} M = 0 \quad \begin{array}{ll} \mathbb{R} & \mathfrak{o}(p, q) \\ \mathbb{C} & \mathfrak{u}(p, q) \\ \mathbb{Q} & \mathfrak{sp}(p, q) \end{array} \quad (5.15)$$

The algebras of groups that leave invariant a nonsingular symmetric indefinite metric are most simply treated by determining their block diagonal structure. For example, the algebra $\mathfrak{so}(2, 1)$ for the Lorentz group in the plane is

$$\begin{bmatrix} A^t & C^t \\ B^t & D \end{bmatrix} \begin{bmatrix} I_2 & 0 \\ 0 & -1 \end{bmatrix} + \begin{bmatrix} I_2 & 0 \\ 0 & -1 \end{bmatrix} \begin{bmatrix} A & B \\ C & D \end{bmatrix} = 0 \quad (5.16)$$

From this, we conclude

$$\begin{aligned} A &= -A^t \\ D &= -D = 0 \\ B^t &= C \end{aligned} \quad (5.17)$$

5.4 Bilinear and quadratic constraints

The matrix algebra of $\mathfrak{so}(2, 1)$ is explicitly

$$\begin{bmatrix} 0 & \theta & v_1 \\ -\theta & 0 & v_2 \\ v_1 & v_2 & 0 \end{bmatrix} \tag{5.18}$$

By an identical argument the matrix Lie algebra for the Lorentz group $\mathfrak{so}(3, 1)$ is

$$\begin{bmatrix} 0 & \theta_3 & -\theta_2 & v_1 \\ -\theta_3 & 0 & \theta_1 & v_2 \\ \theta_2 & -\theta_1 & 0 & v_3 \\ v_1 & v_2 & v_3 & 0 \end{bmatrix} \tag{5.19}$$

10. Antisymmetric nonsingular metric-preserving groups Matrices M for the algebras of these groups satisfy

$$M^\dagger G + GM = 0 \qquad G^t = -G \qquad \mathbb{F} = \begin{cases} \mathbb{R} & \mathfrak{sp}(G; \mathbb{R}) \\ \mathbb{C} & \mathfrak{sp}(G; \mathbb{C}) \end{cases} \tag{5.20}$$

Since G is nonsingular, $M = -G^{-1}M^\dagger G$ and

$$\mathrm{tr}\, M = -\mathrm{tr}\, G^{-1}M^\dagger G = -\mathrm{tr}\, M^\dagger = -\mathrm{tr}\, M^* = 0 \tag{5.21}$$

Therefore the trace of these matrices is imaginary.

11. Singular metric-preserving groups Matrices M for the algebras of these groups satisfy

$$MG + GM^\dagger = 0 \qquad \begin{matrix} \mathbb{R} & \mathfrak{o}(n; G) \\ \mathbb{C} & \mathfrak{u}(n; G) \\ \mathbb{Q} & \mathfrak{sp}(n; G) \end{matrix} \tag{5.22}$$

In the case that the $(p+q) \times (p+q)$ matrix G has singular block diagonal structure $\begin{bmatrix} g & 0 \\ 0 & 0 \end{bmatrix}$ with $\det(g) \neq 0$, this constraint reduces to

$$\begin{bmatrix} A & B \\ C & D \end{bmatrix} \begin{bmatrix} g & 0 \\ 0 & 0 \end{bmatrix} + \begin{bmatrix} g & 0 \\ 0 & 0 \end{bmatrix} \begin{bmatrix} A^\dagger & C^\dagger \\ B^\dagger & D^\dagger \end{bmatrix} = \begin{bmatrix} 0 & 0 \\ 0 & 0 \end{bmatrix}$$

$$\Rightarrow Ag + gA^\dagger = 0 \quad C = C^\dagger = 0 \quad B, D \text{ arbitrary} \tag{5.23}$$

In particular, in the case of real 4×4 matrices with singular symmetric metric diag$(1, 1, 1, 0)$ the Lie algebra is

$$\begin{bmatrix} 0 & \theta_3 & -\theta_2 & t_1 \\ -\theta_3 & 0 & \theta_1 & t_2 \\ \theta_2 & -\theta_1 & 0 & t_3 \\ 0 & 0 & 0 & s_4 \end{bmatrix} \tag{5.24}$$

Here the parameters θ_i describe rotations about the ith coordinate axis and the t_i describe displacements of the origin along the ith coordinate direction. The parameter s_4 describes "scaling" of the time axis: $t' = e^{s_4} t$. If s_4 is set to zero (traceless condition, see the following Section 5.5) the Lie algebra is that of the Euclidean group $E(3) = ISO(3)$ (inhomogeneous rotation group in R^3).

The matrix Lie algebra for the Poincaré group $ISO(3, 1)$ (3.26) is obtained by similar arguments using a singular 5×5 metric $G = \text{diag}(1, 1, 1, -1, 0)$. The Lie algebra is (setting the trace equal to zero):

$$\begin{bmatrix} 0 & \theta_3 & -\theta_2 & v_1 & t_1 \\ -\theta_3 & 0 & \theta_1 & v_2 & t_2 \\ \theta_2 & -\theta_1 & 0 & v_3 & t_3 \\ v_1 & v_2 & v_3 & 0 & t_4 \\ 0 & 0 & 0 & 0 & 0 \end{bmatrix} \tag{5.25}$$

The Galilei group (3.27) has the following 5×5 matrix Lie algebra, obtained by "contraction" (cf., Chapter 13) from the Lie algebra of $ISO(3, 1)$:

$$\begin{bmatrix} 0 & \theta_3 & -\theta_2 & v_1 & t_1 \\ -\theta_3 & 0 & \theta_1 & v_2 & t_2 \\ \theta_2 & -\theta_1 & 0 & v_3 & t_3 \\ 0 & 0 & 0 & 0 & t_4 \\ 0 & 0 & 0 & 0 & 0 \end{bmatrix} \tag{5.26}$$

5.5 Multilinear constraints

12. Special linear groups have algebras that satisfy the zero trace condition

$$\text{tr } M = 0 \qquad \mathbb{F} = \begin{cases} \mathbb{R} & \mathfrak{sl}(n, \mathbb{R}) \\ \mathbb{C} & \mathfrak{sl}(n, \mathbb{C}) \\ \mathbb{Q} & \mathfrak{sl}(n, \mathbb{Q}) \end{cases} \tag{5.27}$$

The exponential of a matrix with zero trace is a matrix with determinant $+1$:

$$\det\left(e^M\right) = e^{\text{tr} M} \tag{5.28}$$

5.6 Intersections of groups

The Lie algebra for the intersection of two groups is the intersection of the groups' Lie algebras. The important algebra $\mathfrak{su}(n)$ is obtained from the intersection of $\mathfrak{u}(n)$ and $\mathfrak{sl}(n; \mathbb{C})$ (cf. (5.14)). For example

$$\mathfrak{su}(2) = \mathfrak{u}(2) \cap \mathfrak{sl}(2; \mathbb{C}) = \frac{i}{2} \begin{bmatrix} x_3 & x_1 - ix_2 \\ x_1 + ix_2 & -x_3 \end{bmatrix} \tag{5.29}$$

5.7 Algebras of embedded groups

The Lie algebras of the embedded groups are constructed in a straightforward way.
The Lie algebra of $U(n)$ consists of $n \times n$ antihermitian matrices M:

$$M \in \mathfrak{u}(n) \Rightarrow (M^\dagger)_{ij} = -M^*_{ji} \tag{5.30}$$

The Lie algebra of $OU(2n)$ is obtained from the Lie algebra of $U(n)$ by replacing each of the $n(n-1)/2$ complex matrix elements M_{ij} ($i < j$) ($M \in \mathfrak{u}(n)$) above the diagonal of M by a 2×2 real matrix, and each of the diagonal matrix elements M_{ii} by a real 2×2 matrix representing an imaginary complex number ($a = 0$, b arbitrary in Eq. (3.3)). The matrix elements M_{ij} below the diagonal ($i > j$) are obtained from the antihermiticity condition. The result is a real antisymmetric $2n \times 2n$ matrix with the property $\mathfrak{u}(n) \to \mathfrak{ou}(2n) \subset \mathfrak{o}(2n)$. The dimension of $\mathfrak{ou}(2n)$ is the dimension of $\mathfrak{u}(n)$: $2 \times n(n-1)/2 + 1 \times n = n^2$.

The Lie algebra of $Sp(n)$ consists of $n \times n$ antihermitian matrices M over $\mathbb{Q}$:

$$M \in \mathfrak{sp}(n) \Rightarrow (M^\dagger)_{ij} = -M^*_{ji} \tag{5.31}$$

The adjoint is taken over the quaternion field. The Lie algebra of $USp(2n)$ is obtained from the Lie algebra of $Sp(n)$ by replacing each of the $n(n-1)/2$ quaternion matrix elements M_{ij} ($i < j$) ($M \in \mathfrak{sp}(n)$) above the diagonal of M by a 2×2 complex matrix, and each of the diagonal matrix elements M_{ii} by a complex 2×2 matrix representing an imaginary quaternion ($q_0 = 0$, q_i arbitrary in Eq. (3.4)). The matrix elements M_{ij} below the diagonal ($i > j$) are obtained from the antihermiticity condition. The result is a real antihermitian $2n \times 2n$ matrix with the property $\mathfrak{sp}(n) \to \mathfrak{usp}(2n) \subset \mathfrak{su}(2n)$. The dimension of $\mathfrak{usp}(2n)$ is the dimension of $\mathfrak{sp}(n)$: $4 \times n(n-1)/2 + n \times 3 = 2n(2n+1)/2$.

5.8 Modular groups

The modular group $GL(n; \mathbb{Z})$ has no Lie algebra because it is not a continuous group.

5.9 Basis vectors

In each of these matrix algebras there is usually a clear choice of basis vectors. A useful choice is made by choosing a basis set that is orthogonal with respect to some inner product on the space of square matrices. In (5.6), (5.7), (5.9), (5.14), (5.18)–(5.19), and (5.24)–(5.26) the infinitesimal generators have been chosen to be orthogonal with respect to a convenient inner product.

As discussed in Section 4.8, the Hilbert–Schmidt inner product on rectangular matrices

$$(X, Y) = \operatorname{tr} X^\dagger Y \tag{5.32}$$

is usually very useful. This inner product is positive–definite: $(X, X) = 0 \Rightarrow X = 0$. For example, for the algebra $\mathfrak{so}(2, 1)$ (Eq. (5.18)), if X, X' are two 3×3 matrices in the algebra

$$(X', X) = \operatorname{tr} X'^\dagger X = 2(+\theta'\theta + v'_1 v_1 + v'_2 v_2) \tag{5.33}$$

There is a yet more useful inner product that can be defined on matrix Lie algebras. This is an analog of the Cartan–Killing inner product

$$(X, Y) = \operatorname{tr} XY \tag{5.34}$$

For $\mathfrak{so}(2, 1)$ this inner product is

$$(X', X) = \operatorname{tr} X'X = 2(-\theta'\theta + v'_1 v_1 + v'_2 v_2) \tag{5.35}$$

This inner product is not positive-definite. For giving up positive-definiteness we gain information of both an algebraic and a topological nature. At the algebraic level, the subspace on which this inner product is identically zero is the largest nilpotent invariant subalgebra (subalgebra of matrices equivalent to upper triangular matrices) in the original algebra. The subspace on which the inner product is negative-definite consists of compact group generators, and the subspace on which it is positive-definite consists of noncompact generators. When appropriate measures are taken (in Chapter 11), the negative-definite subspace closes under commutation, and so describes a compact Lie group.

The Cartan–Killing inner product is defined in terms of the structure constants of a Lie algebra. These are incorporated into the regular matrix representation of the Lie algebra. The Cartan–Killing inner product (X, Y) is specifically defined as the trace of the product of the regular matrix representatives of X and Y. Other inner products are conveniently defined when other matrix representations are used. In many instances it is very convenient to use the defining matrix representation of the Lie algebra: this representation certainly contains no less information than the regular matrix representation. For a large class of Lie algebras (simple Lie algebras) these two different inner products are strictly proportional.

It is remarkable that this metric contains information of both a topological and an algebraic nature. To illustrate the difference between the compact and noncompact cases, we consider 2×2 matrices

$$X = \begin{bmatrix} 0 & +1 \\ -1 & 0 \end{bmatrix} \quad (X, X) = -2 \quad e^{\theta X} = \begin{bmatrix} \cos\theta & \sin\theta \\ -\sin\theta & \cos\theta \end{bmatrix}$$

$$Y = \begin{bmatrix} 0 & +1 \\ +1 & 0 \end{bmatrix} \quad (Y, Y) = +2 \quad e^{\theta X} = \begin{bmatrix} \cosh\theta & \sinh\theta \\ \sinh\theta & \cosh\theta \end{bmatrix} \tag{5.36}$$

In the compact case, the group element $\exp(\theta X)$ periodically returns to the identity as θ increases. Therefore the group can be parameterized by a finite range of parameter values: $-\pi \leq \theta \leq +\pi$, with $-\pi$ and $+\pi$ identified. On the other hand, in the noncompact case the group is parameterized by the entire line $-\infty < \theta < +\infty$. The underlying manifolds for the two groups are the circle S^1 and the line R^1.

In the compact case the simplification of parameterizing the group with a bounded subset of the Lie algebra ($-\pi \leq \theta \leq +\pi$) is somewhat offset by the complication of matching boundary conditions – identifying the group operations parameterized by $-\pi$ and $+\pi$. In the noncompact case the simplification of not having to worry about matching boundary conditions is somewhat offset by the fact that it takes the entire subspace in the Lie algebra, R^k, where k is the number of noncompact generators, to parameterize this piece of the group. This piece of the group is topologically identical to R^k, that is, it is Euclidean. These remarks will be clarified and elaborated on in Chapter 7.

5.10 Conclusion

In this chapter we have constructed the Lie algebras for all the matrix Lie groups defined in Chapter 3. This is done by linearizing the constraints that define the original matrix Lie groups in the neighborhood of the identity. For the general linear groups which are defined by no constraints, the Lie algebras $\mathfrak{gl}(n; \mathbb{F})$ are also defined by no constraints. For the Lie groups defined by linear constraints, linearization is trivial and produces a matrix Lie algebra having structure identical to that of the parent Lie group. Transition from the Lie group to the Lie algebra replaces nonlinear constraints by linear conditions defining the Lie algebras of the metric-preserving groups ($G = I_n$, $I_{p,q}$, nonsingular antisymmetric, general nonsingular) and the unimodular groups. One natural way to choose basis vectors in these Lie algebras has been described.

5.11 Problems

1. The Lie group $UT(1, 1)$ has Lie algebra of the form
$$A = \begin{bmatrix} a & b \\ 0 & c \end{bmatrix} = aX_a + bX_b + cX_c$$
Show that in this matrix Lie algebra an inner product can be defined by $(A, A) = \text{tr}(A)^2 = a^2 + c^2$.

2. Show that the regular representation of the matrix Lie algebra $\mathfrak{ut}(1, 1)$ given in Problem 1 is
$$R(A) = \begin{bmatrix} 0 & 0 & -b \\ 0 & 0 & +b \\ 0 & 0 & a-c \end{bmatrix} \begin{matrix} X_a \\ X_c \\ X_b \end{matrix}$$

with the ordering of the basis vectors given on the right. Show that the Cartan–Killing inner product in the regular representation is $(A, A) = \text{tr} R(A)^2 = (a - c)^2$. The inner product in the regular representation suggests that the linear combination $X_a + X_c$ commutes with all operators in the Lie algebra. Is this true?

3. Write down the algebra inclusions $\mathfrak{gl}(1; \mathbb{R}) \subset \mathfrak{gl}(1; \mathbb{C}) \subset \mathfrak{gl}(1; \mathbb{Q})$ explicitly in terms of the 2×2 complex matrices as defined in (3.3) and (3.4).

4. Construct the table (analogous to (5.8)) giving the commutation relations for the photon energy operator $\hat{n} = a^\dagger a + \frac{1}{2}$, creation and annihilation operators $a^\dagger$ and a, and the identity operator I. Compare with a table for the commutation relations of the matrices $X_\eta, X_r, X_l, X_\delta$ defined in (5.9). Show that the two Lie algebras are isomorphic. The photon number operator is $\hat{n} = a^\dagger a$ and the photon energy operator is $\hat{E} = (a^\dagger a + \frac{1}{2})\hbar\omega \to a^\dagger a + \frac{1}{2}$ for $\hbar\omega = 1$.

5. **Cartan decomposition** Assume a matrix Lie algebra has a block diagonal structure given by

$$Z = \begin{bmatrix} A & B \\ C & D \end{bmatrix} = \begin{bmatrix} A & 0 \\ 0 & D \end{bmatrix} + \begin{bmatrix} 0 & B \\ C & 0 \end{bmatrix}$$

$$\mathfrak{g} = \mathfrak{h} + \mathfrak{p}$$

Show that this decomposition satisfies the commutation relations

$$[\mathfrak{h}, \mathfrak{h}] \subseteq \mathfrak{h}$$
$$[\mathfrak{h}, \mathfrak{p}] \subseteq \mathfrak{p}$$
$$[\mathfrak{p}, \mathfrak{p}] \subseteq \mathfrak{h}$$

This means that if $X, X' \in \mathfrak{h}$ and $Y, Y' \in \mathfrak{p}$, then $[X, X'] \in \mathfrak{h}, [X, Y] \in \mathfrak{p}, [Y, Y'] \in \mathfrak{h}$. Conclude that the subspace $\mathfrak{h}$ is a subalgebra of $\mathfrak{g}$. Is $\mathfrak{p}$ a subalgebra (under what conditions is $\mathfrak{p}$ a subalgebra)?

6. Show that an inner product for the Cartan decomposition given in the previous problem is

$$(Z, Z) = \text{tr} Z^2 = \text{tr} A^2 + \text{tr} BC + \text{tr} CB + \text{tr} D^2$$

If $X = \begin{bmatrix} A & 0 \\ 0 & D \end{bmatrix} \in \mathfrak{h}$ and $Y = \begin{bmatrix} 0 & B \\ C & 0 \end{bmatrix} \in \mathfrak{p}$, then

$$(X, X) = \text{tr}(A^2 + D^2) \qquad (Y, Y) = \text{tr}(BC + CB)$$

Show that X and Y are orthogonal under this inner product: $(X, Y) = 0$.

7. The Lie algebra $\mathfrak{so}(p, q)$ has the structure $\begin{bmatrix} A & B \\ B^t & C \end{bmatrix}$ where the $p \times p$ and $q \times q$ matrices A and C satisfy $A^t = -A$ and $C^t = -C$. If $X = \begin{bmatrix} A & 0 \\ 0 & C \end{bmatrix} \in \mathfrak{h}$ and $Y = \begin{bmatrix} 0 & B \\ B^t & 0 \end{bmatrix} \in \mathfrak{p}$, show
   - $(X, Y) = 0$
   - $(X, X) \leq 0, \quad (X, X) = 0 \Rightarrow X = 0$
   - $(Y, Y) \geq 0, \quad (Y, Y) = 0 \Rightarrow Y = 0$

5.11 Problems

These results are summarized by

$$(\mathfrak{h}, \mathfrak{h}) \leq 0$$
$$(\mathfrak{h}, \mathfrak{p}) = 0$$
$$(\mathfrak{p}, \mathfrak{p}) \geq 0$$

8. The Lie algebra $\mathfrak{su}(p, q)$ has the structure $\begin{bmatrix} A & B \\ B^\dagger & C \end{bmatrix}$ where the $p \times p$ and $q \times q$ matrices A and C satisfy $A^\dagger = -A$, $C^\dagger = -C$, and $\mathrm{tr}(A + C) = 0$. If $X = \begin{bmatrix} A & 0 \\ 0 & C \end{bmatrix} \in \mathfrak{h}$ and $Y = \begin{bmatrix} 0 & B \\ B^\dagger & 0 \end{bmatrix} \in \mathfrak{p}$, then show once again that

$$(\mathfrak{h}, \mathfrak{h}) \leq 0$$
$$(\mathfrak{h}, \mathfrak{p}) = 0$$
$$(\mathfrak{p}, \mathfrak{p}) \geq 0$$

Show that $(X, X) = 0 \Rightarrow X = 0$ and $(Y, Y) = 0 \Rightarrow Y = 0$.

9. The Lie algebra for $\mathfrak{sl}(n; \mathbb{R})$ has a decomposition in terms of real antisymmetric and traceless symmetric matrices $A^t = -A$ and $B^t = B$ with $\mathrm{tr}\, B = 0$:

$$\mathfrak{sl}(n; \mathbb{R}) = A + B$$
$$\mathfrak{g} = \mathfrak{h} + \mathfrak{p}$$

Show

$$[\mathfrak{h}, \mathfrak{h}] \subseteq \mathfrak{h} \qquad (\mathfrak{h}, \mathfrak{h}) \leq 0$$
$$[\mathfrak{h}, \mathfrak{p}] \subseteq \mathfrak{p} \quad \text{and} \quad (\mathfrak{h}, \mathfrak{p}) = 0$$
$$[\mathfrak{p}, \mathfrak{p}] \subseteq \mathfrak{h} \qquad (\mathfrak{p}, \mathfrak{p}) \geq 0$$

Show that $(A, A) = 0 \Rightarrow A = 0$ and $(B, B) = 0 \Rightarrow B = 0$.

10. The Lie algebra for $\mathfrak{sl}(n; \mathbb{C})$ has a decomposition in terms of traceless antihermitian matrices $A^\dagger = -A$ and traceless hermitian matrices $H^\dagger = H$:

$$\mathfrak{sl}(n; \mathbb{C}) = \text{antihermitian} + \text{hermitian}$$
$$\mathfrak{g} = \mathfrak{h} + \mathfrak{p}$$

Show

$$[\mathfrak{h}, \mathfrak{h}] \subseteq \mathfrak{h} \qquad (\mathfrak{h}, \mathfrak{h}) \leq 0$$
$$[\mathfrak{h}, \mathfrak{p}] \subseteq \mathfrak{p} \quad \text{and} \quad (\mathfrak{h}, \mathfrak{p}) = 0$$
$$[\mathfrak{p}, \mathfrak{p}] \subseteq \mathfrak{h} \qquad (\mathfrak{p}, \mathfrak{p}) \geq 0$$

Show that $(A, A) = 0 \Rightarrow A = 0$ and $(H, H) = 0 \Rightarrow H = 0$.

11. Assume that $\mathfrak{g}$ is a Lie algebra with a Cartan decomposition $\mathfrak{g} = \mathfrak{h} + \mathfrak{p}$, with commutation relations and inner product properties given by

$$[\mathfrak{h}, \mathfrak{h}] \subseteq \mathfrak{h} \qquad (\mathfrak{h}, \mathfrak{h}) \leq 0$$
$$[\mathfrak{h}, \mathfrak{p}] \subseteq \mathfrak{p} \quad \text{and} \quad (\mathfrak{h}, \mathfrak{p}) = 0$$
$$[\mathfrak{p}, \mathfrak{p}] \subseteq \mathfrak{h} \qquad (\mathfrak{p}, \mathfrak{p}) \geq 0$$

Show that if every $n \times n$ matrix B in $\mathfrak{p}$ is multiplied by i and the resulting algebra is defined by $\mathfrak{g}' = \mathfrak{h} + i\mathfrak{p} = \mathfrak{h} + \mathfrak{p}'$ then

$$[\mathfrak{h}, \mathfrak{h}] \subseteq \mathfrak{h} \qquad (\mathfrak{h}, \mathfrak{h}) \leq 0$$
$$[\mathfrak{h}, \mathfrak{p}'] \subseteq \mathfrak{p}' \quad \text{and} \quad (\mathfrak{h}, \mathfrak{p}') = 0$$
$$[\mathfrak{p}', \mathfrak{p}'] \subseteq \mathfrak{h} \qquad (\mathfrak{p}', \mathfrak{p}') \leq 0$$

In short, noncompact algebras that satisfy a Cartan decomposition can be analytically continued to compact algebras.

12. Extend the Cartan decomposition and analytic continuation arguments to the quaternion algebra $\mathfrak{g} = \mathfrak{sl}(n; \mathbb{Q})$ with respect to the subalgebra $\mathfrak{h} = \mathfrak{sl}(n; \mathbb{C})$.

13. A matrix Lie algebra has the form

$$A = \begin{bmatrix} 0 & \theta_3 & -\theta_2 & b_1 & t_1 \\ -\theta_3 & 0 & \theta_1 & b_2 & t_2 \\ \theta_2 & -\theta_1 & 0 & b_3 & t_3 \\ \mu b_1 & \mu b_2 & \mu b_3 & 0 & t_4 \\ \sigma t_1 & \sigma t_2 & \sigma t_3 & -\mu \sigma t_4 & 0 \end{bmatrix}$$

$$\frac{1}{2}(A, A) = -(\theta \cdot \theta) + \mu(\mathbf{b} \cdot \mathbf{b}) + \sigma(\mathbf{t} \cdot \mathbf{t}) - \mu\sigma t_4^2$$

Show

μ	σ	Algebra	Singular subspace
$+1$	$+1$	$\mathfrak{so}(3, 2)$	
-1	$+1$	$\mathfrak{so}(4, 1)$	
-1	-1	$\mathfrak{so}(5)$	
$+1$	0	Poincaré	translations t_μ
0	0	Galilei	translations t_μ, boosts $\mathbf{b}$

14. Assume that $\mathfrak{g} = A$, where A is a Lie algebra of $n \times n$ matrices on which the inner product is negative-definite: $\operatorname{tr} A^2 \leq 0, = 0 \Rightarrow A = 0$. Then show that $\operatorname{EXP}(tA)$ returns to any neighborhood of the identity I_n if t becomes large enough. If the eigenvalues of A are rationally related ($\lambda_i = \gamma n_i$, n_i are integers, $\gamma \neq 0$ is rational or irrational), $\operatorname{EXP}(tA)$ returns periodically to I_n. What is t_0, the minimum period in t?

15. Use the parameterization of $\mathfrak{so}(3)$ given in Problem 3.14. Show that the differentials (dx, dy, dz) of a point in the neighborhood of (x, y, z) are related to the displacements $(\delta x, \delta y, \delta z)$ in the neighborhood of the identity by

$$\begin{bmatrix} dx \\ dy \\ dz \end{bmatrix} = \begin{bmatrix} m_{11} & 0 & -y \\ 0 & m_{11} & x \\ 0 & m_{21} & m_{22} \end{bmatrix} \begin{bmatrix} \delta x \\ \delta y \\ \delta z \end{bmatrix}$$

Use the values you constructed for the matrix elements m_{ij} to construct explicitly the metric tensor $g(x, y, z)$ and the invariant measure $d\mu(x, y, z)$ on the group $SO(3)$ with this parameterization.

16. $\mathfrak{g}$ is a matrix Lie algebra. Show that if the matrix subspaces $\mathfrak{h}$ and $\mathfrak{p}$ defined below exist in the algebra ($\mathfrak{g} \cap \mathfrak{h} = \mathfrak{h}$, $\mathfrak{g} \cap \mathfrak{p} = \mathfrak{p}$),

$\mathfrak{h}$	$\mathfrak{p}$
$\mathfrak{g} + \mathfrak{g}^*$	$\mathfrak{g} - \mathfrak{g}^*$
$\mathfrak{g} - \mathfrak{g}'$	$\mathfrak{g} + \mathfrak{g}'$
$\mathfrak{g} - \mathfrak{g}^\dagger$	$\mathfrak{g} + \mathfrak{g}^\dagger$

then the following commutation relations are satisfied:

$$[\mathfrak{h}, \mathfrak{h}] \subseteq \mathfrak{h} \quad [\mathfrak{h}, \mathfrak{p}] \subseteq \mathfrak{p} \quad [\mathfrak{p}, \mathfrak{p}] \subseteq \mathfrak{h}$$

6
Operator algebras

Lie algebras of matrices can be mapped onto Lie algebras of operators in a number of different ways. Three useful matrix algebra to operator algebra mappings are described in this chapter.

6.1 Boson operator algebras

It is possible to construct useful operator algebras from Lie algebras. An operator Lie algebra can be constructed from a Lie algebra of $n \times n$ matrices by introducing a set of n independent boson creation ($b_i^\dagger$) and annihilation (b_j) operators that obey the commutation relations

$$[b_i, b_j^\dagger] = I\delta_{ij} \qquad (6.1)$$

with all other commutators (e.g., $[b_i, b_j]$, $[b_i^\dagger, b_j^\dagger]$, $[b_i, I]$, $[b_j^\dagger, I]$) equal to zero. The operator algebra is constructed from the matrix algebra by associating to each matrix A the operator $\mathcal{A}$ that is a linear combination of creation and annihilation operators:

$$A \to \mathcal{A} = b^\dagger A b = \sum_i \sum_j b_i^\dagger A_{ij} b_j \qquad (6.2)$$

The matrices and their associated operators have isomorphic commutation relations

$$\begin{aligned}
[\mathcal{A}, \mathcal{B}] &= [b_i^\dagger A_{ij} b_j, b_r^\dagger B_{rs} b_s] \\
&= A_{ij} B_{rs} [b_i^\dagger b_j, b_r^\dagger b_s] \\
&= A_{ij} B_{rs} (b_i^\dagger \delta_{jr} b_s - b_r^\dagger \delta_{si} b_j) \\
&= b_i^\dagger A_{ij} B_{js} b_s - b_r^\dagger B_{rs} A_{sj} b_j \\
&= b_i^\dagger [A, B]_{ij} b_j \\
&= \mathcal{C}
\end{aligned} \qquad (6.3)$$

where $[A, B] = C$. This argument is invertible. An algebra of operators bilinear in boson creation and annihilation operators for n independent modes has an isomorphic $n \times n$ matrix algebra (or matrix representation)

$$[A, B] = C \Leftrightarrow [\mathcal{A}, \mathcal{B}] = \mathcal{C} \qquad \mathcal{A} = \sum_{ij} b_i^\dagger A_{ij} b_j \qquad (6.4)$$

Remark The $2n + 1$ operators $b_i, b_j^\dagger, I$ ($1 \leq i, j \leq n$) span the Heisenberg algebra.

6.2 Fermion operator algebras

The success of the calculation above does not depend on the boson commutation relations (6.1). It depends, rather, on the commutation relations of bilinear products of these operators

$$[b_i^\dagger b_j, b_r^\dagger b_s] = b_i^\dagger b_s \delta_{jr} - b_r^\dagger b_j \delta_{si} \qquad (6.5)$$

Any set of operators X_{ij} that satisfies isomorphic commutation relations

$$[X_{ij}, X_{rs}] = X_{is} \delta_{jr} - X_{rj} \delta_{si} \qquad (6.6)$$

can be used in place of the bilinear combinations $b_i^\dagger b_j$:

$$\mathcal{A} \to \mathcal{A} = \sum_{ij} A_{ij} X_{ij} \qquad (6.7)$$

Another useful set of operators with this property is obtained from the fermion creation ($f_i^\dagger$) and annihilation (f_j) operators for n independent modes. These operators do not even satisfy commutation relations. Rather, they satisfy **anticommutation relations**

$$\{f_i, f_j^\dagger\} = f_i f_j^\dagger + f_j^\dagger f_i = I \delta_{ij} \qquad (6.8)$$

with all other bilinear anticommutators (e.g., $\{f_i, f_j\}$, $\{f_i^\dagger, f_j^\dagger\}$) equal to zero. Bilinear combinations of fermion operators satisfy **commutation relations** of the form (6.6), for

$$\begin{aligned}
[f_i^\dagger f_j, f_r^\dagger f_s] &= f_i^\dagger f_j f_r^\dagger f_s - f_r^\dagger f_s f_i^\dagger f_j \\
&= f_i^\dagger (\delta_{jr} - f_r^\dagger f_j) f_s - f_r^\dagger (\delta_{is} - f_i^\dagger f_s) f_j \\
&= f_i^\dagger f_s \delta_{jr} - f_r^\dagger f_j \delta_{si}
\end{aligned} \qquad (6.9)$$

As a result, matrix Lie algebras can be associated with bilinear products of either boson or fermion operators:

$$[A, B] = C \Leftrightarrow [\mathcal{A}, \mathcal{B}] = \mathcal{C} \qquad \mathcal{A} = \sum_{ij} f_i^\dagger A_{ij} f_j \qquad (6.10)$$

These two matrix algebra → operator algebra mappings are useful for constructing particular classes of representations for the unitary group $U(n)$ and its subgroup $SU(n)$. The mapping to a boson operator algebra greatly simplifies the construction of the symmetric representations of $U(n)$. The mapping to a fermion operator algebra greatly simplifies the construction of the antisymmetric representations of $U(n)$. A closely related mapping allows an elegant construction of the spin representations of the orthogonal groups.

6.3 First order differential operator algebras

Yet another useful set of operators that satisfies the commutation relations (6.6) are the first order differential operators

$$X_{ij} \to x_i \partial_j = x_i \frac{\partial}{\partial x_j} \qquad (6.11)$$

Then

$$[A, B] = C \Leftrightarrow [\mathcal{A}, \mathcal{B}] = \mathcal{C} \qquad \mathcal{A} = \sum_{ij} x_i A_{ij} \partial_j = \sum_{ij} A_{ij} X_{ij} \qquad (6.12)$$

To illustrate the use of this operator combination, we treat the matrix algebra $\mathfrak{so}(3)$ of the orthogonal group $SO(3)$

$$\mathfrak{so}(3) = \begin{bmatrix} 0 & \theta_3 & -\theta_2 \\ -\theta_3 & 0 & \theta_1 \\ \theta_2 & -\theta_1 & 0 \end{bmatrix} = \theta \cdot \mathbf{L} \qquad (6.13)$$

The operator algebra is

$$(x_1 \ x_2 \ x_3) \begin{bmatrix} 0 & \theta_3 & -\theta_2 \\ -\theta_3 & 0 & \theta_1 \\ \theta_2 & -\theta_1 & 0 \end{bmatrix} \begin{bmatrix} \partial_1 \\ \partial_2 \\ \partial_3 \end{bmatrix} = \theta \cdot \mathcal{L} \qquad (6.14)$$

where $\mathcal{L}_1 = x_2 \partial_3 - x_3 \partial_2$, $\mathcal{L}_2 = x_3 \partial_1 - x_1 \partial_3$, $\mathcal{L}_3 = x_1 \partial_2 - x_2 \partial_1$. The two algebras have isomorphic commutation relations

$$[L_i, L_j] = -\epsilon_{ijk} L_k \qquad [\mathcal{L}_i, \mathcal{L}_j] = -\epsilon_{ijk} \mathcal{L}_k \qquad (6.15)$$

where L_i are 3×3 matrices and $\mathcal{L}_i$ are first order differential operators.

6.3 First order differential operator algebras

As another example, we treat the Lie algebra for the group $E(2) = ISO(2)$ of rigid motions (translations and rotations) in the x–y plane, whose matrix algebra may be taken in the form

$$\begin{bmatrix} 0 & \theta & 0 \\ -\theta & 0 & 0 \\ t_1 & t_2 & 0 \end{bmatrix} = \theta L_z + t_i T_i \qquad (6.16)$$

This describes rotations about an axis perpendicular to the x–y plane through an angle θ and displacements in the x and y directions by t_1 and t_2. The associated operator algebra is

$$(x_1 \; x_2 \; 1) \begin{bmatrix} 0 & \theta & 0 \\ -\theta & 0 & 0 \\ t_1 & t_2 & 0 \end{bmatrix} \begin{bmatrix} \partial_1 \\ \partial_2 \\ 1 \end{bmatrix} = \theta L_z + t_i T_i \qquad (6.17)$$

where $L_z = x_1 \partial_2 - x_2 \partial_1$ and $T_i = \partial_i$. The matrix algebra and operator algebra have isomorphic commutation relations.

Differential operator realizations of Lie algebras come about in a natural way. This is illustrated by two simple examples. The general procedure can easily be inferred from these examples. Both involve the group of affine transformations of the real line parameterized by points (a, b) in R^2 as follows

$$(a, b) \rightarrow \begin{bmatrix} e^a & b \\ 0 & 1 \end{bmatrix} \qquad (6.18)$$

Imagine a function defined for every point p in R^1. Once a coordinate system S is chosen a coordinate, $x(p)$, can be introduced and the function can be written explicitly as a function of x

$$\begin{array}{ccc} f(p) & & \\ \downarrow & \downarrow & \\ f_S[x(p)] & = & f_{S'}[x'(p)] \end{array} \qquad (6.19)$$

If a new coordinate system S' is chosen, the value of the function at p remains unchanged but the new coordinate of p, $x'(p)$, is different. Therefore the functions f_S and $f_{S'}$ must be different. We ask: how is $f_{S'}$ related to f_S?

To answer this question, assume $x'(p)$ and $x(p)$ are related by an infinitesimal group transformation

$$\begin{bmatrix} x' \\ 1 \end{bmatrix} = \begin{bmatrix} 1+da & db \\ 0 & 1 \end{bmatrix} \begin{bmatrix} x \\ 1 \end{bmatrix} \qquad (6.20)$$

Then
$$f_{S'}[x'(p)] = f_S[x(x'(p))] \tag{6.21}$$

We solve for x in terms of x' by inverting the linear relation (6.20)

$$\begin{aligned} f_{S'}[x'(p)] &= f_S[x'(1-da) - db] \\ &= f_S[x'] - da\, x' \frac{\partial f_S}{\partial x'} - db\, \frac{\partial f_S}{\partial x'} \end{aligned} \tag{6.22}$$

The infinitesimal generators that transform the function at p are

$$\mathcal{X}_a = -x' \frac{\partial}{\partial x'} \qquad \mathcal{X}_b = -\frac{\partial}{\partial x'} \tag{6.23}$$

These operators have commutation relations that are isomorphic with those of the original matrix group

$$[X_a, X_b] = X_b \Leftrightarrow [\mathcal{X}_a, \mathcal{X}_b] = \mathcal{X}_b \tag{6.24}$$

As a second example we consider functions $G(x, y)$ defined on the plane R^2 that parameterizes the affine group. By repeating the arguments above

$$G_{S'}(x', y') = G_S(x, y) \tag{6.25}$$

where (x', y') and (x, y) are related by

$$\begin{bmatrix} x' & y' \\ 0 & 1 \end{bmatrix} = \begin{bmatrix} 1+da & db \\ 0 & 1 \end{bmatrix} \begin{bmatrix} x & y \\ 0 & 1 \end{bmatrix} \tag{6.26}$$

Inverting the infinitesimal transformation, we have

$$\begin{aligned} G_{S'}(x', y') &= G_S[x = (1-da)x', y = (1-d)y' - db] \\ &= G_S(x', y') + \left\{ da \left(-x' \frac{\partial}{\partial x'} - y' \frac{\partial}{\partial y'} \right) + db \left(-\frac{\partial}{\partial y'} \right) \right\} G_S(x', y') \end{aligned} \tag{6.27}$$

The two infinitesimal generators are

$$\begin{aligned} \mathcal{X}_a &= -x' \partial/\partial x' - y' \partial/\partial y' \\ \mathcal{X}_b &= -\partial/\partial y' \end{aligned} \tag{6.28}$$

The commutation relations are preserved

$$[X_a, X_b] = X_b \Leftrightarrow [\mathcal{X}_x, \mathcal{X}_b] = \mathcal{X}_b \tag{6.29}$$

These two examples serve to demonstrate that a single matrix algebra can have many different operator realizations.

Remark In the example above we have adopted the "passive" interpretation of group action. That is, the coordinates of a point changed by virtue of a choice of a different coordinate system, but the value of the function did not. Therefore the particular form of the function was required to change. There is another interpretation of the group action – the "active" interpretation. In this interpretation the group operation defines a new function at the initial point in accordance with (see Eq. (6.19))

$$f_{S'}[x(p)] = f_S[x'(p)] \quad (6.30)$$

Infinitesimal generators for changes in the function under the active interpretation can be computed. They are exactly the same as those computed for the passive interpretation, except for a sign change. This sign difference is encountered in the theory of rotating bodies as the difference in commutation relations for the generators of rotation in a laboratory-fixed frame and a body-fixed frame.

The "active" and "passive" interpretations of group operations are related by the equivalence principle (see Section 14.2).

6.4 Conclusion

Matrix algebra to operator algebra isomorphisms are easily constructed by associating to each matrix A in a matrix Lie algebra an operator $\mathcal{A} = \sum_i \sum_j A_{ij} X_{ij}$. If the operators X_{ij} obey the simple commutation relations (6.6), the commutation relations of the matrix Lie algebra and the operator algebra are isomorphic: $[A, B] = C \Leftrightarrow [\mathcal{A}, \mathcal{B}] = \mathcal{C}$. Under these conditions, complicated commutators in an operator algebra can be replaced by simpler commutators in the matrix algebra. These results extend to the respective Lie groups. Products of exponentials of operators can be replaced by products of exponentials of the corresponding matrices with a little care: $e^A e^B = e^D \Leftrightarrow e^{\mathcal{A}} e^{\mathcal{B}} = e^{\mathcal{D}}$.

6.5 Problems

1. Bilinear products involving one creation and one annihilation operator for two modes generate a four-dimensional Lie algebra with basis vectors $a_i^\dagger a_j$, $1 \leq i, j \leq 2$.
   a. Show that $\hat{n} = a_1^\dagger a_1 + a_2^\dagger a_2$ commutes with all the operators in this set.
   b. If $\hat{n}$ is chosen as one basis vector in this four-dimensional space, the remaining three operators can be chosen as $a_1^\dagger a_1 - a_2^\dagger a_2$, $a_1^\dagger a_2$, and $a_2^\dagger a_1$. Construct their commutation relations.
   c. These calculations simplify considerably under the operator to matrix mapping

$$\begin{array}{cccc} a_1^\dagger a_1 + a_2^\dagger a_2 & a_1^\dagger a_1 - a_2^\dagger a_2 & a_1^\dagger a_2 & a_2^\dagger a_1 \\ \downarrow & \downarrow & \downarrow & \downarrow \\ \begin{bmatrix} 1 & 0 \\ 0 & 1 \end{bmatrix} & \begin{bmatrix} 1 & 0 \\ 0 & -1 \end{bmatrix} & \begin{bmatrix} 0 & 1 \\ 0 & 0 \end{bmatrix} & \begin{bmatrix} 0 & 0 \\ 1 & 0 \end{bmatrix} \end{array}$$

d. Show that the three operators $\frac{1}{2}(a_1^\dagger a_1 - a_2^\dagger a_2)$, $a_1^\dagger a_2$, and $a_2^\dagger a_1$ satisfy commutation relations isomorphic to the comutation relations of the angular momentum algebra J_z, $J_\pm$. In particular, show

$$J_z = \tfrac{1}{2}(a_1^\dagger a_1 - a_2^\dagger a_2)$$
$$J_x = \tfrac{1}{2}(J_+ + J_-) = \tfrac{1}{2}(a_1^\dagger a_2 + a_2^\dagger a_1)$$
$$J_y = \tfrac{1}{2i}(J_+ - J_-) = \tfrac{1}{2i}(a_1^\dagger a_2 - a_2^\dagger a_1)$$

   e. Evaluate $J^2 = J_x^2 + J_y^2 + J_z^2$ in terms of the creation and annihilation operators, and show

$$J^2 = \left(\frac{1}{2}\hat{n}\right)\left(\frac{1}{2}\hat{n} + 1\right)$$

2. **Schwinger representation of angular momentum** Introduce two independent modes. Assume that the quantum state of mode i ($i = 1, 2$) is $|n_i\rangle$, where n_i is the number of excitations in mode i. Assume also that the creation and annihilation operators $a_i^\dagger$ and a_i act on state $|n_i\rangle$ in the usual way:

$$a_i^\dagger |n_i\rangle = \sqrt{n_i + 1}\, |n_i + 1\rangle \qquad a_i |n_i\rangle = \sqrt{n_i}\, |n_i - 1\rangle$$

   Choose as a set of basis vectors the direct product states $|n_1\rangle \otimes |n_2\rangle = |n_1, n_2\rangle$. Define

$$\left|\begin{array}{c}j\\m\end{array}\right\rangle = |n_1, n_2\rangle \qquad j = \frac{1}{2}(n_1 + n_2), \quad m = \frac{1}{2}(n_1 - n_2)$$

   a. Identify the lattice sites in Fig. 6.1 with the states $|n_1, n_2\rangle = |jm\rangle$, the diagonal operator $\frac{1}{2}(a_1^\dagger a_1 - a_2^\dagger a_2)$ with the operator J_z, and the shift operators $a_1^\dagger a_2$, $a_2^\dagger a_1$ with J_+ and J_-.
   b. Show that the four operators $a_i^\dagger a_j$ leave invariant the sum $n_1 + n_2$.
   c. $J^2 |jm\rangle = j(j+1) |jm\rangle$.
   d. $J_z |jm\rangle = m |jm\rangle$.
   e. $J_+ |jm\rangle = a_1^\dagger a_2 |n_1, n_2\rangle = \sqrt{n_1 + 1}\sqrt{n_2}\, |n_1 + 1, n_2 - 1\rangle = |j, m+1\rangle\sqrt{j + m + 1}\sqrt{j - m}$.
   f. $J_- |jm\rangle = a_2^\dagger a_1 |n_1, n_2\rangle = \sqrt{n_1}\sqrt{n_2 + 1}\, |n_1 - 1, n_2 + 1\rangle = |j, m-1\rangle\sqrt{j + m}\sqrt{j - m + 1}$.
   g. $J_\pm |jm\rangle = |j, m \pm 1\rangle\sqrt{(j \pm m + 1)(j \mp m)}$. Note that $J_+ |j, j\rangle = 0$, $J_- |j, -j\rangle = 0$.
   h. $\langle j'm'|J_\pm|jm\rangle = \sqrt{(j' \pm m')(j \mp m)}\, \delta_{j'j}\, \delta_{m', m \pm 1}$.

3. Basis vectors in the Lie algebra u(3) for the group $U(3)$ have commutation relations that are isomorphic to the commutation relations of the nine boson operators $a_i^\dagger a_j$, $1 \le i, j \le 3$. Choose a set of basis vectors for a matrix representation of this algebra of the form $|n_1, n_2, n_3\rangle = |n_1\rangle \otimes |n_2\rangle \otimes |n_3\rangle$, where for example $b_i |n_i\rangle = |n_i - 1\rangle\sqrt{n_i}$, etc.

6.5 Problems

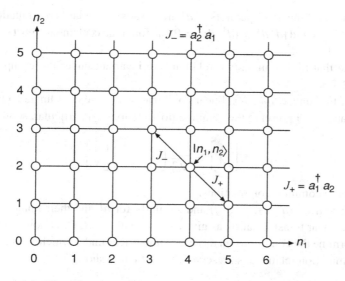

Figure 6.1. Identification of the angular momentum operators with operators for two boson modes simplifies computation of the angular momentum matrix elements.

a. Show $N = \sum_{i=1}^{3} n_i$ is not changed by the action of any of the nine operators in this set.

b. Show that the dimension, D, of this representation is $D = (N + 3 - 1)!/N!(3 - 1)!$. This is the number of ways three nonnegative integers can be chosen whose sum is N (Bose–Einstein counting problem). In higher dimensions (n) replace 3 by n. D is also the number of monomials of degree N in the Taylor series expansion of a function $f(x_1, x_2, \ldots, x_n)$ of n variables.

c. Compute the matrix elements of all operators $b_i^\dagger b_j$ in this representation:

$$\langle n_1', n_2', n_3' | b_i^\dagger b_j | n_1, n_2, n_3 \rangle \tag{6.31}$$

d. Is there some operator in the Lie algebra that maps to the identity matrix, I_D, in this representation?

$$\langle n_1', n_2', n_3' | \mathcal{O} | n_1, n_2, n_3 \rangle = I_D \delta_{n_1', n_1} \delta_{n_2', n_2} \delta_{n_3', n_3} \tag{6.32}$$

What is $\mathcal{O}$?

4. Repeat the steps of Problem 3, replacing the boson operators $b_i^\dagger b_j$ by Fermion operators $f_i^\dagger f_j$. What is now the dimension of this representation?

5. Construct operators $d, d^\dagger$ defined formally from the standard creation and annihilation operators $a, a^\dagger$ as follows:

$$\begin{bmatrix} d \\ d^\dagger \end{bmatrix} = \begin{bmatrix} A & B \\ C & D \end{bmatrix} \begin{bmatrix} a \\ a^\dagger \end{bmatrix}$$

a. Show that if the new operators $d, d^\dagger$ are to satisfy standard commutation relations $[d, d^\dagger] = 1$ and $[d, d] = [d^\dagger, d^\dagger] = 0$, the four matrix elements must satisfy $AD - BC = 1$.
b. Argue that the commutation relations are invariant under the group $Sp(2; \mathbb{R}) = SL(2; \mathbb{R})$.
c. Show that under $Sp(2; \mathbb{R})$, linear combinations of the coordinate and differential operators x, ∂ preserve the commutation relations. In particular, show that

$$\begin{bmatrix} a \\ a^\dagger \end{bmatrix} = \frac{1}{\sqrt{2}} \begin{bmatrix} 1 & 1 \\ -1 & 1 \end{bmatrix} \begin{bmatrix} \partial \\ x \end{bmatrix}$$

preserve commutation relations.
d. Replace a by $(a_1, a_2, \ldots, a_n)$ and similarly for $a^\dagger$ and their images $d, d^\dagger$ under some linear transformation as given above, with A, B, C, D now $n \times n$ matrices. Determine the conditions on these $n \times n$ matrices under which the structure of the commutation relations is preserved. In particular, show

$$AD^t - BC^t = I_n \qquad AB^t = BA^t \qquad CD^t = DC^t$$

Show that these transformations belong to the Lie group $Sp(2n; \mathbb{R})$.

6. The N-dimensional isotropic harmonic oscillator has hamiltonian

$$\mathcal{H} = \hbar\omega \sum_{i=1}^{N} \left(a_i^\dagger a_i + \frac{1}{2} \right)$$

and eigenstates $|n_1, n_2, \ldots, n_N\rangle$.
a. Show that the degeneracy of the multiplet containing n quanta, with energy $\hbar\omega(n + \frac{N}{2})$ is $\deg(N, n) = (n + N - 1)!/n!(N - 1)!$. This solution to the Bose–Einstein counting problem is exactly equal to the number of coefficients of degree n in the Taylor series expansion of a function of N variables: $f(x_1, x_2, \ldots, x_N)$.
b. Show that the symmetry group of this hamiltonian has Lie algebra spanned by the N^2 operators $a_i^\dagger a_j$. This is isomorphic to the Lie algebra $\mathfrak{u}(N)$. Since $[\mathcal{H}, a_i^\dagger a_j] = 0$, this algebra is a direct sum of a simple Lie algebra, $\mathfrak{su}(N)$, plus the one-dimensional algebra spanned by $\mathcal{H}$.
c. If the generators $a_i^\dagger a_j$ that span the invariance algebra are supplemented with the single creation and annihilation operators $a_i^\dagger$ and a_j, as well as their commutator I, the resulting set of operators closes to form an $(N + 1)^2$ dimensional Lie algebra that is nonsemisimple. This is called the **spectrum generating algebra** of the isotropic harmonic oscillator. Show that there is a sequence of operations drawn from this algebra that transform any state in a multiplet with n excitations to any state in a multiplet with n' excitations.

7. The set of matrices $R, S, T, U, \ldots$ belong to a Lie algebra of $n \times n$ matrices, $a^\dagger = (a_1^\dagger, a_2^\dagger, \ldots, a_n^\dagger)$ is a row vector of creation operators for n boson modes, and a is its

adjoint, a column vector of annihilation operators. Define $\mathcal{R} = a^\dagger R a = a_i^\dagger R_{ij} a_j$, and similarly for $\mathcal{S}, \mathcal{T}, \mathcal{U}, \ldots$.
 a. $[\mathcal{R}, \mathcal{S}] = \mathcal{T} \Leftrightarrow [R, S] = T$
 b. $e^{\mathcal{R}} e^{\mathcal{S}} = e^{\mathcal{U}} \Leftrightarrow e^R e^S = e^U$

8. The Rodriguez formula is often used to generate the Hermite polynomials:

$$H_n(x) = e^{x^2} \left(-\frac{d}{dx}\right)^n e^{-x^2}$$

a. Show $[\frac{d}{dx}, e^{-x^2/2}] = -xe^{-x^2/2}$.
b. Use this result to show

$$\left(-\frac{d}{dx}\right) e^{-x^2} = e^{-x^2/2} \left(x - \frac{d}{dx}\right) e^{-x^2/2}$$

$$\left(-\frac{d}{dx}\right)^n e^{-x^2} = e^{-x^2/2} \left(x - \frac{d}{dx}\right)^n e^{-x^2/2}$$

c. As a result

$$H_n(x) e^{-x^2/2} = e^{+x^2/2} \left(-\frac{d}{dx}\right)^n e^{-x^2} = \left(x - \frac{d}{dx}\right)^n e^{-x^2/2}$$

d. Introduce the annihilation operator $a = \frac{1}{\sqrt{2}}(x + \frac{d}{dx})$, define the normalized ground state $\langle x|0\rangle$ by $a\langle x|0\rangle = 0$. Solve this equation, normalize the solution, and show $\langle x|0\rangle = e^{-x^2/2}/\sqrt{1\sqrt{\pi}}$.

e. Introduce the creation operator $a^\dagger = \frac{1}{\sqrt{2}}(x - \frac{d}{dx})$ and show

$$\langle x|n\rangle = \frac{(\sqrt{2}a^\dagger)^n}{\sqrt{2^n n!}} \langle x|0\rangle = \frac{H_n(x) e^{-x^2/2}}{\sqrt{2^n n!\sqrt{\pi}}} = \psi_n(x) \quad (6.33)$$

where $\psi_n(x)$ is the nth normalized harmonic oscillator eigenstate $\langle x|n\rangle = \frac{(a^\dagger)^n}{\sqrt{n!}} \langle x|0\rangle$.

9. Assume a set of n harmonic oscillators interact through an angular momentum term ($L_{ij} = a_i^\dagger a_j - a_j^\dagger a_i$) and a quadrupole interaction ($Q_{ij} = a_i^\dagger a_j + a_j^\dagger a_i$).
a. Show that the hamiltonian for this system is

$$H = \sum_{i=1}^n \hbar\omega_i \left(a_i^\dagger a_i + \frac{1}{2}\right) + i \sum_{i<j} \theta_{ij}\left(a_i^\dagger a_j - a_j^\dagger a_i\right) + \sum_{i\leq j} q_{ij}\left(a_i^\dagger a_j + a_j^\dagger a_i\right)$$

b. Show that this hamiltonian can be represented by a hermitian matrix. Show that for $i \leq j$ the matrix elements are

$$\Gamma_{ij} = \hbar\omega_i \delta_{ij} + (q + i\theta)_{ij}$$

with $\Gamma_{ji}^* = \Gamma_{ij}$.

c. Show that an orthogonal transformation can be constructed so that the hamiltonian can be expressed in terms of n independent oscillators represented by creation and annihilation operators $b_i = m_{ij}a_j$: $H = \sum_{i=1}^{n} \hbar\omega'_i(b_i^\dagger b_i + \frac{1}{2}) +$ constant. Express the amplitudes m_{ij} in terms of the eigenvectors of $\Gamma(H)$.

d. Compute the shift in the zero point energy ("constant").

7
EXPonentiation

Linearization of a Lie group to form a Lie algebra introduces an enormous simplification in the study of Lie groups. The inverse process, reconstructing the Lie group from the Lie algebra, is carried out by the EXPonential map. We return to a more thorough study of the exponential map in this chapter. In particular, we address the three problems raised in Chapter 4. Does the EXPonential operation map the Lie algebra back onto the Lie group? Are Lie groups with isomorphic Lie algebras themselves isomorphic? Are there natural ways to parameterize Lie groups? We close this chapter with a spectrum of applications of the EXPonential mapping in physics. Applications include computing the dynamical evolution of quantum systems and their thermal expectation values.

7.1 Preliminaries

In Chapter 4 we saw how the linearization and EXPonentiation operations relate Lie groups and Lie algebras

$$\text{Lie groups} \underset{\text{EXP}}{\overset{\ln}{\rightleftharpoons}} \text{Lie algebras} \tag{7.1}$$

At that time three questions, and their answers, were briefly raised about the EXPonential mapping. These questions are more thoroughly explored in this chapter.

The three questions, and their answers, are now presented.

Question 1 Does EXP map the Lie algebra onto the entire group?
Answer 1 No, but with some effort and insight, Yes.
Question 2 Are Lie groups with isomorphic Lie algebras isomorphic?
Answer 2 No, but there is a unique Lie group (covering group) and all others with the same Lie algebra are simply related to this unique simply connected Lie group.

Question 3 Are all mappings of the Lie algebra onto the Lie group identical?
Answer 3 No, but with care they are all analytically related to each other (by Baker–Campbell–Hausdorff formulas).

Each question is now discussed in more detail.

7.2 The covering problem

Cartan gave a simple example which showed that it is not always possible to map a Lie algebra onto the entire Lie group through a single mapping of the form EXP(X). We consider the Lie group $SL(2; \mathbb{R})$ with Lie algebra $\mathfrak{sl}(2; \mathbb{R})$:

$$X = \begin{bmatrix} a & b+c \\ b-c & -a \end{bmatrix} \in \mathfrak{sl}(2; \mathbb{R}) \tag{7.2}$$

For this matrix algebra

$$\text{tr EXP}(X) \geq -2 \tag{7.3}$$

Since $SL(2; \mathbb{R})$ contains group operations of the form

$$\begin{bmatrix} -\lambda & 0 \\ 0 & -1/\lambda \end{bmatrix} \qquad \lambda > 1 \tag{7.4}$$

with trace less than -2, a single exponential cannot map the Lie algebra onto the entire group.

The lower bound (-2) on the trace of the exponential can be seen as follows. Trace is an invariant under similarity transformation, so

$$\text{tr } e^X = \text{tr } S\, e^X\, S^{-1} = \text{tr } e^{SXS^{-1}} \tag{7.5}$$

Now choose S to diagonalize (7.2). Since Tr $X = 0$, the eigenvalues λ can only have the form $\pm\theta$ or $\pm i\theta$ (θ real)

$$\text{tr } e^{SXS^{-1}} \longrightarrow \begin{cases} 2\cosh\theta \geq 2 & \text{real eigenvalues} \\ 2\cos\theta \geq -2 & \text{imaginary eigenvalues} \end{cases} \tag{7.6}$$

The problem in attempting to parameterize the Lie group with a single exponential map lies with the compact generators. The compact generators "go around" in circles, while the noncompact generators "go on forever." Furthermore, the compact generators always form a subgroup in the Lie group while the noncompact generators do not.

7.2 The covering problem

To make these cryptic statements less mysterious, we compute EXP(X), with X given in (7.2), and find

$$\text{EXP}\begin{bmatrix} a & b+c \\ b-c & -a \end{bmatrix}$$

$$= \begin{bmatrix} \cosh r + a \sinh r/r & (b+c) \sinh r/r \\ (b-c) \sinh r/r & \cosh r - a \sinh r/r \end{bmatrix} \quad r^2 = a^2 + b^2 - c^2 > 0$$

$$= \begin{bmatrix} 1+a & b+c \\ b-c & 1-a \end{bmatrix} \qquad\qquad a^2 + b^2 - c^2 = 0 \quad (7.7)$$

$$= \begin{bmatrix} \cos r + a \sin r/r & (b+c) \sin r/r \\ (b-c) \sin r/r & \cos r - a \sin r/r \end{bmatrix} \quad -r^2 = a^2 + b^2 - c^2 < 0$$

The "light cone" structure of the (a, b, c) coordinate space of the Lie algebra is shown in Fig. 7.1. Points inside this cone map onto 2×2 rotation matrices in the group $SO(2)$. Points outside this cone map onto noncompact group elements. Points on the cone itself map onto some interesting group operations.

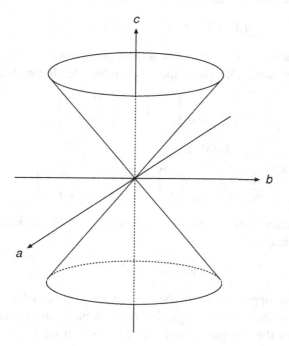

Figure 7.1. "Light cone" for $SL(2; \mathbb{R})$.

Many points inside the cone map onto the same operation in the subgroup $SO(2)$. To see this most easily set $a = b = 0$. Points on the c-axis map onto

$$(0, 0, c) \longrightarrow \begin{bmatrix} \cos c & \sin c \\ -\sin c & \cos c \end{bmatrix} \tag{7.8}$$

and therefore points separated by $2\pi n$ along the c-axis map onto the same group operation in $SO(2) \subset SL(2; R)$. The complementary subspace $(a, b, 0)$ maps onto noncompact group operations in $SL(2; \mathbb{R})$

$$(a, b, 0) \longrightarrow \begin{bmatrix} \cosh r + (a/r)\sinh r & (b/r)\sinh r \\ (b/r)\sinh r & \cosh r - (a/r)\sinh r \end{bmatrix} \quad r^2 = a^2 + b^2 \tag{7.9}$$

that are not recurrent. In fact, this two-parameter set of group operations has the same topology as the subspace $(a, b, 0)$ in the Lie algebra. We show this below.

In addition to providing an example that shows that $\text{EXP}(X)$ may not map onto the group when the group is noncompact, Cartan provided a theorem that a succession of mappings would always do the job. For simple groups (Chapter 9) the product of two exponential mappings – one of the compact generators, the other of the noncompact generators – will map the algebra onto the group. To separate compact and noncompact generators we use the Cartan–Killing inner product (4.43) computed in the defining matrix representation (7.2)

$$(X, X) = \text{tr } X^2 = 2(a^2 + b^2 - c^2) \tag{7.10}$$

The metric is positive-definite on noncompact generators and negative-definite on noncompact generators. This decomposition in the Lie algebra leads to

$$\begin{bmatrix} a & b \\ b & -a \end{bmatrix} + \begin{bmatrix} 0 & c \\ -c & 0 \end{bmatrix}$$

$$\text{EXP} \downarrow \qquad \downarrow \qquad \downarrow \text{EXP} \tag{7.11}$$

$$\begin{bmatrix} z+y & x \\ x & z-y \end{bmatrix} \times \begin{bmatrix} \cos c & \sin c \\ -\sin c & \cos c \end{bmatrix}$$

For simplicity we have set $z = \cosh r \geq 1$ and $(x, y) = (b, a)\sinh(r)/r$, $r^2 = a^2 + b^2$. We observe that

$$z^2 - x^2 - y^2 = 1 \tag{7.12}$$

which is just the upper sheet of the two-sheeted hyperboloid H^2_{2+}, shown in Fig. 7.2(a). This sheet is topologically equivalent to the space R^2, the plane that it covers. For the compact generator only a small range of parameter values $-\pi \leq c \leq +\pi$ is required to map the subalgebra onto the subgroup $SO(2)$.

7.2 The covering problem

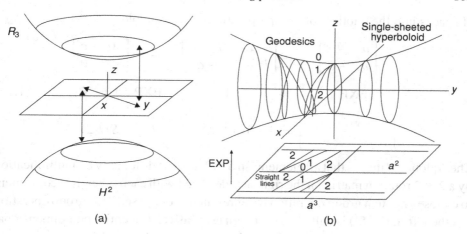

Figure 7.2. (a) Two-sheeted and (b) single-sheeted hyperboloids. Both are quotients (coset spaces) of $SL(2; \mathbb{R})$ by one of its two inequivalent types of subgroups, $SO(2)$ and $SO(1, 1)$.

The connection of $SL(2; \mathbb{R})$ with geometry may be unexpected, but it is not unique to $SL(2; \mathbb{R})$. Moreover, other geometric structures are obtained by exponentiating different subspaces of the algebra $\mathfrak{sl}(2; \mathbb{R})$. For example

$$\begin{bmatrix} a & c \\ -c & -a \end{bmatrix} + \begin{bmatrix} 0 & b \\ b & 0 \end{bmatrix}$$

$$\text{EXP} \downarrow \qquad \downarrow \qquad \downarrow \text{EXP} \qquad (7.13)$$

$$\begin{bmatrix} z+y & x \\ -x & z-y \end{bmatrix} \times \begin{bmatrix} \cosh b & \sinh b \\ \sinh b & \cosh b \end{bmatrix}$$

In this expression for the coset representatives (recall the definition of cosets, or quotients of a group by a subgroup, given in Chapter 1) the three real parameters (x, y, z) obey

$$z^2 + x^2 - y^2 = 1 \qquad (7.14)$$

This equation describes the surface of the single-sheeted hyperboloid H_1^2, shown in Fig. 7.2(b). Many other algebraic surfaces can be obtained from Lie algebras in this way.

We point out that the EXPonential function maps the sum of two subspaces in the algebra into the product of the associated group operations (cf. (7.11) and (7.13)). We can regard one of the subspaces as the difference between the full space (Lie algebra) and the other subspace (subalgebra). The EXPonential maps the difference of spaces into the quotient of group operations. For example

$$\begin{bmatrix} a & b \\ b & -a \end{bmatrix} = \begin{bmatrix} a & b+c \\ b-c & -a \end{bmatrix} - \begin{bmatrix} 0 & c \\ -c & 0 \end{bmatrix}$$

EXP ↓   EXP ↓   ↓ EXP   ↓ EXP   (7.15)

$$\begin{bmatrix} z+y & x \\ x & z-y \end{bmatrix} = \quad SL(2;\mathbb{R}) \quad / \quad SO(2)$$

The "quotient" means that all elements in $SL(2;\mathbb{R})$ that differ only by multiplication by a 2×2 rotation matrix on the right are identified with each other. It is convenient to choose one such group operation to represent this entire set. This group operation (on the left in (7.15)) is called a **coset representative**. The entire one-dimensional set parameterized by c, $0 \leq c < 2\pi$, is the coset. In the theory of Lie groups, cosets and coset representatives are usually interesting spaces.

From this discussion we conclude that the group $SL(2;\mathbb{R})$ can be viewed in various different ways involving coset decompositions. In the parameterization (7.11) obtained from the coset decomposition $SL(2;\mathbb{R})/SO(2)$, the manifold parameterizing the group is the direct product of the upper sheet of the two-sheeted hyperboloid with a circle. Since the upper sheet of a two-sheeted hyperboloid is topologically (but not geometrically!) equivalent to R^2, the manifold that parameterizes $SL(2;\mathbb{R})$ is the direct product $R^2 \times S^1$. A different parameterization (7.13) based on the coset decomposition $[SL(2;\mathbb{R})/SO(1,1)] \times SO(1,1)$ ($SO(1,1) \simeq R^1$) shows that the manifold underlying $SL(2;\mathbb{R})$ is the direct product of the single-sheeted hyperboloid (equivalent to $R^1 \times S^1$) with R^1. This product is once again $R^2 \times S^1$.

Since matrix Lie groups are defined by algebraic constraints, so are their subgroups and quotient spaces. This means that the underlying manifold for each matrix Lie group is an algebraic manifold. For example, for subgroups of $GL(n;\mathbb{R})$ the underlying manifold is a subset of R^N, $N = n^2$, that is defined by algebraic constraints. This manifold can be expressed as products of algebraic submanifolds, each parameterizing a subgroup or coset.

We conclude this discussion of the covering problem by stating a theorem due to Cartan. It is always possible to map a Lie algebra onto its Lie group with a product of exponential mappings. In fact, if the algebra can be written in the form

algebra = noncompact generators + compact generators

EXP ↓   EXP ↓   ↓   ↓ EXP   (7.16)

group = coset representatives × compact subgroup

then the product of two exponential maps, one of the noncompact generators, the other of the compact generators (which form a subalgebra), maps onto the entire Lie

7.3 The isomorphism problem and the covering group

Isomorphic Lie groups have isomorphic Lie algebras, but two Lie groups with isomorphic Lie algebras need not be isomorphic. To illustrate this point, we treat the groups $SO(2, 1)$ and $SU(1, 1)$ with Lie algebras

$$\mathfrak{so}(2,1) = \begin{bmatrix} 0 & a_3 & a_2 \\ -a_3 & 0 & a_1 \\ a_2 & a_1 & 0 \end{bmatrix} \quad \mathfrak{su}(1,1) = \frac{i}{2}\begin{bmatrix} b_3 & ib_1+b_2 \\ ib_1-b_2 & -b_3 \end{bmatrix} \quad (7.17)$$

The Lie algebras are isomorphic but the Lie groups are not. The group $SO(2,1)$ is covered by the map

$$\begin{bmatrix} 0 & 0 & a_2 \\ 0 & 0 & a_1 \\ a_2 & a_1 & 0 \end{bmatrix} + \begin{bmatrix} 0 & a_3 & 0 \\ -a_3 & 0 & 0 \\ 0 & 0 & 0 \end{bmatrix}$$

$$\text{EXP} \downarrow \qquad \downarrow \qquad \downarrow \text{EXP} \qquad\qquad (7.18)$$

$$[SO(2,1)/SO(2)] \times \begin{bmatrix} \cos a_3 & \sin a_3 & 0 \\ -\sin a_3 & \cos a_3 & 0 \\ 0 & 0 & 1 \end{bmatrix}$$

The group $SU(1,1)$ is similarly covered by

$$\frac{i}{2}\begin{bmatrix} 0 & ib_1+b_2 \\ ib_1-b_2 & 0 \end{bmatrix} + \frac{i}{2}\begin{bmatrix} b_3 & 0 \\ 0 & -b_3 \end{bmatrix}$$

$$\text{EXP} \downarrow \qquad \downarrow \qquad \downarrow \text{EXP} \qquad\qquad (7.19)$$

$$[SU(1,1)/U(1)] \quad \times \quad \begin{bmatrix} e^{+ib_3/2} & 0 \\ 0 & e^{-ib_3/2} \end{bmatrix}$$

The cosets $SO(2,1)/SO(2)$ and $SU(1,1)/U(1)$ are both isomorphic to R^2 and have a 1:1 correspondence. The subgroups $SO(2)$ and $U(1)$ have a 2:1 correspondence. This can be seen by increasing b_3 by 2π and noticing that the 2×2 unitary matrix in (7.19) goes to its negative: $U(b_3 + 2\pi) = -U(b_3)$. However, increasing a_3 by 2π does not change the 3×3 rotation matrix in (7.18). The 2:1 correspondence can be seen in a better and simpler way. One can ask: how far along a straight line through the origin does one have to go to return to the identity? For the subgroup

$U(1) \subset SU(1, 1)$ the result is 4π; for the subgroup $SO(2) \subset SO(2, 1)$ the result is 2π. Therefore, $SU(1, 1)$ is "twice as large" as $SO(2, 1)$. More formally, there is a $2 \to 1$ homomorphism of $SU(1, 1)$ onto $SO(2, 1)$.

Once again there is a result due to Cartan that is useful for comparing Lie groups that have isomorphic Lie algebras. Since the noncompact parts of the Lie algebras map to elements of the group with the topology of a Euclidean space, a comparison of the largest compact subgroups of the two groups is sufficient to determine whether the groups are isomorphic.

The most familiar example of nonisomorphic groups with isomorphic Lie algebras is the pair $SO(3)$ and $SU(2)$ with algebras

$$\mathfrak{so}(3) = \begin{bmatrix} 0 & a_3 & -a_2 \\ -a_3 & 0 & a_1 \\ a_2 & -a_1 & 0 \end{bmatrix} \quad \mathfrak{su}(2) = \frac{i}{2}\begin{bmatrix} b_3 & b_1 - ib_2 \\ b_1 + ib_2 & -b_3 \end{bmatrix} \quad (7.20)$$

It can be checked that all points in the interior of a sphere of radius $\sqrt{a_1^2 + a_2^2 + a_3^2} \leq \pi$) map onto $SO(3)$ provided antipodal points at $|\mathbf{a}| = \pi$ are identified

$$\pi(\sin\theta\cos\phi, \sin\theta\sin\phi, \cos\theta) \sim -\pi(\sin\theta\cos\phi, \sin\theta\sin\phi, \cos\theta)$$

with θ the latitude, and ϕ the longitude on a sphere. For $SU(2)$ all points within a sphere of radius 2π ($\sqrt{b_1^2 + b_2^2 + b_3^2} < 2\pi$) are mapped onto distinct elements of $SU(2)$ and all points at a radius of 2π are mapped onto $-I_2$. There is an easier way to verify the $2 \to 1$ nature of the map $SU(2)$ to $SO(3)$. All straight lines through the origin of the Lie algebra are equivalent (since the algebra has rank 1, see Chapter 8). Therefore, we can compare how a convenient line (z-axis) maps onto the two groups. This has already been done for the comparison of $SU(1, 1)$ with $SO(2, 1)$.

Another convenient parameterization of $SO(3)$ and $SU(2)$ can be used to show the 2:1 map. This is analogous to (7.18)

$$\mathfrak{so}(3) = \begin{bmatrix} 0 & 0 & -a_2 \\ 0 & 0 & a_1 \\ a_2 & -a_1 & 0 \end{bmatrix} + \begin{bmatrix} 0 & a_3 & 0 \\ -a_3 & 0 & 0 \\ 0 & 0 & 0 \end{bmatrix}$$

$$\text{EXP} \downarrow \qquad \downarrow \text{EXP} \qquad \downarrow \text{EXP} \qquad (7.21)$$

$$\begin{bmatrix} * & * & -x \\ * & * & y \\ x & -y & z \end{bmatrix} \qquad \times \qquad \begin{bmatrix} \cos a_3 & \sin a_3 & 0 \\ -\sin a_3 & \cos a_3 & 0 \\ 0 & 0 & 1 \end{bmatrix}$$

7.3 The isomorphism problem and the covering group

A similar parameterization for $SU(2)$ gives

$$\mathfrak{su}(2) = \frac{i}{2}\begin{bmatrix} 0 & b_1 - ib_2 \\ b_1 + ib_2 & 0 \end{bmatrix} + \frac{i}{2}\begin{bmatrix} b_3 & 0 \\ 0 & -b_3 \end{bmatrix}$$

$$\text{EXP} \downarrow \qquad\qquad \downarrow \text{EXP} \qquad\qquad \downarrow \text{EXP} \qquad\qquad (7.22)$$

$$\begin{bmatrix} z' & i(x' - iy') \\ i(x' + iy') & z' \end{bmatrix} \times \begin{bmatrix} e^{ib_3/2} & 0 \\ 0 & e^{-ib_3/2} \end{bmatrix}$$

The coset representatives $SO(3)/SO(2)$, parameterized by the real numbers (x, y, z) subject to $x^2 + y^2 + z^2 = 1$, and $SU(2)/U(1)$, parameterized by the real numbers (x', y', z') subject to $x'^2 + y'^2 + z'^2 = 1$, are in 1:1 correspondence with points in the same geometric space – a sphere in this case. As a result, the 2:1 nature of the mapping $SU(2) \to SO(3)$ can be seen from the 2:1 nature of the rotations around the "3" axis.

Yet another result of Cartan establishes a unique connection between Lie groups and Lie algebras. There is a unique Lie algebra for every Lie group. For each Lie algebra there may be many inequivalent Lie groups. But there is a unique Lie group, $\overline{G}$, called the **universal covering group**. This group is simply connected: every loop starting and ending at the identity can be continuously deformed to the identity. Moveover, every other Lie group with this Lie algebra is either identical to this simply connected Lie group, or else has the form of a quotient $\overline{G}/D$, where D is a discrete invariant subgroup of $\overline{G}$ whose elements commute with $\overline{G}$: $gd_i = d_i g$ for $d_i \in D$ and $g \in G$. If $\overline{G}$ is compact it is useful to determine the largest such subgroup, D_{MAX}, of $\overline{G}$. Then all compact Lie groups with the same Lie algebra as $\overline{G}$ are obtained by "dividing" $\overline{G}$ by all possible subgroups of D_{MAX}, as shown in Fig. 7.3.

For simple matrix Lie groups G, computation of the discrete invariant subgroup D is a simple matter. The only discrete group operations d_i that commute with all $g \in G$ are multiples of the identity, by Schur's lemma

$$g \in G, \quad d_i \in D, \quad G \text{ simple}, \quad gd_i = d_i g \Rightarrow d_i = \lambda I_n \qquad (7.23)$$

Two Lie groups with isomorphic Lie algebras are **locally isomorphic**. If G_1 and G_2 have the same Lie algebra, $G_1 = \overline{G}/D_1$ and G_1 is locally isomorphic with $\overline{G}$. By the same argument G_2 is locally isomorphic with $\overline{G}$, and therefore also with G_1. If $\overline{G}$ is compact, G_1 and G_2 are also locally isomorphic with $\overline{G}/D_{\text{MAX}}$, which is a universal image Lie group.

$$G_1 = \overline{G}/D_1 \to \overline{G}/D_{\text{MAX}} \leftarrow \overline{G}/D_2 = G_2$$

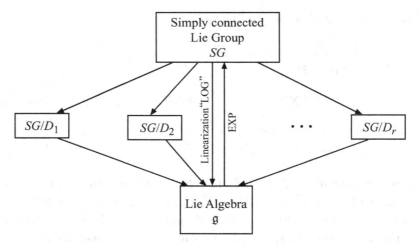

Figure 7.3. Cartan's covering theorem. There is a unique correspondence between Lie algebras $\mathfrak{g}$ and simply connected Lie groups $SG = \overline{G}$. Every other Lie group with this Lie algebra is a quotient of the universal covering group by one of the discrete invariant subgroups D_i of $\overline{G}$.

Example The maximal discrete invariant subgroup of $SU(2)$ consists of matrices λI_2 that obey $\lambda^*\lambda = 1$ and $\det(\lambda I_2) = +1$, so that $\lambda = \pm 1$. D is the two-element subgroup $D = \{I_2, -I_2\}$. For the locally isomorphic Lie group $SO(3)$, $D = \lambda I_3$ with $\lambda = +1$. As a result $SU(2)/\{I_2, -I_2\} = SO(3)/I_3 = SO(3)$. For each group operation in $SO(3)$ there are two matrices in $SU(2)$ that differ in sign.

Remark The maximal compact subgroups $SO(2)$ of $SO(2, 1)$ and $U(1)$ of $SU(1, 1)$ are not simply connected. Their simply connected covering group is R^1, the group of translations of the line. The covering group $\overline{SO(2,1)} = \overline{SU(1,1)}$ has no compact subgroup at all. Its underlying group manifold is $\overline{SO(2,1)/SO(2)} \times \overline{SO(2)} = \overline{SU(1,1)/U(1)} \times \overline{U(1)} = [SO(2,1)/SO(2)] \times SO(2) = SU(1,1)/U(1) \times \overline{U(1)} = R^2 \times R^1$. It is the only group we will encounter in this book that is not a matrix group. The covering group $\overline{SO(2,1)} = \overline{SU(1,1)}$ has many discrete invariant subgroups but does not have a maximal discrete invariant subgroup.

7.4 The parameterization problem and BCH formulas

A Lie algebra can be mapped onto a Lie group in many different ways. More generally, points in the underlying topological space can be identified with group operations in an unlimited number of ways. These different parameterizations of a Lie group can be related to each other by analytic transformations in a way that

7.4 The parameterization problem and BCH formulas

can often be used to simplify computations. Reparameterization formulas involving products of exponentials of operators are called **Baker–Campbell–Hausdorff** (BCH) formulas for historical reasons. Once again we illustrate by example rather than present a general theory.

As a first example we consider the affine group of transformations of the line, and two different parameterizations of this group. One maps a point (x, y) in the right half-plane R_+^2 into the group operator

$$(x, y) \to \begin{bmatrix} x & y \\ 0 & 1 \end{bmatrix} \qquad x > 0 \tag{7.24}$$

The second maps a point (w, z) in R^2 into the group under the EXPonential map

$$(w, z) = \text{EXP} \begin{bmatrix} w & z \\ 0 & 1 \end{bmatrix} = \begin{bmatrix} e^w & (e^w - 1)z/w \\ 0 & 1 \end{bmatrix} \tag{7.25}$$

We ask: is there some mapping of the half-plane R_+^2 ($x > 0$, y) into R^2 (w, z) that makes these two group operations, and the group multiplication laws derived from them, equivalent? The transformation between these two parameterizations is obtained by identifying matrix elements:

$$(x, y) \to \begin{bmatrix} x & y \\ 0 & 1 \end{bmatrix} = \begin{bmatrix} e^w & (e^w - 1)z/w \\ 0 & 1 \end{bmatrix} \leftarrow (w, z) \tag{7.26}$$

The mapping ("diffeomorphism") between the half-plane R_+^2 and the plane R^2, or the coordinates (x, y) and (w, z), is

$$\begin{aligned} x &= e^w \\ y &= (e^w - 1)z/w = z\left(1 + \frac{w}{2!} + \frac{w^2}{3!} + \cdots\right) \end{aligned} \tag{7.27}$$

and the inverse transformation is

$$\begin{aligned} w &= \ln x \\ z &= y \ln(x)/(x - 1) \qquad z = 0 \text{ for } x = 1 \end{aligned} \tag{7.28}$$

These transformations are analytic for $x > 0$.

As a second example we treat the algebra of upper triangular 3×3 matrices

$$\begin{bmatrix} 0 & l & \delta \\ 0 & 0 & r \\ 0 & 0 & 0 \end{bmatrix} = l X_l + r X_r + \delta X_\delta \tag{7.29}$$

The commutation relations of these three generators are

$$[X_l, X_r] = X_\delta \qquad [X_l, X_\delta] = [X_r, X_\delta] = 0 \tag{7.30}$$

The single-mode photon operators a, $a^\dagger$, I obey isomorphic commutation relations

$$[a, a^\dagger] = I \qquad [a, I] = [a^\dagger, I] = 0 \tag{7.31}$$

The two Lie algebras are isomorphic under

$$\begin{aligned} X_l &\to a \\ X_r &\to a^\dagger \\ X_\delta &\to I \end{aligned} \tag{7.32}$$

For many quantum computations it is convenient to relate several different parameterizations of the Lie group. For example, the following "disentangling" results are useful

$$e^{ra^\dagger + la + \delta I}$$

$$\parallel \qquad\qquad \parallel \tag{7.33}$$

$$e^{r'a^\dagger} e^{\delta' I} e^{l'a} \;=\; e^{l''a} e^{\delta'' I} e^{r''a^\dagger}$$

This reparameterization computation can be carried out using 3×3 matrices

$$\mathrm{EXP} \begin{bmatrix} 0 & l & \delta \\ 0 & 0 & r \\ 0 & 0 & 0 \end{bmatrix} \;=\; \begin{bmatrix} 1 & l & \delta + \tfrac{1}{2}lr \\ 0 & 1 & r \\ 0 & 0 & 1 \end{bmatrix}$$

$$\parallel \qquad\qquad\qquad \parallel \tag{7.34}$$

$$e^{r'a^\dagger} e^{\delta' I} e^{l'a} \to \begin{bmatrix} 1 & l' & \delta' \\ 0 & 1 & r' \\ 0 & 0 & 1 \end{bmatrix} = \begin{bmatrix} 1 & l'' & \delta'' + l''r'' \\ 0 & 1 & r'' \\ 0 & 0 & 1 \end{bmatrix} \leftarrow e^{l''a} e^{\delta'' I} e^{r''a^\dagger}$$

We see immediately that $l = l' = l''$, $r = r' = r''$, $\delta' = \delta + \tfrac{1}{2}lr = \delta'' + l''r''$, and obtain the Heisenberg identity (for $\delta = 0$)

$$e^{ra^\dagger} e^{+\tfrac{1}{2}lrI} e^{la} = e^{ra^\dagger + la} = e^{la} e^{-\tfrac{1}{2}lrI} e^{ra^\dagger} \tag{7.35}$$

As a third example we treat the four-parameter Lie group of solvable 3×3 matrices with Lie algebra

$$\begin{bmatrix} 0 & l & \delta \\ 0 & \eta & r \\ 0 & 0 & 0 \end{bmatrix} = \eta X_\eta + l X_l + r X_r + \delta X_\delta \tag{7.36}$$

7.4 The parameterization problem and BCH formulas

This Lie algebra is isomorphic with the Lie algebra spanned by the four single-mode photon operators $\hat{n} = a^\dagger a, a, a^\dagger, I$ under the identification

$$\begin{aligned} X_n &\to \hat{n} \\ X_l &\to a \\ X_r &\to a^\dagger \\ X_s &\to I \end{aligned} \tag{7.37}$$

If for some reason $\text{EXP}(\eta a^\dagger a + r a^\dagger + l a)$ needed to be rewritten in the more conveniently ordered form $\text{EXP}(r' a^\dagger)\text{EXP}(\eta' a^\dagger a + \delta' I)\text{EXP}(l' a)$, then the reparameterization computation could be carried out in the 3×3 matrix representation

$$\text{EXP}(\eta a^\dagger a + r a^\dagger + l a) = \text{EXP}(r' a^\dagger)\text{EXP}(\eta' a^\dagger a + \delta' I)\text{EXP}(l' a)$$

$$\parallel \qquad\qquad \parallel \tag{7.38}$$

$$\begin{bmatrix} 1 & (e^\eta - 1)l/\eta & (e^\eta - 1 - \eta)lr/\eta^2 \\ 0 & e^\eta & (e^\eta - 1)r/\eta \\ 0 & 0 & 1 \end{bmatrix} = \begin{bmatrix} 1 & l' & \delta' \\ 0 & e^{\eta'} & r' \\ 0 & 0 & 1 \end{bmatrix}$$

By inspection, we obtain

$$\begin{aligned} \eta' &= \eta & l' &= (e^\eta - 1)l/\eta \\ \delta' &= (e^\eta - 1 - \eta)lr/\eta^2 & r' &= (e^\eta - 1)r/\eta \end{aligned} \tag{7.39}$$

If it is necessary to compute the expectation value of $\text{EXP}(\eta a^\dagger a + r a^\dagger + l a)$ in the ground state of the harmonic oscillator, then

$$\langle 0| e^{\eta a^\dagger a + r a^\dagger + l a} |0\rangle = \langle 0| e^{r' a^\dagger} e^{\eta' a^\dagger a + \delta' I} e^{l' a} |0\rangle \tag{7.40}$$

Since $e^{l'a}|0\rangle = |0\rangle$, $\langle 0|e^{r'a^\dagger} = \langle 0|$ and $e^{\eta' a^\dagger a}|0\rangle = |0\rangle$, the expectation value is

$$\langle 0| e^{\eta a^\dagger a + r a^\dagger + l a} |0\rangle = e^{\delta'} = \text{EXP}\left(\frac{(e^\eta - 1 - \eta)lr}{\eta^2}\right) \tag{7.41}$$

This result is not easy to derive by other techniques.

As a final example we treat the Lie algebra $\mathfrak{su}(2)$. First, we show how to compute the matrix element of an arbitrary rotation between "ground state" wavefunctions $(|j, -j\rangle)$

$$\left\langle \begin{matrix} j \\ -j \end{matrix} \middle| e^{i\theta \cdot \mathbf{J}} \middle| \begin{matrix} j \\ -j \end{matrix} \right\rangle \tag{7.42}$$

This expectation would be easy to compute if the exponential were written in a "normally ordered form"

$$\left\langle \begin{array}{c} j \\ -j \end{array} \right| e^{i\theta \cdot \mathbf{J}} \left| \begin{array}{c} j \\ -j \end{array} \right\rangle = \left\langle \begin{array}{c} j \\ -j \end{array} \right| e^{i\theta'_+ J_+} e^{i\theta'_z J_z} e^{i\theta'_- J_-} \left| \begin{array}{c} j \\ -j \end{array} \right\rangle \tag{7.43}$$

Since

$$e^{i\theta'_- J_-} \left| \begin{array}{c} j \\ -j \end{array} \right\rangle = (I + i\theta'_- J_- + \cdots) \left| \begin{array}{c} j \\ -j \end{array} \right\rangle = \left| \begin{array}{c} j \\ -j \end{array} \right\rangle \tag{7.44}$$

with a similar result for J_+ acting on the left, we find

$$\left\langle \begin{array}{c} j \\ -j \end{array} \right| e^{i\theta \cdot \mathbf{J}} \left| \begin{array}{c} j \\ -j \end{array} \right\rangle = \left\langle \begin{array}{c} j \\ -j \end{array} \right| e^{i\theta'_z J_z} \left| \begin{array}{c} j \\ -j \end{array} \right\rangle = e^{-ij\theta'_z} \tag{7.45}$$

The only problem that remains is to compute θ'_z as a function of θ. To do this we carry out the operator disentangling calculations in the faithful 2×2 matrix representation $\mathbf{J} \to \frac{1}{2}\sigma$, where σ are the Pauli spin matrices (5.14):

$$e^{i\theta \cdot \mathbf{J}} \to \mathrm{EXP} \frac{i}{2} \begin{bmatrix} \theta_z & \theta_x - i\theta_y \\ \theta_x + i\theta_y & -\theta_z \end{bmatrix}$$

$$= \begin{bmatrix} \cos(\theta/2) + i(\theta_z/\theta)\sin(\theta/2) & i[(\theta_x - i\theta_y)/\theta]\sin(\theta/2) \\ i[(\theta_x + i\theta_y)/\theta]\sin(\theta/2) & \cos(\theta/2) - i(\theta_z/\theta)\sin(\theta/2) \end{bmatrix} \tag{7.46}$$

In a similar way we find

$$\begin{array}{ccc} \mathrm{EXP}(i\theta'_+ J_+) & \mathrm{EXP}(i\theta'_z J_z) & \mathrm{EXP}(i\theta'_- J_-) \\ \downarrow & \downarrow & \downarrow \end{array}$$

$$\begin{bmatrix} 1 & i\theta'_+ \\ 0 & 1 \end{bmatrix} \begin{bmatrix} e^{i\theta'_z/2} & 0 \\ 0 & e^{-i\theta'_z/2} \end{bmatrix} \begin{bmatrix} 1 & 0 \\ i\theta'_- & 1 \end{bmatrix}$$

$$= \begin{bmatrix} e^{i\theta'_z/2} - \theta'_+ \theta'_- e^{-i\theta'_z/2} & i\theta'_+ e^{-i\theta'_z/2} \\ i\theta'_- e^{-i\theta'_z/2} & e^{-i\theta'_z/2} \end{bmatrix} \tag{7.47}$$

where $\theta_\pm = \theta_1 \pm i\theta_2$. Comparison of the two matrices gives immediately

$$e^{-i\theta'_z/2} = \cos(\theta/2) - i(\theta_z/\theta)\sin(\theta/2) \tag{7.48}$$

As a result, we find

$$\left\langle \begin{array}{c} j \\ -j \end{array} \right| e^{i\theta \cdot \mathbf{J}} \left| \begin{array}{c} j \\ -j \end{array} \right\rangle = e^{-ij\theta'_z} = (e^{-i\theta'_z/2})^{2j} = [\cos(\theta/2) - i(\theta_z/\theta)\sin(\theta/2)]^{2j} \tag{7.49}$$

This result is useful in the field of quantum optics but is not easy to compute by other means.

7.4 The parameterization problem and BCH formulas

To illustrate the use of Baker–Campbell–Hausdorff formulas in another situation we compute the matrix elements

$$\left\langle \begin{matrix} j \\ j \end{matrix} \middle| J_+^k J_-^k \middle| \begin{matrix} j \\ j \end{matrix} \right\rangle \tag{7.50}$$

To do this we construct a generating function

$$\left\langle \begin{matrix} j \\ j \end{matrix} \middle| e^{\alpha J_+} e^{\beta J_-} \middle| \begin{matrix} j \\ j \end{matrix} \right\rangle = \sum_{rs} \frac{\alpha^r \beta^s}{r! s!} \left\langle \begin{matrix} j \\ j \end{matrix} \middle| J_+^r J_-^s \middle| \begin{matrix} j \\ j \end{matrix} \right\rangle \tag{7.51}$$

The operator product $e^{\alpha J_+} e^{\beta J_-}$ is written in normally ordered form $\mathrm{EXP}(\beta' J_-) \mathrm{EXP}(n' J_z) \mathrm{EXP}(\alpha' J_+)$ and the parameters α', β', n' computed. We find

$$\left\langle \begin{matrix} j \\ j \end{matrix} \middle| e^{\beta' J_-} e^{n' J_z} e^{\alpha' J_+} \middle| \begin{matrix} j \\ j \end{matrix} \right\rangle = e^{jn'} = (1 + \alpha\beta)^{2j} \tag{7.52}$$

By expanding $(1 + \alpha\beta)^{2j}$ and invoking analyticity, we find

$$\left\langle \begin{matrix} j \\ j \end{matrix} \middle| J_+^r J_-^s \middle| \begin{matrix} j \\ j \end{matrix} \right\rangle = \frac{(2j)! r!}{(2j - r)!} \delta_{rs} \tag{7.53}$$

Other matrix elements of products of angular momentum operators can be constructed similarly from appropriate generating functions.

The general computational procedure should now be clear. Given a Lie algebra of operators and the associated group operations that are exponentials of the elements in the Lie algebra, it is possible to carry out all calculations in either the algebra or the group using a faithful matrix representation of the operator algebra. In general, the smaller the size of the matrices, the easier the computation.

For example, if operators $\mathcal{A}, \mathcal{B}$ belong to two complementary subspaces in some operator Lie algebra $\mathfrak{g}$ then the operator product $e^{\mathcal{A}} e^{\mathcal{B}}$ can be reparameterized as $e^{\mathcal{B}'} e^{\mathcal{A}'}$ ($\mathcal{A}', \mathcal{B}'$ different operators in the same subspaces as $\mathcal{A}, \mathcal{B}$) by

(i) finding a faithful matrix representation of the operator algebra,
(ii) identifying the operators $\mathcal{A}, \mathcal{B}$ with matrices A, B,
(iii) Carrying out the matrix calculations $e^A e^B$ and $e^{B'} e^{A'}$,
(iv) determining the matrices A', B' by comparing matrix elements; and
(v) using the isomorphism $A' \leftrightarrow \mathcal{A}'\ B' \leftrightarrow \mathcal{B}'$.

This procedure will produce a local analytic reparameterization $(\mathcal{A}, \mathcal{B}) \leftrightarrow (\mathcal{A}', \mathcal{B}')$. If the matrix group used to construct this reparameterization is simply connected (the covering group) the analytic reparameterization will be global. Otherwise, some care must be taken to compare the maximal discrete invariant subgroups of the operator group and the matrix group. When the operators $\mathcal{A}, \mathcal{B}, \ldots$ are related to matrices A, B, ... by a matrix–operator mapping (see Chapter 6) $\mathcal{A} \leftrightarrow A$, the disentangling formulas can be constructed using the matrices A, B, ...

7.5 EXPonentials and physics

By the greatest good fortune – or perhaps by the deepest possible connections between mathematics and physics – the exponential function also plays a most fundamental role in physics. In fact, it plays two roles: one in dynamics and another in equilibrium statics (thermo"dynamics"). More fundamental yet, these two roles are related by analytic continuation ("Wick rotation"). We describe both roles in this section, in terms of two examples, one related to fermions, the other related to bosons.

7.5.1 Dynamics

The dynamics of quantum systems is governed by the time-dependent Schrödinger equation:

$$H|\psi\rangle = i\hbar \frac{\partial}{\partial t}|\psi\rangle \tag{7.54}$$

The state of the system at time $t + \delta t$ is related to the state at time t by

$$|\psi(t+\delta t)\rangle = \left(I - \frac{i}{\hbar} H \delta t\right)|\psi(t)\rangle = e^{-\frac{i}{\hbar} H \delta t}|\psi(t)\rangle \tag{7.55}$$

The exponential is unitary since the hamiltonian operator H is hermitian. The state $|\psi(t_f)\rangle$ at some final time t_f is related to the state at initial time t_i by $|\psi(t_f)\rangle = U(t_f, t_i)|\psi(t_i)\rangle$. The finite time unitary operator is built up from small displacements

$$U(t_f, t_i) = U(t_f, t_f - \delta t) \cdots U(t_i + 2\delta t, t_i + \delta t) U(t_i + \delta t, t_i)$$

$$= \prod U(t_i + (n+1)\delta t, t_i + n\delta t) = \text{``} \int_{t_i}^{t_f} \text{''} U(\tau) d\tau$$

$$= T \int_{t_i}^{t_f} e^{-\frac{i}{\hbar} H(t)} dt \tag{7.56}$$

Care must be taken with the formal integration in this equation, as in general $H(t')$ does not commute with $H(t)$, $t' \neq t$. It is for this reason that the symbol "T" precedes the integral: this signifies a time-ordered product. If the hamiltonian is not explicitly time dependent then the integral in Eq. (7.56) reduces to an everyday Riemann integral.

7.5 EXPonentials and physics

Expression of the time dependence in terms of a unitary evolution operator is useful for two very different reasons.

(i) The evolution is decoupled from the initial state.
(ii) In special cases it is very simple to construct this unitary evolution operator when it would be much more difficult to construct the evolution of a specific state.

The second case becomes important when the hamiltonian is a linear superposition of operators that exist in a Lie algebra. In that case the unitary operator is a group operation, and it may be possible to find some shortcuts for its computation. We give two examples.

Example 1. A Hamiltonian acts in a $2j+1$ dimensional space through a set of three operators J_z, $J_\pm$ that obey angular momentum commutation relations. We wish to determine the evolution of some particular state $|j, m_j\rangle$. The Hamiltonian is

$$H = \epsilon(t)J_z + \alpha(t)J_+ + \alpha^*(t)J_- \xrightarrow{j \to \frac{1}{2}} \begin{bmatrix} \frac{1}{2}\epsilon(t) & \alpha(t) \\ \alpha^*(t) & -\frac{1}{2}\epsilon(t) \end{bmatrix} \quad (7.57)$$

The unitary operator acting in the $2j+1$ dimensional space is a unitary representation of some operation in the group $SU(2)$. It is simpler to determine how $g(t) \in SU(2)$ evolves, and then construct its unitary representation, than it is to determine the time evolution of the $(2j+1) \times (2j+1)$ unitary matrix. Specifically, the equation of motion *in the group* is

$$\frac{d}{dt}\begin{bmatrix} a(t) & b(t) \\ -b^*(t) & a^*(t) \end{bmatrix} = -\frac{i}{\hbar}\begin{bmatrix} \frac{1}{2}\epsilon(t) & \alpha(t) \\ \alpha^*(t) & -\frac{1}{2}\epsilon(t) \end{bmatrix}\begin{bmatrix} a(t) & b(t) \\ -b^*(t) & a^*(t) \end{bmatrix} \quad (7.58)$$

After some algebraic manipulation this matrix equation reduces to two equations for the complex coefficients $a(t)$ and $b(t)$ or three equations for the real coefficients of the Pauli spin matrices $\sigma_1, \sigma_2, \sigma_3$. These are first order equations and can be solved by standard integration methods (e.g., RK4). The initial conditions are $a(t_i) = 1$, $b(t_i) = 0$. The final 2×2 unitary matrix is determined by $a(t_f), b(t_f)$. This is a group operation in $SU(2)$ that can subsequently be mapped into the $(2j+1) \times (2j+1)$ unitary irreducible representation of this group. At this point the problem is solved, independent of the initial state $|\psi(t_i)\rangle$.

Example 2. As a second example we treat a hamiltonain that is a linear combination of the boson number, creation, and annihilation operators (and their commutator):

$$H = \omega(t)a^\dagger a + \alpha(t)a^\dagger + \alpha^*(t)a + \delta(t)I \to \begin{bmatrix} 0 & \alpha^*(t) & \delta(t) \\ 0 & \omega(t) & \alpha(t) \\ 0 & 0 & 0 \end{bmatrix} \quad (7.59)$$

The boson operators act as a hermitian superposition in an infinite-dimensional space with basis vectors $|n\rangle$, $n = 0, 1, 2, \ldots$. The matrix on the right is a faithful finite-dimensional *nonhermitian* representation of these operators. The most general unitary operator that can be constructed from these operators is $U = \text{EXP}(i[n(t)a^\dagger a + r(t)a^\dagger + r^*(t)a + d(t)I])$. This exponential is easy to compute in the faithful 3×3 nonunitary representation. The matrix equation of motion analogous to Eq. (7.58) is explicitly

$$\frac{d}{dt}\begin{bmatrix} 1 & r^*\frac{(e^{in}-1)}{(in)} & r^*r\left(\frac{e^{in}-1-in}{(in)^2}\right)+id \\ 0 & e^{in} & r\frac{(e^{in}-1)}{(in)} \\ 0 & 0 & 1 \end{bmatrix}$$

$$= -\frac{i}{\hbar}\begin{bmatrix} 0 & \alpha^*(t) & \delta(t) \\ 0 & \omega(t) & \alpha(t) \\ 0 & 0 & 0 \end{bmatrix}\begin{bmatrix} 1 & r^*\frac{(e^{in}-1)}{(in)} & r^*r\left(\frac{e^{in}-1-in}{(in)^2}\right)+id \\ 0 & e^{in} & r\frac{(e^{in}-1)}{(in)} \\ 0 & 0 & 1 \end{bmatrix} \quad (7.60)$$

This matrix equation leads to an ugly but manageable set of coupled nonlinear equations in four real variables (n, r, r^*, d) that can be integrated by standard methods. In the case that $d\omega(t)/dt = 0$ the equations simplify considerably, and can almost be solved by inspection.

7.5.2 Equilibrium thermodynamics

In classical and quantum physics expectation values are expressed in terms of a density operator ρ

$$\langle \mathcal{O} \rangle = \text{tr}\,\rho \mathcal{O} \quad (7.61)$$

In thermodynamic equilibrium the density operator is expressed in terms of the hamiltonian describing the system as $\rho = e^{-\beta H}/Z$, where the normalization constant, or partition function, is $Z = \text{tr}\,e^{-\beta H}$ and $\beta = 1/k_B T$, k_B is the Boltzmann constant and T is the absolute temperature. When H is an element in a finite-dimensional Lie algebra, many simplifications in the computation of thermal expectation values occur. Again, we give two examples.

Example 1. We choose a hamiltonian constructed from angular momentum operators

$$H = \epsilon J_z + \alpha J_+ + \alpha^* J_- \xrightarrow{j \to \frac{1}{2}} \begin{bmatrix} \frac{1}{2}\epsilon(t) & \alpha(t) \\ \alpha^*(t) & -\frac{1}{2}\epsilon(t) \end{bmatrix} \quad (7.62)$$

7.5 EXPonentials and physics

We would like to be able to compute thermal expectation values of various moments of the angular momentum operators. The simplest way to go about this is to compute *generating functions* for these expectation values. To do this we compute $\langle e^\Lambda \rangle$, where $\Lambda = \lambda \cdot \mathbf{J}$. All symmetric moments can be constructed by taking derivatives of this generating function. We first compute this generating function in the smallest faithful matrix representation:

$$e^{-\beta H} e^\Lambda \to \left(I_2 \cosh(\beta|H|) - \beta \begin{bmatrix} \epsilon/2 & \alpha \\ \alpha^* & -\epsilon/2 \end{bmatrix} \frac{\sinh(\beta|H|)}{\beta|H|} \right)$$
$$\times \left(I_2 \cosh(|\Lambda|) + \begin{bmatrix} \lambda_3/2 & \lambda \\ \lambda^* & -\lambda_3/2 \end{bmatrix} \right) \frac{\sinh(|\Lambda|)}{|\Lambda|} \quad (7.63)$$

The trace of this expression is

$$\text{tr } e^{-\beta H} e^\Lambda \to$$
$$2\cosh(\beta|H|)\cosh(|\Lambda|) - 2 \frac{H \cdot \Lambda}{\sqrt{H \cdot H}\sqrt{\Lambda \cdot \Lambda}} \sinh(\beta|H|)\sinh(|\Lambda|) \quad (7.64)$$

In these expressions $H \cdot \Lambda = (H, \Lambda) = \frac{1}{2} \text{tr } H\Lambda$, and similarly for $|H| = \sqrt{(H, H)}$ and $|\Lambda| = \sqrt{(\Lambda, \Lambda)}$.

The trace of this 2×2 matrix can be written in another useful way after a similarity transform that diagonalizes it:

$$\text{tr } e^{-\beta H} e^{\lambda \cdot \mathbf{J}} = \text{tr } \begin{bmatrix} e^{+\mu(H,\Lambda)/2} & 0 \\ 0 & e^{-\mu(H,\Lambda)/2} \end{bmatrix} = 2\cosh(\mu(H,\Lambda)/2) \quad (7.65)$$

If N two-level atoms are acting incoherently, the trace over the 2^N states of all N atoms is the Nth power of the trace expressed in (7.65). On the other hand, if all N atoms are acting coherently, there are $2J+1$ states, where $N = 2J$. The trace over these states is (Arecchi *et al.*, 1972)

$$\chi(H, \Lambda, J) = \frac{\sinh(J + \frac{1}{2})\mu(H, \Lambda)}{\sinh(\frac{1}{2})\mu(H, \Lambda)} \quad (7.66)$$

where $\mu(H, \Lambda, T)$ is determined from Eq. (7.65). The thermodynamic generating function is

$$\langle e^\Lambda \rangle = \frac{\chi(H, \Lambda, J)}{\chi(H, 0, J)} \quad (7.67)$$

To construct explicit expectation values (e.g., $\langle J_- \rangle$) it is sufficient to differentiate the generating function (e.g., $\frac{\partial}{\partial \lambda^*} \langle e^\Lambda \rangle / \langle e^0 \rangle$) and evaluate the result at $\Lambda = 0$. It is even more convenient to differentiate the logarithm and evaluate at $\Lambda = 0$: $\frac{\partial}{\partial \lambda^*} \log(\langle e^\Lambda \rangle)|_{\Lambda=0}$.

Example 2. As a second example we treat a harmonic oscillator described by a time-independent hamiltonian of the form (7.68) in thermodynamic equilibrium at temperature T

$$H = \hbar\omega a^\dagger a + \alpha a^\dagger + \alpha^* a + \delta I \rightarrow \begin{bmatrix} 0 & \alpha^* & \delta \\ 0 & \hbar\omega & \alpha \\ 0 & 0 & 0 \end{bmatrix} \quad (7.68)$$

The density operator is $\rho = e^{-\beta(\hbar\omega a^\dagger a + \alpha a^\dagger + \alpha^* a + \delta I)}/Z$. The generating function for operator expectation values is $\chi(H, \Lambda, T) = \mathrm{tr}\, e^{-\beta H} e^{\lambda_n a^\dagger a + \lambda a^\dagger + \lambda^* a + dI}/Z = \langle e^\Lambda \rangle$. The trace is taken in the infinite-dimensional Hilbert space with Fock basis $|0\rangle, |1\rangle, |2\rangle, \ldots$. It would be insane to attempt to compute this expectation value without exploiting opportunities allowed by choice of a smaller, more convenient faithful matrix representation M of the group. The calculation proceeds according to the following steps.

(i) Write each of the operators H, Λ in the 3×3 matrix representation M (cf., Eq. (7.59));

(ii) Compute the exponential of each. For example

$$e^{-\beta M(H)} = \mathrm{EXP} - \beta \begin{bmatrix} 0 & \alpha^* & \delta \\ 0 & \hbar\omega & \alpha \\ 0 & 0 & 0 \end{bmatrix} = \begin{bmatrix} 1 & \alpha^* \frac{e^{-\beta\hbar\omega}-1}{\hbar\omega} & \frac{e^{-\beta\hbar\omega}-1+\beta\hbar\omega}{(\hbar\omega)^2}\alpha^*\alpha - \beta\delta \\ 0 & e^{-\beta\hbar\omega} & \alpha\frac{e^{-\beta\hbar\omega}-1}{\hbar\omega} \\ 0 & 0 & 1 \end{bmatrix}$$

(7.69)

(iii) Multiply the group operations together:

$$e^{-\beta M(H)} e^{M(\Lambda)} = \begin{bmatrix} 1 & Z_l & * \\ 0 & * & Z_r \\ 0 & 0 & 1 \end{bmatrix}$$

(iv) Find a similarity transformation, S, that zeroes out Z_l and Z_r:

$$M(S) \begin{bmatrix} 1 & Z_l & * \\ 0 & * & Z_r \\ 0 & 0 & 1 \end{bmatrix} M(S^{-1}) = \begin{bmatrix} 1 & 0 & B \\ 0 & A & 0 \\ 0 & 0 & 1 \end{bmatrix}$$

(v) Map this group operation to the infinite-dimensional matrix representation acting on the Fock space

$$\begin{bmatrix} 1 & 0 & B \\ 0 & A & 0 \\ 0 & 0 & 1 \end{bmatrix} \rightarrow e^{Aa^\dagger a + BI}$$

(vi) Take the trace. Assuming $A < 0$ the sum converges to

$$\text{tr } e^{Aa^\dagger a + BI} = \frac{e^B}{1 - e^A}$$

(vii) Take the logarithm to find

$$\log(\chi(H, \Lambda, T)) = B - A - \log(e^{-A} - 1)$$

(viii) These steps can be implemented easily using symbol manipulation codes. The result is

$$-A = \beta\hbar\omega - \lambda_n$$

$$B = \frac{e^{-\beta\hbar\omega} - 1 + \beta\hbar\omega}{(\hbar\omega)^2}\alpha^*\alpha - \beta\delta + d + \frac{e^{\lambda_n} - 1 - \lambda_n}{\lambda_n^2}\lambda^*\lambda$$

$$+ \frac{e^{-\beta\hbar\omega} - 1}{\hbar\omega}\frac{e^{\lambda_n} - 1}{\lambda_n}(\alpha^*\lambda + \alpha\lambda^*)/(1 - e^{-(\beta\hbar\omega - \lambda_n)})$$

$$+ \left[e^{-\beta\hbar\omega}\left(\frac{e^{\lambda_n} - 1}{\lambda_n}\right)^2 \lambda^*\lambda + e^{\lambda_n}\left(\frac{e^{-\beta\hbar\omega} - 1}{\hbar\omega}\right)^2 \alpha^*\alpha\right]\bigg/\left[1 - e^{-(\beta\hbar\omega - \lambda_n)}\right]$$

(7.70)

The generating function for only the creation and annihilation operators ($\lambda_n = d = 0$) is considerably simpler.

7.6 Conclusion

The EXPonential mapping from a Lie algebra to a Lie group is generally not onto. It is not in general possible to recover the entire Lie group by taking a single exponential of the Lie algebra. However, a sequence of exponential mappings from various linear vector subspaces in the Lie algebra can be found that covers the Lie group. This sequence of exponential mappings can be used to determine the structure of the underlying manifold of the Lie group. It also provides a useful parameterization for the Lie group.

Associated with every Lie algebra $\mathfrak{g}$ is a unique Lie group $\overline{G}$ that is simply connected. Every matrix group with this Lie algebra is locally isomorphic to this covering group. Every Lie group G with Lie algebra $\mathfrak{g}$ has the structure $\overline{G}/D$, where D is a discrete invariant subgroup of $\overline{G}$. If $D = \text{Id}$, G is isomorphic to $\overline{G}$, otherwise it is a homomorphic image of $\overline{G}$. For simple matrix groups, D consists of multiples of the identity matrix, λI_n, and is simple to compute. If G_1 and G_2 have isomorphic Lie algebras they are locally isomorphic with the universal covering group and with each other.

Many different parameterizations of a Lie group are possible. The most useful ones typically involve a sequence of exponential mappings of linear vector subspaces of the Lie algebra into the Lie group. These are "linear" in the sense that the coordinates parameterizing elements in the Lie group are components of a vector in a linear vector space (the Lie algebra). Different parameterizations are related by analytic reparameterization formulas, called Baker–Campbell–Hausdorff formulas for historical reasons. These BCH formulas can be constructed by finding a faithful matrix representation of the Lie algebra, then carrying out the reparameterization computation using products of exponentials of these matrices.

Exponentials play a fundamental role in physics as well as mathematics. We have explored two of the most useful applications of the exponential function in physics. These describe dynamics and statics. The dynamical evolution of a quantum system is governed by a unitary transformation that can be written as a time-ordered exponential. If the hamiltonian is a linear superposition of basis vectors in a finite dimensional Lie algebra many useful computational methods are available for its simple computation. We have provided two illustrations of the methods that are available. If the physical system is in thermodynamic equilibrium, the density operator is also the exponential of the hamiltonian. The two (dynamics and statics) are related by a "Wick rotation": $it/\hbar \leftrightarrow 1/k_B T$. We have used the same two physical systems as vehicles to illustrate how the exponential mapping, and suitable stepping back and forth through large and small unitary or nonunitary but faithful representations, has been used to simplify computation of partition functions and generating functions for symmetrized operator expectation values.

7.7 Problems

1. Construct the analytic group mapping $\phi((x_1, y_1), (x_2, y_2))$ for the parameterization (7.24) of the affine group. Construct the mapping $\phi((w_1, z_1), (w_2, z_2))$ for the parameterization (7.25) of this group.

2. Show that a straight line through the origin of the parameter space (a, b, c) that is inside the light cone $a^2 + b^2 - c^2 < 0$ (Eq. (7.7)) maps onto the subgroup $SO(2) \subset SL(2; \mathbb{R})$. Show that if $a = b = 0$, the basic "repetition period" in the c-direction, c_T, in the subgroup is 2π but if $a^2 + b^2 > 0$ ($\sqrt{a^2 + b^2} = \beta \times c$, $|\beta| < 1$), the basic repetition period in the c-direction is increased to $2\pi \gamma$, where $\gamma = 1/\sqrt{1 - \beta^2}$ and $\beta^2 = (a^2 + b^2)/c^2$. Compare this renormalization of periodicity with "time dilation."

3. Compute the maximal discrete invariant subgroup D_{MAX} of $SU(3)$ and show that it is $\{I_3, \lambda I_3, \lambda^2 I_3\}$, where $\lambda = e^{2\pi i/3}$. Next, show that $SU(3)/D_{MAX}$ is isomorphic to the group of real 8×8 matrices EXP[$\mathfrak{Reg}(\mathfrak{su}(3))$] ("eight-fold way").

4. Compute the maximal discrete invariant subgroup for the special unitary groups $SU(n)$ and show that it is the cyclic group of order n generated by ϵI_n, $\epsilon = e^{2\pi i/n}$. What real matrix group is $SU(n)/D_{MAX}$ equivalent to?

7.7 Problems

5. Show that the covering group $\overline{SU(1,1)}$ does not have a maximum discrete invariant subgroup.

6. It is convenient to introduce the creation and annihilation operators $a^\dagger$, a to study the one-dimensional quantum oscillator. These two operators are defined by

$$a^\dagger = \frac{1}{\sqrt{2}}\left(x - \frac{d}{dx}\right) \qquad a = \frac{1}{\sqrt{2}}\left(x + \frac{d}{dx}\right)$$

Computation of the matrix elements of the moments of x in the harmonic oscillator basis, $\langle n'|x^k|n\rangle$, can be simplified using disentangling theorems. This problem indicates how.

a. The function $e^{\lambda x}$ is a generating function for matrix elements of x^k. Show that

$$\langle n'|x^k|n\rangle = \frac{d^k}{d\lambda^k}\langle n'|e^{\lambda x}|n\rangle_{\lambda=0}$$

b. Use the 3×3 matrix representation for the photon creation and annihilation operators and their commutator $[a, a^\dagger] = I$ to show

$$e^{\lambda x} = e^{\lambda(a^\dagger + a)/\sqrt{2}} = \text{EXP}\left(\frac{\lambda}{\sqrt{2}}\begin{bmatrix} 0 & 1 & 0 \\ 0 & 0 & 1 \\ 0 & 0 & 0 \end{bmatrix}\right) = \begin{bmatrix} 1 & \lambda/\sqrt{2} & \lambda^2/4 \\ 0 & 1 & \lambda/\sqrt{2} \\ 0 & 0 & 1 \end{bmatrix}$$

c. Construct a disentangling theorem that expresses this group operator in the form $e^{ra^\dagger}e^{\delta I}e^{la}$ by constructing the matrix product of these three operators:

$$e^{ra^\dagger}e^{\delta I}e^{la} = \begin{bmatrix} 1 & 0 & 0 \\ 0 & 1 & r \\ 0 & 0 & 1 \end{bmatrix}\begin{bmatrix} 1 & 0 & \delta \\ 0 & 1 & 0 \\ 0 & 0 & 1 \end{bmatrix}\begin{bmatrix} 1 & l & 0 \\ 0 & 1 & 0 \\ 0 & 0 & 1 \end{bmatrix} = \begin{bmatrix} 1 & l & \delta \\ 0 & 1 & r \\ 0 & 0 & 1 \end{bmatrix}$$

d. By comparing the matrices in **b** and **c**, conclude

$$e^{\lambda(a^\dagger + a)/\sqrt{2}} = e^{\lambda a^\dagger/\sqrt{2}} e^{\lambda^2/4} e^{\lambda a/\sqrt{2}}$$

e. Use the disentangling theorem in **d** to compute $\langle n'|x^4|n\rangle$. In particular, show

$$\langle n'|x^4|n\rangle = \frac{d^4}{d\lambda^4}\sum_{p,q,r}\frac{\lambda^{p+2q+r}}{p!q!r!} 2^{-(p/2+2q+r/2)}\langle n'|(a^\dagger)^p a^r|n\rangle_{\lambda=0}$$

$$\to \sum_{p+2q+r=4}\frac{4!}{p!q!r!}\frac{\langle n'|(a^\dagger)^p(a)^r|n\rangle}{2^{(p/2+q+r/2)}}$$

The point of this exercise is that the computation of the matrix elements is simplified because the operators are in **normally ordered** form (all annihilation operators first, on the right and all creation operators last, on the left). As a result, the calculation reduces to summing a descending series with no more than three nonzero terms.

7. In order to describe the scattering of X-rays from an atom moving in a harmonic potential it is necessary to compute a structure factor $\langle e^{ikx}\rangle$. The expectation value is thermal: $P_n \simeq e^{-n\beta\hbar\omega}$. This expectation value can be written in algebraic form as

$$\langle e^{ikx} \rangle = \frac{\text{tr } e^{ikx} e^{-\beta\mathcal{H}}}{\text{tr } e^{-\beta\mathcal{H}}} \tag{7.71}$$

We concentrate on the numerator, as the denominator is obtained in the limit $k \to 0$.

a. Show

$$\text{tr } e^{ikx} e^{-\beta\mathcal{H}} = \sum_{n=0}^{\infty} \langle n|e^{ikx}|n\rangle e^{-n\beta\hbar\omega} = \sum_{n=0}^{\infty} \langle n|e^{ikx} e^{-n\beta\hbar\omega}|n\rangle \tag{7.72}$$

b. The trace is invariant under similarity transform (the operator is bounded). Show that

$$\text{tr } e^{ikx} e^{-\beta\hbar\omega a^\dagger a} = \text{tr } e^{-\beta\hbar\omega a^\dagger a} e^\delta = e^\delta \text{ tr } e^{-\beta\hbar\omega a^\dagger a} \tag{7.73}$$

As a result $\langle e^{ikx} \rangle = e^\delta$.

c. Compute δ using 3×3 nonunitary matrix multiplications to carry out multiplications *in the group* rather than in an $\infty \times \infty$ unitary representation of the group.

$$M(S) \, M(e^{ikx}) \, M(e^{-\beta\hbar\omega\hat{n}}) \, M(S^{-1}) = M(e^{-\beta\hbar\omega'\hat{n}}) \, M(e^\delta) \tag{7.74}$$

$$\begin{array}{cccc}
S & e^{ikx} & e^{-\beta\hbar\omega a^\dagger a} & S^{-1} \\
\downarrow & \downarrow & \downarrow & \downarrow
\end{array}$$

$$\begin{bmatrix} 1 & \alpha & \alpha\beta/2 \\ 0 & 1 & \beta \\ 0 & 0 & 1 \end{bmatrix} \begin{bmatrix} 1 & ik/\sqrt{2} & -k^2/4 \\ 0 & 1 & ik/\sqrt{2} \\ 0 & 0 & 1 \end{bmatrix} \begin{bmatrix} 1 & 0 & 0 \\ 0 & e^{-\beta\hbar\omega} & 0 \\ 0 & 0 & 1 \end{bmatrix} \begin{bmatrix} 1 & -\alpha & \alpha\beta/2 \\ 0 & 1 & -\beta \\ 0 & 0 & 1 \end{bmatrix}$$

$$\tag{7.75}$$

Carry out the multiplication of 3×3 matrices in this nonunitary representation M. Show that $\omega' = \omega$. Determine α, β, and compute γ. Show

$$\langle e^{ikx} \rangle = e^\delta \qquad \delta = -\tfrac{1}{2}k^2 \coth\left(\tfrac{1}{2}\beta\hbar\omega\right) \tag{7.76}$$

8. A finite set of operators X_i closes under commutation: $[X_i, X_j] = \sum_{k=1}^{N} C_{ij}{}^k X_k$. These operators span a finite-dimensional Lie algebra $\mathfrak{g}$ of Lie group G. Assume that this set of operators has two representations R and S with the following properties:
   - R is hermitian: $\left(R(a^i X_i)\right)^\dagger = \left(a^i R(X_i)\right)^\dagger = (a^i)^* R^\dagger(X_i)$.
   - S is faithful: $S(a^i X_i) = 0 \Rightarrow a^i = 0$.

   We require S to be finite dimensional so that simple matrix computations are possible. We require R to be hermitian to make an immediate connection with quantum mechanics.

   a. It happens frequently that $\mathcal{H} = R(a^i X_i)$ describes the physics of some quantum mechanical system. Show that if $H_1, H_2, \ldots, H_r \in \mathfrak{g}$ span a maximal commutative subspace, so that $[H_i, H_j] = 0$, $1 \leq i, j \leq r$, then the hermitian operators $R(H_i)$ are mutually commutative and can all be made diagonal simultaneously in this representation: $[R(H_i)]_{\alpha\beta} = r_\alpha(i)\delta_{\alpha\beta}$.

b. Show that $[S(H_i), S(H_j)] = 0$, but show by example that the r matrices $S(H_i)$ cannot always be simultaneously diagonal.
   c. Show the time evolution of the quantum system is given by the unitary operator $U(t) = R(e^{-\frac{i}{\hbar}\mathcal{H}t}) = e^{-\frac{i}{\hbar}R(\mathcal{H})t}$.
   d. Show that the density operator for thermal expectation values is $\rho(T) = e^{-\beta\mathcal{H}}/Z = R(e^{-\beta\mathcal{H}})/Z = e^{-\beta R(\mathcal{H})}/Z$. What is Z?
   e. Show that the unitary time evolution operator $U(t)$ and the hermitian density operator $\rho(T)$ are related by a Wick rotation $it/\hbar \leftrightarrow \beta = 1/k_B T$.
   f. A generating function for thermal expectation values has the form

   $$\langle e^{x^i X_i} \rangle = \frac{\text{tr } e^{R(x^i X_i)} e^{-\beta\mathcal{H}}}{\text{tr } e^{-\beta\mathcal{H}}} \rightarrow \frac{\text{tr } R(e^{x^i X_i}) R(e^{-\beta a^i X_i})}{\text{tr } R(e^{-\beta a^i X_i})} \quad (7.77)$$

   g. The operator product in the numerator is in the group $G = e^\mathfrak{g}$ or its complex extension. If this operator product can be transformed to "diagonal" form (i.e., expressed in terms of the operators H_i) the trace can easily be constructed. Show that for x^i sufficiently small it is always possible to construct a similarity transformation $S = e^{y^k X_k}$ with the property

   $$S e^{x^j X_j} e^{-\beta a^i X_i} S^{-1} = e^{-\beta d^i(x,a) H_i} \quad (7.78)$$

   h. The thermal expectation value then reduces to

   $$\langle e^{x^i X_i} \rangle = \frac{\text{tr } R(e^{-\beta d^i(x,a) H_i})}{\text{tr } R(e^{-\beta d^i(0,a) H_i})} \quad (7.79)$$

   Since the H_i are diagonal in the representation R, the sums are straightforward.
   i. Relate the steps in the algorithm described in this problem to the steps followed in the previous problem for computing the result derived in Eq. (7.76). In particular, identify the operators X_i, the "diagonal" operators H_i, the hermitian representation R (it is invisible), the faithful representation S (it is given explicitly), the generating function $e^{x^i X_i}$, and the Wick rotation.

9. Coherent states were first discussed by Schrödinger in 1926. For many purposes it is useful to apply a unitary transformation to the harmonic oscillator ground state. The unitary transformation has the form $U(\alpha) = e^{(\alpha a^\dagger - \alpha^* a)}$, where $a^\dagger$ and a are the usual photon creation and annihilation operators. This unitary operator, acting on the ground state, is relatively simple to compute if it can be disentangled as follows

$$U(\alpha)|0\rangle = e^{(\alpha a^\dagger - \alpha^* a)}|0\rangle = e^{\beta a^\dagger} e^{\delta I} e^{\beta' a}|0\rangle \quad (7.80)$$

This disentangling theorem can be worked out easily in the 3×3 *nonunitary* representation. (It is the group multiplication property that we are after; unitarity is an additional structure that is applied to the *representation* of the group.)

a. Show that the left-hand side of Eq. (7.80) simplifies to

$$\text{EXP}\begin{bmatrix} 1 & -\alpha^* & 0 \\ 0 & 1 & \alpha \\ 0 & 0 & 1 \end{bmatrix} = \begin{bmatrix} 1 & -\alpha^* & -\alpha^*\alpha/2 \\ 0 & 1 & \alpha \\ 0 & 0 & 1 \end{bmatrix} \quad (7.81)$$

b. Show that the right-hand side of Eq. (7.80) becomes

$$\begin{bmatrix} 1 & \beta' & \delta \\ 0 & 1 & \beta \\ 0 & 0 & 1 \end{bmatrix} \quad (7.82)$$

c. Use this result to compute

$$e^{(\alpha a^\dagger - \alpha^* a)}|0\rangle = e^{\alpha a^\dagger} e^{-\alpha^* \alpha I/2} e^{-\alpha^* a}|0\rangle = \sum \frac{(\alpha a^\dagger)^n}{n!}|0\rangle e^{-\alpha^* \alpha/2} \quad (7.83)$$

d. Use a further property of the creation operators (this is a representation-dependent property, so the calculation has now moved back into the infinite-dimensional Hilbert space and out of the nonunitary 3×3 matrix representation), $a^\dagger|n\rangle = |n+1\rangle\sqrt{n+1}$ to conclude

$$|\alpha\rangle = U(\alpha)|0\rangle = e^{-\alpha^*\alpha/2} \sum_{n=0}^{\infty} \frac{\alpha^n}{\sqrt{n!}}|n\rangle \quad (7.84)$$

e. Compute the inner product $\langle\beta|\alpha\rangle$ and show $\langle\alpha|\alpha\rangle = 1$.
f. Show $a|\alpha\rangle = \alpha|\alpha\rangle$.
g. Show $\langle\alpha|x|\alpha\rangle = (\alpha^* + \alpha)/\sqrt{2}$.

9. An $SU(2)$ coherent state (also called atomic coherent state) is constructed by the action of an arbitrary $SU(2)$ group operation on the ground state, or lowest lying state, in a $2j+1$ dimensional invariant space (Arecchi et al., 1972; Gilmore, 1974b):

$$\left|\begin{matrix} j \\ \theta \end{matrix}\right\rangle = SU(2)\left|\begin{matrix} j \\ -j \end{matrix}\right\rangle \quad (7.85)$$

a. Show that rotations by ϕ around the z-axis serve only to multiply the fiducial state by a phase angle: $e^{i\phi J_z}|\begin{smallmatrix}j\\-j\end{smallmatrix}\rangle = |\begin{smallmatrix}j\\-j\end{smallmatrix}\rangle e^{-ij\phi}$. This simply "renormalizes" the fiducial state, and is generally not important.

b. Rotations about an axis in the x–y plane produce a two-parameter family of coherent states parameterized by coset representatives in $SU(2)/U(1)$:

$$\left|\begin{matrix} j \\ \theta \end{matrix}\right\rangle = e^{i(\theta_x J_x + \theta_y J_y)}\left|\begin{matrix} j \\ -j \end{matrix}\right\rangle \quad i(\theta_x J_x + \theta_y J_y) = \frac{i}{2}\begin{bmatrix} 0 & \theta_x - i\theta_y \\ \theta_x + i\theta_y & 0 \end{bmatrix} \quad (7.86)$$

c. Rewrite $e^{i(\theta_x J_x + \theta_y J_y)}$ in the form $e^{i\alpha_+ J_+} e^{i\alpha_z J_z} e^{i\alpha_- J_-}$ and compute the analytic relation between the angles θ and the parameters α.
d. Show $e^{i\alpha_- J_-}|\begin{smallmatrix}j\\-j\end{smallmatrix}\rangle = |\begin{smallmatrix}j\\-j\end{smallmatrix}\rangle$.
e. Show $e^{i\alpha_z J_z}|\begin{smallmatrix}j\\-j\end{smallmatrix}\rangle = |\begin{smallmatrix}j\\-j\end{smallmatrix}\rangle e^{-ij\alpha_z}$.

f. Compute finally

$$U(\alpha)|0\rangle = e^{i\alpha_+ J_+}\left|\begin{matrix}j\\-j\end{matrix}\right\rangle e^{-ij\alpha_z} = \sum_{m=-j}^{m=+j}\frac{(i\alpha_+ J_+)^{j+m}}{(j+m)!}\left|\begin{matrix}j\\-j\end{matrix}\right\rangle e^{-ij\alpha_z} \quad (7.87)$$

g.
Show that $J_-|^{\ j}_{\theta_x\theta_y}\rangle$ cannot be proportional to $|^{\ j}_{\theta_x\theta_y}\rangle$ because the state $|^{\ j}_{+j}\rangle$ is not occupied. This is different from the harmonic oscillator (photon operator) case. The difference arises because $SU(2)$ is compact with finite-dimensional unitary irreducible representations and the harmonic oscillator group H_4 is not compact with only an infinite-dimensional unitary irreducible representation of interest.

h. Compute the inner product and show

$$\left\langle\begin{matrix}j\\\theta'_x\theta'_y\end{matrix}\bigg|\begin{matrix}j\\\theta_x\theta_y\end{matrix}\right\rangle = \left[\cos\left(\frac{\theta'}{2}\right)\cos\left(\frac{\theta}{2}\right) + e^{i(\phi'-\phi)}\sin\left(\frac{\theta'}{2}\right)\sin\left(\frac{\theta}{2}\right)\right]^{2j} \quad (7.88)$$

where $e^{-i\phi} = (\theta_x - i\theta_y)/\theta$, and similarly for θ' (cf., Eq. (7.46)).

10. A number of important quantum eigenvalue equations can be expressed in algebraic format. A toy example is

$$(EJ_3 + pJ_1 - Z)|u\rangle = 0$$

Here E is an energy eigenvalue, p is some sort of coupling strength, Z could (and sometimes does) represent a charge, and $|u\rangle$ is an eigenfunction. In this toy example, the operators J_3 and J_1 are assumed to belong to the Lie algebra $\mathfrak{su}(2)$ and the equation applies to half-integer spin spaces ($(2j+1)$ is even).

   a. Show that a unitary transformation U transforms this equation to the diagonal form $(E'J_3 - Z)|v\rangle = 0$, where $E' = \sqrt{E^2 + p^2}$ and $|v\rangle = U|u\rangle$.
   b. Show that $E = \pm\sqrt{(Z/m)^2 - p^2}$.
   c. Compare this spectrum with the unperturbed spectrum ($p \to 0$).
   d. Under what conditions on j, p, Z are these solutions valid?
   e. Construct the unitary transformation that diagonalizes the eigenvalue equation, and show that $|u\rangle = e^{i\theta J_2}|^{\ j}_m\rangle$. Compute θ for each E.

11. Compute the matrix elements of the rotation matrices in the $2j+1$ unitary irreducible representations of $SU(2)$ and show

$$\text{EXP}(i\beta J_y)_{mn} = D^j_{mn}(\beta) = P^j_{mn}(z) = \frac{(-)^{j-m}}{2^j(j-n)!}\left[\frac{(j-n)!(j+m)!}{(j+n)!(j-m)!}\right]^{1/2}$$

$$\times (1+z)^{-(m+n)/2}(1-z)^{-(m-n)/2}\left(\frac{d}{dz}\right)^{j-m}\left[(1-z)^{j-n}(1+z)^{j+n}\right]$$

where $z = \cos(\beta)$. The Wigner matrix elements D^j_{mn} are related to the Jacobi polynomials when $j = l$, where l is an integer.

12. Use the decompositions (7.21) for $SO(3)$ and (7.22) for $SU(2)$ to show the following.
    a. Geodesics through $I_2 \in SU(2)$ focus at $-I_2$ and geodesics through $I_3 \in SO(3)$ focus at I_3. Conclude that $SU(2)$ is a two-fold covering group of $SO(3)$.
    b. Geodesics through the "north pole" of $SU(2)/U(1)$ ($z = 1, x = y = 0$) focus at its "south pole" ($z = -1, x = y = 0$) and geodesics through the north pole of $SO(3)/SO(2)$ ($z = 1, x = y = 0$) focus at its south pole ($z = -1, x = y = 0$).
    c. Conclude that $SU(2)/U(1) = S^2 = SO(3)/SO(2)$ and the $2 \to 1$ nature of the covering $SU(2) \downarrow SO(3)$ is contained in the subgroup of rotations about the z-axis $U(1) \downarrow SO(2)$:
    $$\begin{bmatrix} e^{i\theta/2} & 0 \\ 0 & e^{-i\theta/2} \end{bmatrix} \xrightarrow{2 \to 1} \begin{bmatrix} \cos\theta & \sin\theta \\ -\sin\theta & \cos\theta \end{bmatrix}$$

13. Show that the discrete invariant subgroups of $SU(n)$ are all commutative groups of order r, with group elements $e^{2\pi i k/r} I_n$, with n/r integer. Compute the foci in $SU(n)$. How are the foci related to the group operations of the form $e^{2\pi i k/n} I_n$?

14. Show that the matrix $\begin{bmatrix} -\lambda & 0 \\ 0 & -1/\lambda \end{bmatrix}$ in $SL(2; \mathbb{R})$ cannot be reached by exponentiating any element in the Lie algebra if $\lambda > 1$. Show that it can be reached by following a "broken geodesic" $e^A e^B$. Find matrices A and B that do this. (Hint: do not work too hard.)

15. A simple model has been introduced to describe the interaction of light with matter. In this model (Dicke model) N atoms interact with a single mode of the electromagnetic field. Each atom is modeled as a two-level system, with energy separation ϵ. A single photon has energy $\hbar\omega$. The hamiltonian is chosen as

    $$\mathcal{H} = \sum_{i=1}^{N} \frac{\epsilon}{2} \sigma_z^{(i)} + \hbar\omega a^\dagger a + \frac{\lambda}{\sqrt{N}} \sum_{i=1}^{N} \sigma_+^{(i)} a + \sigma_-^{(i)} a^\dagger$$

    The operator $\sigma_z^{(i)}$ describes the two states of atom i and the operator $a^\dagger a$ describes the number of photons in the field mode. The operator $\sigma_+^{(j)}$ ($\sigma_\pm^{(j)} = \frac{1}{2}(\sigma_x^{(j)} \pm i\sigma_y^{(j)})$) describes transitions of the jth atom from the ground to its excited state. This atomic transition is accompanied by the absorption (annihilation) of a single photon. The operator $\sigma_-^{(j)} a^\dagger$, describes deexcitation of an atom with emission (creation, $a^\dagger$) of a photon. The strength of interaction of the atom with the electromagnetic field (the dipole moment) is parameterized by λ.
    a. Assume the atoms are independent and show
    $$[\sigma_z^{(i)}, \sigma_\pm^{(j)}] = \pm\sigma_\pm^{(i)} \delta_{ij} \quad [\sigma_+^{(i)}, \sigma_-^{(j)}] = \sigma_z^{(i)} \delta_{ij}$$
    b. If all the atoms behave cooperatively it is possible to replace $\sum_{i=1}^{N} \frac{1}{2}\sigma_z^{(i)} \to J_z$, $\sum_{i=1}^{N} \sigma_\pm^{(i)} \to J_\pm$. Show that the operators J_z, $J_\pm$ satisfy the usual su(2) commutation relations.
    c. Assume the atoms "behave classically." This means that the quantum mechanical operators J_z, $J_\pm$ can be replaced by their c-number expectation values:

7.7 Problems

$J_z \to \langle J_z(t)\rangle$, $J_+ \to \langle J_+(t)\rangle$, $J_- \to \langle J_-(t)\rangle = \langle J_+(t)\rangle^*$. Show that this semiclassical hamiltonian

$$\mathcal{H}_{\text{field}} = \epsilon \langle J_z(t)\rangle + \hbar\omega a^\dagger a + \frac{\lambda}{\sqrt{N}}(\langle J_+(t)\rangle a + \langle J_-(t)\rangle a^\dagger)$$

maps the ground state of the field (the state with no photons) into a coherent state of the electromagnetic field: $|\alpha(t)\rangle = U(\alpha(t))|0\rangle = e^{\alpha a^\dagger - \alpha^* a}|0\rangle$. Use the disentangling theorems to compute the relation between the coherent state parameter $\alpha(t)$ and the classical driving fields $\langle J_z(t)\rangle$ and $\langle J_+(t)\rangle = \langle J_-(t)\rangle^*$.

d. Show that if the initial state of the field is not the ground state, but rather a coherent state $|\beta\rangle$, the state obtained by the action of the classical current is still a coherent state. How are the parameters β, describing the initial condition, and α, describing the unitary evolution of the field, related?

e. Suppose now that the atoms are considered quantum mechanically but the field is considered classically. Show that this amounts to the substitutions $a^\dagger \to \langle a(t)\rangle^*$, $a \to \langle a(t)\rangle$, and $a^\dagger a \to \langle a(t)\rangle^* \langle a(t)\rangle$.

f. Show that the resulting semiclassical hamiltonian is

$$\mathcal{H}_{\text{atoms}} = \epsilon J_z + \frac{\lambda}{\sqrt{N}}(J_+ \langle a(t)\rangle + J_- \langle a(t)\rangle^*)$$

Show that under this semiclassical hamiltonian, if the atoms are in their collective ground state ($m = -\frac{1}{2}$ for each atom, or $M = -J$, $J = N/2$ for the ensemble of N atoms) the ground state will evolve into a coherent state of the group $SU(2)$ parameterized by a point in the coset $SU(2)/U(1)$.

g. Show that, under the action of this semiclassical Hamiltonian a coherent state will evolve into a coherent state: $|\theta(t)\rangle = e^{i\theta(t)\cdot\mathbf{J}}|J, -J\rangle$, where $J = N/2$. How are the angles $\theta(t)$ related to the classical field variables $\langle a(t)\rangle$ and $\langle a(t)\rangle^*$?

h. Conclude that there is a duality between the atoms and the field in this model: a classical current will produce a coherent state of the electromagnetic field; a classical electromagnetic field will produce a coherent atomic state.

i. The semiclassical hamiltonian for the field can be used to construct time-dependent field expectation values $\langle a\rangle$ and $\langle a\rangle^*$. Conversely, the semiclassical hamiltonian for the atoms can be used to construct time-dependent atomic expectation values $\langle J_+\rangle = \langle J_-\rangle^*$. Construct a self-consistent model by requiring that both sets of time-dependent quantities are equal.

16. The thermodynamic properties of the Dicke model can be studied in a similar fashion. Assume N identical atoms interacting with a single field mode are in thermodynamic equilibrium at temperature T ($\beta = 1/k_B T$).

a. Assume $\langle \sigma_+^{(i)}\rangle_T$ has some fixed unknown value, and similarly for the other atomic thermal expectation values. Use these values in the semiclassical approximation for the field hamiltonian to compute the density operator. Compute the thermal expectation values for the operators $a^\dagger, a, a^\dagger a$.

b. Dualize. Assume the field operators have fixed but unknown expectation values. Use these values in the semiclassical approximation for the atomic hamiltonian to compute the density operator. Compute the thermal expectation values for the operators $\sigma_z, \sigma_+, \sigma_-$.

c. Impose self-consistency. Require that if a set of field thermal expectation values produces specific atomic expectation values, these atomic expectation values produce the same set of field expectation values. This leads to a nonlinear set of self-consistency equations. These self-consistent equations may have more than one solution.

d. To lift the self-consistent solution degeneracy, construct the thermal expectation value for $\mathcal{H}$. Choose the minimum energy solution. Under what conditions on $\epsilon, \hbar\omega, \lambda, N$ is there a nontrivial solution (e.g., $\langle J_+ \rangle_T \neq 0$)?

e. Show that a thermodynamic phase transition occurs as $\lambda^2/\epsilon\hbar\omega$ increases through $+1$. Is this a first or second order phase transition?

17. The two complex parameters $a(t), b(t)$ in the evolution equation (7.58) can be expressed in terms of their real and imaginary parts. These obey $a_r^2 + a_i^2 + b_r^2 + b_i^2 = 1$ (unitarity condition). This condition simply reflects that the state of the system is given by a unit quaternion. As numerical integration proceeds, imprecisions may cause these parameters to depart slightly from the unitarity condition. Devise a self-correcting integration procedure to correct for this type of error. After N small integration steps, compute the length of the vector (a_r, a_i, b_r, b_i) and scale this length back to $+1$.

18. The thermodynamic generating functions for $SU(2)$ and H_4 given by expressions (7.67) and (7.70) simplify considerably if the "diagonal operator" is not included. Simplify (7.67) by taking the limit $\lambda_3 \to 0$. Simplify (7.70) by taking the limit $\lambda_n \to 0$ and setting $d = \delta = 0$.

19. For many reasons it is less desirable to compute thermal expectation values for *symmetric* operator products such as $\langle J_+ J_- + J_- J_+ \rangle$ or $\langle aa^\dagger + a^\dagger a \rangle$ than it is to construct generating functions for *ordered* products of operators such as $\langle J_+ J_- \rangle$ or $\langle a^\dagger a \rangle$. Show how to use disentangling theorems to transform the generating functions for symmetric operator products in (7.67) and (7.70), or their simplified forms constructed in the previous problem, into generating functions for ordered products of operators.

8
Structure theory for Lie algebras

In this chapter we discuss the structure of Lie algebras. A typical Lie algebra is a semidirect sum of a semisimple Lie algebra and a solvable subalgebra that is invariant. By inspection of the regular representation "in suitable form," we are able to determine the maximal nilpotent and solvable invariant subalgebras of the Lie algebra and its semisimple part. We show how to use the Cartan–Killing inner product to determine which subalgebras in the Lie algebra are nilpotent, solvable, semisimple, and compact.

8.1 Regular representation

A Lie algebra is defined by its commutation relations. The commutation relations are completely encapsulated by the structure constants. These are conveniently summarized in the regular representation

$$[Z, X_i] = R(Z)_i{}^j X_j \qquad (8.1)$$

Under a change of basis $X_j = A_j{}^s Y_s$ this $n \times n$ matrix undergoes a similarity transformation

$$S(Z)_r{}^s = (A^{-1})_r{}^i R(Z)_i{}^j A_j{}^s \qquad (8.2)$$

It is very useful to find a basis, or construct a similarity transformation, that brings the regular representation of every operator in the Lie algebra *simultaneously* to some standard form. The structure of the Lie algebra can be decided by inspecting this standard form.

8.2 Some standard forms for the regular representation

We summarize in Fig. 8.1 the standard forms that the regular representation can assume. We also provide an example for each.

1. Zero In this case all structure constants vanish and the algebra is commutative.

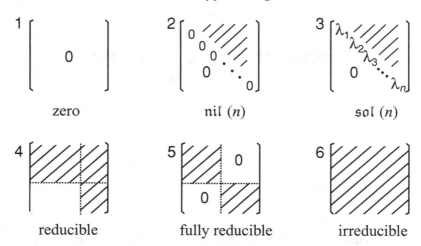

Figure 8.1. Structure of the regular representation for different types of Lie algebras.

Example The Lie algebra $a(p, q)$ consists of matrices of the form

$$\left[\begin{array}{c|c} 0 & A \\ \hline 0 & 0 \end{array}\right] \begin{array}{c} \uparrow \\ p \\ \downarrow \\ \uparrow \\ q \\ \downarrow \end{array} \qquad (8.3)$$

This is an $n \times n$ ($n = p + q$) matrix algebra which is $N = pq$ dimensional. The independent degrees of freedom are the N independent matrix elements of the $p \times q$ matrix A. The $n \times n$ matrices all commute under matrix multiplication. The group operation is equivalent to addition of the $p \times q$ matrices. The regular representation consists of $N \times N$ matrices, all N of them are zero.

2. $\mathfrak{nil}(n)$ Strictly upper triangular In this case the Lie algebra is nilpotent.

Example We consider the Lie algebra spanned by the photon operators a, $a^\dagger$, and $I = [a, a^\dagger]$ or the isomorphic 3×3 matrix algebra (5.11). The regular representation is a 3×3 matrix

$$\mathfrak{Reg}(la + ra^\dagger + \delta I) = \begin{bmatrix} 0 & 0 & l \\ 0 & 0 & -r \\ 0 & 0 & 0 \end{bmatrix} \begin{array}{c} a^\dagger \\ a \\ I \end{array} \qquad (8.4)$$

3. $\mathfrak{sol}(n)$ Upper triangular In this case nonzero elements occur on and above the diagonal. The algebra is solvable.

8.2 Some standard forms for the regular representation

Example The algebra spanned by the photon number operator $\hat{n} = a^\dagger a$, creation and annihilation operators $a^\dagger$ and a, and their commutator $I = [a, a^\dagger]$ is isomorphic to the algebra described by the 3×3 matrices (5.9). The regular representation is a 4×4 matrix

$$\mathfrak{Reg}(\eta\hat{n} + la + ra^\dagger + \delta I) = \begin{bmatrix} 0 & -r & l & 0 \\ 0 & \eta & 0 & l \\ 0 & 0 & -\eta & -r \\ 0 & 0 & 0 & 0 \end{bmatrix} \begin{matrix} \hat{n} \\ a^\dagger \\ a \\ I \end{matrix} \qquad (8.5)$$

4. ut(p, q) In this case the regular representation is reducible and the Lie algebra is nonsemisimple.

Example We consider the algebra consisting of the six photon operators $\hat{n} = \frac{1}{2}\{a, a^\dagger\} = a^\dagger a + \frac{1}{2}$, $a^{\dagger 2}$, a^2, $a^\dagger$, a, $I = [a, a^\dagger]$. This is isomorphic to the algebra of six 4×4 matrices presented in (5.7). The algebra of 4×4 matrices (the "defining" representation) and the regular representation of this algebra are given below:

$$\eta(\hat{n} + \tfrac{1}{2}) + Ra^{\dagger 2} + La^2 + ra^\dagger + la + \delta I$$

$$\mathfrak{def} = \begin{bmatrix} 0 & l & r & -2\delta \\ 0 & \eta & 2R & -r \\ 0 & -2L & -\eta & l \\ 0 & 0 & 0 & 0 \end{bmatrix}$$

$$\mathfrak{Reg} = \left[\begin{array}{ccc|ccc|c} 0 & -2R & 2L & -r & l & 0 & \hat{n} + \tfrac{1}{2} \\ 4L & 2\eta & 0 & 2l & 0 & 0 & a^{\dagger 2} \\ -4R & 0 & -2\eta & 0 & -2r & 0 & a^2 \\ \hline & & & \eta & 2L & l & a^\dagger \\ & & & -2R & -\eta & -r & a \\ & & & 0 & 0 & 0 & I \end{array}\right]$$

(8.6)

The subspace spanned by the three operators $a^\dagger$, a, I is invariant, as is shown by the structure of the regular representation.

5. Block diagonal In this case the regular representation is fully reducible and the Lie algebra is semisimple.

Example The Lie algebra $\mathfrak{so}(4)$ has six generators $X_{ij} = -X_{ji}$, $1 \leq i, j \leq 4$ and commutation relations

$$[X_{ij}, X_{rs}] = X_{is}\delta_{jr} + X_{jr}\delta_{is} - X_{ir}\delta_{js} - X_{js}\delta_{ir} \qquad (8.7)$$

The following two linear combinations of generators

$$Y_i = \tfrac{1}{2}(X_{i4} + \tfrac{1}{2}\epsilon_{irs}X_{rs}) \qquad X_{i4} = Y_i + Z_i$$
$$Z_i = \tfrac{1}{2}(X_{i4} - \tfrac{1}{2}\epsilon_{irs}X_{rs}) \qquad X_{ij} = \epsilon_{ijk}(Y_k - Z_k)$$
(8.8)

obey the commutation relations

$$\begin{aligned}[Y_i, Y_j] &= -\epsilon_{ijk}Y_k \\ [Z_i, Z_j] &= +\epsilon_{ijk}Z_k \\ [Y_i, Z_j] &= 0\end{aligned}$$
(8.9)

The 4×4 defining matrix representation and the 6×6 regular representation have the structure

$$X = \sum y_i Y_i + \sum z_j Z_j$$

$$\mathfrak{def}(X) \to \begin{bmatrix} 0 & +(y_3 - z_3) & -(y_2 - z_2) & +(y_1 + z_1) \\ -(y_3 - z_3) & 0 & +(y_1 - z_1) & +(y_2 + z_2) \\ +(y_2 - z_2) & -(y_1 - z_1) & 0 & +(y_3 + z_3) \\ -(y_1 + z_1) & -(y_2 + z_2) & -(y_3 + z_3) & 0 \end{bmatrix}$$

(8.10)

$$\mathfrak{Reg}(X) \to \left[\begin{array}{ccc|ccc} 0 & -y_3 & +y_2 & & & \\ +y_3 & 0 & -y_1 & & & \\ -y_2 & +y_1 & 0 & & & \\ \hline & & & 0 & +z_3 & -z_2 \\ & & & -z_3 & 0 & +z_1 \\ & & & +z_2 & -z_1 & 0 \end{array}\right] \begin{array}{c} Y_1 \\ Y_2 \\ Y_3 \\ Z_1 \\ Z_2 \\ Z_3 \end{array}$$

Since the regular representation has a block diagonal structure, the algebra is semisimple. It is not at all obvious that the Lie algebra $\mathfrak{so}(4)$ is semisimple and can be written as the direct sum of two simple algebras. This is not true for the other orthogonal Lie algebras, $\mathfrak{so}(n)$, $n > 4$. We will have to wait until Chapter 10 to be able to see easily that $\mathfrak{so}(4)$ is semisimple, not simple.

6. Irreducible In this case the regular representation is irreducible and the Lie algebra is simple.

Example The Lie algebras $\mathfrak{su}(n)$ ($n \geq 2$), $\mathfrak{so}(n)$ ($n > 4$), and $\mathfrak{sp}(n)$ ($n \geq 1$) are all simple. To be concrete, the Lie algebra for $SU(2)$ has defining and regular representations

$$\mathfrak{def} = \frac{i}{2}\begin{bmatrix} a_3 & a_1 - ia_2 \\ a_1 + ia_2 & -a_3 \end{bmatrix} \quad \mathfrak{Reg} = \begin{bmatrix} 0 & -a_3 & +a_2 \\ +a_3 & 0 & -a_1 \\ -a_2 & +a_1 & 0 \end{bmatrix} \begin{matrix} X_1 \\ X_2 \\ X_3 \end{matrix} \quad (8.11)$$

8.3 What these forms mean

Reducing the regular representation to one of the standard forms described in the previous section means that the structure constants, and therefore the commutation relations, have also been reduced to some standard form. We discuss in this section what each of the standard forms implies about the commutation relations and structure of the Lie algebra.

1. Commutative case If all the structure constants are zero, then

$$[X_i, X_j] = 0 \qquad (8.12)$$

for each element in the Lie algebra.

2, 3. Nilpotent and solvable In these cases

$$[Z, X_i] = R(Z)_i^{\ j} X_j$$
$$R(Z)_i^{\ j} = 0 \text{ unless } \begin{matrix} j > i \text{ nilpotent} \\ j \geq i \text{ solvable} \end{matrix} \qquad (8.13)$$

This means that $[Z, X_i]$ can be expressed as a linear combination of basis vectors X_j with $j \geq i$. This in turn means that the basis vectors $X_i, X_{i+1}, \ldots, X_n$ span a subalgebra for each value of $i = 1, 2, \ldots, n$. Since this subalgebra is mapped onto itself by every element Z in the original algebra, each subalgebra is an invariant subalgebra. The result is shown schematically in Fig. 8.2 and is summarized by

$$\begin{matrix} V_1 & \text{spanned by } X_n, X_{n-1}, X_{n-2}, \ldots, X_2, X_1 \\ \cup \\ V_2 & \text{spanned by } X_n, X_{n-1}, X_{n-2}, \ldots, X_2 \\ \cup \\ \vdots & \vdots & \vdots \\ \cup \\ V_{n-2} & \text{spanned by } X_n, X_{n-1}, X_{n-2} \\ \cup \\ V_{n-1} & \text{spanned by } X_n, X_{n-1} \\ \cup \\ V_n & \text{spanned by } X_n \end{matrix} \qquad (8.14)$$

Each V_i is an invariant subalgebra in V_j, $i > j$. The original algebra is V_1.

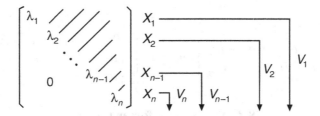

Figure 8.2. Structure of nilpotent and solvable algebras.

4. Reducible or nonsemisimple The block diagonal form for the regular representation requires the commutation relations

$$\left[\begin{array}{c|c} * & * \\ \hline 0 & * \end{array}\right] \begin{array}{c} \uparrow \\ V_1 \\ \downarrow \\ \uparrow \\ V_2 \\ \downarrow \end{array} \quad \Rightarrow \quad [\text{Any}, V_2] \subset V_2 \quad (8.15)$$

Since in particular $[V_2, V_2] \subseteq V_2$, V_2 is a subalgebra in the original algebra. Further, since the commutator of anything in the original algebra with V_2 is in V_2, V_2 is an **invariant** subalgebra. The complementary subspace V_1 is not generally even a subalgebra of the original algebra.

5. Fully reducible or semisimple In this case the block diagonal form for the regular representation requires the commutation relations

$$\left[\begin{array}{c|c} * & 0 \\ \hline 0 & * \end{array}\right] \begin{array}{c} \uparrow \\ V_1 \\ \downarrow \\ \uparrow \\ V_2 \\ \downarrow \end{array} \quad \Rightarrow \quad \begin{array}{c} [V_1, V_1] \subseteq V_1 \\ [V_2, V_2] \subseteq V_2 \\ [V_1, V_2] = 0 \end{array} \quad (8.16)$$

Both V_1 and V_2 are invariant subalgebras. Moreover, every element in V_1 commutes with every operator in V_2. Therefore the two subalgebras V_1 and V_2 can be studied separately and independently.

6. Irreducible or simple In this case every generator X can be written as the commutator of some pair of operators Y and Z in the Lie algebra:

$$X = [Y, Z] \quad (8.17)$$

It is this ability of an algebra to reproduce itself under commutation that distinguishes simple and semisimple Lie algebras from solvable and nilpotent algebras. Nonsemisimple algebras are composed of a semisimple subalgebra and a solvable invariant subalgebra.

8.4 How to make this decomposition

There is a systematic procedure for decomposing a Lie algebra into its semisimple component and its maximal solvable invariant subalgebra. This is a simple two-step procedure. In the first step we identify the subspace of the Lie algebra on which the Cartan–Killing inner product is identically zero. If there is no such subspace the algorithm stops here and the algebra is semisimple. If there is a nontrivial subspace, it forms the maximal nilpotent invariant subalgebra of the algebra. This subspace is "removed" from the algebra, and the commutation relations and Cartan–Killing inner product for the remaining operators are computed. The algorithm stops here, regardless of the outcome. If there is a nontrivial subspace on which the new Cartan–Killing inner product is identically zero, the elements in this subspace, together with the previously identified nilpotent invariant subalgebra, span a solvable algebra. This is the maximal solvable invariant subalgebra in the original Lie algebra. The complementary subspace on which the new Cartan–Killing inner product is nonsingular is the semisimple part of the original Lie algebra.

In small, easy-to-digest steps, this two-step algorithm takes the following form.

(i) From the structure constants of the original Lie algebra $\mathfrak{g}$ form the Cartan–Killing inner product.
(ii) Determine the subspace on which this inner product is positive-definite, negative-definite, and zero:

$$\mathfrak{g} = (V_- + V_+) + V_0 \qquad (8.18)$$

(iii) If V_0 is zero, stop. If not, V_0 is the maximal nilpotent invariant subalgebra in $\mathfrak{g}$.
(iv) Form the difference $\mathfrak{g}' = \mathfrak{g} - V_0$. This is a Lie algebra (under the "mod" operation: set to zero any part of the commutator that is in V_0). Compute the structure constants and Cartan–Killing inner product for $\mathfrak{g}'$.
(v) Effect another decomposition:

$$\mathfrak{g}' = (V'_- + V'_+) + V'_0 \qquad (8.19)$$

(vi) The original Lie algebra has the following structure

$$\mathfrak{g} = \underbrace{\underbrace{\underbrace{V'_-}_{\text{compact subalgebra}} + \underbrace{V'_+}_{\text{noncompact generators}}}_{\text{semisimple}} + \underbrace{\underbrace{V'_0}_{\text{abelian}} + \underbrace{V_0}_{\text{nilpotent}}}_{\text{maximum solvable invariant subalgebra}}}_{\text{nonsemisimple Lie algebra}} \qquad (8.20)$$

8.5 An example

To illustrate this procedure, we compute the structure of the six-dimensional Lie algebra of two photon operators. The regular representation is given in (8.6). The inner product of a vector with itself is

$$(X, X) = -40RL + 10\eta^2 \qquad (8.21)$$

The inner product is identically zero on the subspace V_0 spanned by $a^\dagger, a$ and I. The three remaining operators have regular representation

$$\eta(a^\dagger a + \tfrac{1}{2}) + R a^{\dagger 2} + L a^2 \longrightarrow \begin{bmatrix} 0 & -2R & 2L \\ 4L & 2\eta & 0 \\ -4R & 0 & -2\eta \end{bmatrix} \begin{array}{l} \hat{n} + \tfrac{1}{2} \\ a^{\dagger 2} \\ a^2 \end{array} \qquad (8.22)$$

with inner product

$$(X, X)' = -32RL + 8\eta^2 \qquad (8.23)$$

In this case $V_0' = 0$ and the two photon algebra has the decomposition

$$\mathfrak{g} = \underbrace{(\hat{n} + \tfrac{1}{2}, a^{\dagger 2}, a^2)}_{su(1,1)} + \underbrace{(a^\dagger, a, I)}_{\text{nilpotent invariant subalgebra}} \qquad (8.24)$$

The Cartan–Killing inner product can be diagonalized by choosing two linear combinations of the operators $a^{\dagger 2}$ and a^2. Then $a^{\dagger 2} + a^2$ is a compact generator, since the Cartan–Killing form is negative-definite on it. The other two operators, $a^\dagger a + \tfrac{1}{2}$ and $a^{\dagger 2} - a^2$, are noncompact.

8.6 Conclusion

An arbitrary Lie algebra is a semidirect sum of a semisimple Lie algebra and a solvable invariant subalgebra. The structure of a Lie algebra can be determined by inspecting its regular representation, once this has been brought to suitable form by a similarity transformation. To facilitate constructing this transformation, we have shown how to use the Cartan–Killing inner product to determine the linear vector subspaces in the Lie algebra that are maximal nilpotent invariant subalgebras, the maximal solvable invariant subalgebra, the semisimple subalgebra, and its maximal compact subalgebra.

8.7 Problems

1. Compute the decomposition (8.20) for
   a. The photon algebra $\hat{n}, a^\dagger, a, I$ (Eq. (8.5)).
   b. The algebra $so(3, 1)$.
   c. The algebra for the Poincaré group (Eq. (3.26)).
   d. The algebra for the Galilei group (Eq. (3.27)).

2. Compute the decomposition (8.20) for Lie algebras spanned by various combinations of the boson creation and annihilation operators (a–g below). These satisfy $[b_i, b_j^\dagger] = I\delta_{ij}$, $1 \le i, j \le n$. Commutators involving bilinear (trilinear, ...) products are computed in the usual way.
   a. $b_i, b_j^\dagger, I$.
   b. $b_i^\dagger b_j$.
   c. $b_i^\dagger b_j, b_i, b_j^\dagger, I$.
   d. $b_i^\dagger b_j^\dagger, b_i^\dagger b_j + \frac{1}{2}\delta_{ij}, b_i b_j$.
   e. $b_i^\dagger b_j^\dagger, b_i^\dagger b_j + \frac{1}{2}\delta_{ij}, b_i b_j, b_i, b_j^\dagger, I$.
   f. $b, b^\dagger b, b^\dagger b^\dagger b$.
   g. $b^\dagger, b^\dagger b, b^\dagger b b$.

3. Fermion creation and annihilation operators obey anticommutation relations $\{f_i, f_j^\dagger\} = \delta_{ij}$, but their bilinear combinations close under the same commutation relations as do boson operators. Compute the structure of these fermion algebras:
   a. $f_i^\dagger f_j$.
   b. $f_i^\dagger f_j^\dagger, f_i^\dagger f_j + \frac{1}{2}\delta_{ij}, f_i f_j$.

4. Compute the decomposition (8.20) for Lie algebras spanned by various combinations of the position (x_i) and momentum (∂_j) operators for n independent directions:
   a. x_i, ∂_j, I.
   b. $x_i \partial_j$.
   c. $x_i \partial_j, x_i, \partial_j, I$.
   d. $x_i x_j, x_i \partial_j + \frac{1}{2}I\delta_{ij}, \partial_i \partial_j$.
   e. $x_i x_j, x_i \partial_j, \partial_i \partial_j, x_i, \partial_j, I$.
   f. $\frac{d}{dx}, x\frac{d}{dx}, x^2\frac{d}{dx}$.
   g. $x, x\frac{d}{dx}, x\frac{d^2}{dx^2}$.

5. What is the relation between the Cartan–Killing inner product computed using the defining matrix representation of a matrix Lie algebra and using the regular matrix representation of the Lie algebra?

6. The Lorentz, Poincaré, and Galilei groups in $2 + 1$ dimensions (x, y and t) have Lie algebras with matrix structures:

$$\left[\begin{array}{cc|c} 0 & \theta & v_1 \\ -\theta & 0 & v_2 \\ \hline v_1 & v_2 & 0 \end{array}\right] \quad \left[\begin{array}{cc|cc} 0 & \theta & v_1 & t_1 \\ -\theta & 0 & v_2 & t_2 \\ \hline v_1 & v_2 & 0 & t_3 \\ 0 & 0 & 0 & 0 \end{array}\right] \quad \left[\begin{array}{cc|cc} 0 & \theta & v_1 & t_1 \\ -\theta & 0 & v_2 & t_2 \\ \hline 0 & 0 & 0 & t_3 \\ 0 & 0 & 0 & 0 \end{array}\right] \quad (8.25)$$

$\qquad$ Lorentz $\qquad\qquad$ Poincare $\qquad\qquad$ Galilei

Structure theory for Lie algebras

   a. Compute the matrix infinitesimal generators for each.
   b. Construct their commutation relations.
   c. Decompose each Lie algebra into the standard form (8.20).
   d. For each Lie algebra, express the generators in terms of the operators x_i, ∂_j.
   e. For each Lie algebra, express the generators in terms of the boson operators $b_i^\dagger$, b_j, $1 \le i, j \le 3$.

7. In a semisimple Lie algebra the Cartan–Killing metric $g_{ij} = C_{ir}{}^s C_{js}{}^r$ is nonsingular and therefore the contravariant metric g^{ij} is well defined. Show that the bilinear operator $C^2 = g^{ij} X_i X_j$ satisfies $[C^2, X_k] = 0$. If there is one quadratic Casimir operator, it must therefore be proportional to C^2.

8. Show that $C_{ijk} = C_{ij}{}^r g_{rk}$ is a third order antisymmetric tensor: $C_{ijk} = C_{jki} = C_{kij} = -C_{kji} = -C_{jik} = -C_{ikj}$. (Hint: use the Jacobi identity.)

9. Determine the structure of the Lie algebra defined by the following operators (cf., Eq. (16.57)):

$$X_{ij} = x^i \partial_j - x^j \partial_i$$

$$Y_i = 2t \frac{\partial}{\partial x^i} - x^i u \frac{\partial}{\partial u} \qquad (8.26)$$

$$Z = 2t \frac{\partial}{\partial t} + x^i \frac{\partial}{\partial x^i} - nu \frac{\partial}{\partial u}$$

9
Structure theory for simple Lie algebras

In this chapter we continue the development begun in the previous chapter. These two chapters focus on determining the structure of a Lie algebra and putting it into some canonical form. In the previous chapter we determined the types of subalgebras that every Lie algebra is constructed from. In this chapter we put the commutation relations into a standard form. This can be done for any Lie algebra. For semisimple Lie algebras this standard form has a very rigid structure whose usefulness is surpassed only by its beauty.

9.1 Objectives of this program

In the previous chapter we studied the commutation relations of a Lie algebra through its regular representation. This study was carried out using as a tool the Cartan–Killing inner product. As far as possible, this was the only method used. In the present chapter we introduce a second powerful tool from the theory of linear vector spaces. This is the eigenvalue decomposition. This tool is introduced in an attempt to find standard forms for the commutation relations. If a standard form is available then the properties of a Lie algebra, as well as its identification (classification), can be determined at sight.

The eigenoperator decomposition is effected by computing and studying a secular equation determined from the matrix of the regular (or any other matrix) representation of the Lie algebra. To get the most information from this study we seek the maximum number of independent roots of this equation. The decomposition of the Lie algebra into eigenoperators according to the roots of the secular equation, and the properties of these roots, can also be discussed for any Lie algebra. However, for Lie algebras with a nonsingular Cartan–Killing inner product – semisimple and simple Lie algebras – the properties of the roots are very rigidly prescribed. This leads to a very elegant set of canonical commutation relations.

In introducing an eigenvalue equation it is necessary to extend the field over which the Lie algebra is defined from the real to the complex numbers. Without this extension it is not always possible to find roots of the secular equation. This field extension has the drawback that several different Lie algebras (e.g., su(2) and su(1, 1)) have the same complex extension and have their different commutation relations cast into the same canonical form. We return to this question in Chapter 11, where the problem is resolved.

9.2 Eigenoperator decomposition – secular equation

It would be very useful to find vectors Z, X in the Lie algebra that obeyed the "eigenoperator" commutation relations

$$[Z, X] = \lambda X \tag{9.1}$$

It would be even more useful if we could find a set of basis vectors for the Lie algebra which all *simultaneously* obeyed commutation relations of the eigenoperator type.

To determine operators X for which such commutation relations are possible, we write $X = \sum_{i=1}^{N} a^i X_i$, where X_i form a basis set. Then

$$\left[Z, \sum a^i X_i\right] = \lambda \sum a^j X_j$$
$$\sum \sum a^i (R(Z)_i{}^j - \lambda \delta_i{}^j) X_j = 0 \tag{9.2}$$

This equation has a nonzero solution for the coefficients a^i provided the secular equation

$$\| R(Z) - \lambda I_N \| = 0 \tag{9.3}$$

can be solved. This equation can be expanded as a polynomial in λ

$$\sum_{j=0}^{N} (-\lambda)^{N-j} \phi_j(Z) = 0 \tag{9.4}$$

where N is the dimension of the Lie algebra and its regular representation. The coefficients $\phi_j(Z)$ are homogeneous polynomials of degree j in the coefficients z^i ($Z = \sum z^i X_i$) that describe Z:

$$\phi_j(Z) \to \phi_j(z^i) \tag{9.5}$$

Example The regular representation for the three-dimensional Lie algebra spanned by the photon creation and annihilation operators and their commutator

9.2 Eigenoperator decomposition – secular equation

$a^\dagger, a, I = [a, a^\dagger]$ is

$$ra^\dagger + la + \delta I \xrightarrow[\text{representation}]{\text{regular}} \begin{bmatrix} 0 & l & 0 \\ 0 & 0 & 0 \\ 0 & -r & 0 \end{bmatrix} \begin{matrix} a^\dagger \\ I \\ a \end{matrix} \qquad (9.6)$$

With this ordering of basis vectors the regular representation does not have the structure indicated in (8.4) and Fig. 8.1 for a nilpotent algebra. The secular equation is

$$\| \mathfrak{Reg}(ra^\dagger + la + \delta I) - \lambda I_3 \| = (-\lambda)^3 = 0 \qquad (9.7)$$

Strictly upper (or lower) triangular matrices have a secular equation of this form. The converse is true. If the secular equation of an $N \times N$ matrix is $(-\lambda)^N = 0$, then a basis can be found in which the matrix has strictly upper (or lower) triangular form. Therefore, there is a permutation transformation of the basis vectors that brings the regular representation of this Lie algebra to strictly upper triangular form, and the algebra is nilpotent by inspection.

Example For $X = \sum a_i X_i \in \mathfrak{su}(2)$ the defining 2×2 matrix representation $\mathfrak{def}(X)$ and the regular 3×3 matrix representation $\mathfrak{Reg}(X)$ are

$$\mathfrak{def}(X) = \frac{1}{2} \begin{bmatrix} ia_3 & i(a_1 - ia_2) \\ i(a_1 + ia_2) & -ia_3 \end{bmatrix}$$

$$\mathfrak{Reg}(X) = \begin{bmatrix} 0 & -a_3 & a_2 \\ a_3 & 0 & -a_1 \\ -a_2 & a_1 & 0 \end{bmatrix} \qquad (9.8)$$

The secular equation for the regular representation is

$$\| \mathfrak{Reg}(X) - \lambda I_3 \| = (-\lambda)^3 + (-\lambda)(+a_1^2 + a_2^2 + a_3^2) = 0$$
$$= (-\lambda)(\lambda^2 + \phi_2(\mathbf{a})) \qquad (9.9)$$
$$\phi_2(\mathbf{a}) = +a_1^2 + a_2^2 + a_3^2$$

Since $\phi_2(\mathbf{a}) \geq 0$, this secular equation cannot be solved over the field of real numbers. Extension of the field from the real to the complex numbers allows factorization to find the three (three is the dimension of $\mathfrak{su}(2)$) roots: $\lambda = 0$, $\lambda = \pm ia$, $a^2 = +a_1^2 + a_2^2 + a_3^2$.

Example For $Y = \sum b_i Y_i \in \mathfrak{su}(1, 1)$ the defining 2×2 matrix representation $\mathfrak{def}(Y)$ and the regular 3×3 matrix representation $\mathfrak{Reg}(Y)$ are

$$\mathfrak{def}(Y) = \frac{1}{2}\begin{bmatrix} ib_3 & b_1 - ib_2 \\ b_1 + ib_2 & -ib_3 \end{bmatrix}$$

$$\mathfrak{Reg}(y) = \begin{bmatrix} 0 & -b_3 & -b_2 \\ b_3 & 0 & b_1 \\ -b_2 & b_1 & 0 \end{bmatrix} \quad (9.10)$$

The secular equation for the regular representation is

$$\| \mathfrak{Reg}(Y) - \lambda I_3 \| = (-\lambda)^3 + (-\lambda)(-b_1^2 - b_2^2 + b_3^2) = 0$$
$$= (-\lambda)(\lambda^2 + \phi_2(\mathbf{b})) \quad (9.11)$$
$$\phi_2(\mathbf{b}) = -b_1^2 - b_2^2 + b_3^2$$

By comparing the secular equations for $\mathfrak{su}(1, 1)$ and $\mathfrak{su}(2)$, it is clear that the coefficients of the respective secular equations are "analytic continuations" of each other. That is, under rotation of some coordinates from the real to the imaginary axis, $(a_1, a_2, a_3) \to (ib_1, ib_2, b_3)$, the coefficient $\phi_2(\mathbf{a}) = a_1^2 + a_2^2 + a_3^2$ of the secular equation for $\mathfrak{su}(2)$ maps to $\phi_2(\mathbf{b}) = -b_1^2 - b_2^2 + b_3^2$ for $\mathfrak{su}(1, 1)$. This same rotation of coordinates maps the Lie algebra $\mathfrak{su}(2)$ to the Lie algebra $\mathfrak{su}(1, 1)$.

The secular equation was written down for the regular representation, since it can always be constructed from the Lie algebra. A secular equation could just as easily be written down for any matrix representation of the Lie algebra. We are by and large interested in studying matrix Lie algebras, so secular equations can be written directly for the defining matrix algebras. There is a great deal of utility in this approach. First, the matrices in a matrix algebra are almost always smaller – much smaller – than the matrices of its regular representation. Second, a matrix Lie algebra contains at least as much information (certainly not less) as its regular representation.

Example The secular equation for the defining 2×2 matrix representation of $\mathfrak{su}(2)$ in (9.8) is

$$\| \mathfrak{def}(X) - \lambda I_2 \| = \lambda^2 + \left(\tfrac{1}{2}\right)^2 (+a_1^2 + a_2^2 + a_3^2) = 0 \quad (9.12)$$

Similarly, the secular equation for the defining 2×2 matrix representation of $\mathfrak{su}(1, 1)$ in (9.10) is

$$\| \mathfrak{def}(Y) - \lambda I_2 \| = \lambda^2 + \left(\tfrac{1}{2}\right)^2 (-b_1^2 - b_2^2 + b_3^2) = 0 \quad (9.13)$$

For each algebra the functional forms of the nonzero coefficient ϕ_2 in the secular equation are the same in the defining and the regular matrix representations.

9.3 Rank

The **rank**, l, of a Lie algebra is the number of independent coefficients in the secular equation of its regular representation, $\mathfrak{Reg}$. Since the number of independent roots of the secular equation is equal to the number of independent coefficients $\phi_j(z^i)$, the rank is also the number of independent roots of the secular equation. The rank is always smaller than the dimension of the Lie algebra, since there is always at least one zero root (put $X = Z$ in (9.1)). For simple Lie algebras of dimension N, $l^2 \sim N$, so describing commutation relations in terms of rank rather than dimension effects a big simplification.

9.4 Invariant operators

If $\phi_j(z^i)$ is a coefficient in the secular equation, the operator obtained by the symmetrized substitution

$$z^i \to X_i \qquad \phi_j(z^i) \longrightarrow \phi_j(X_i) \tag{9.14}$$

is an invariant operator: it commutes with all elements of the Lie algebra

$$[\phi_j(X_i), X_k] = 0 \tag{9.15}$$

The number of independent invariant operators ("Casimir invariants") is at least equal to the rank of the algebra, and may be as large as the dimension for a commutative algebra, where all N operators mutually commute.

Example From the secular equation (9.9) for $\mathfrak{su}(2)$ we immediately construct a second order invariant operator that commutes with all operators in $\mathfrak{su}(2)$

$$\phi_2(\mathbf{a}) = +a_1^2 + a_2^2 + a_3^2 \longrightarrow \phi_2(X) = +X_1^2 + X_2^2 + X_3^2 \tag{9.16}$$

A similar calculation for $\mathfrak{su}(1,1)$ gives

$$\phi_2(\mathbf{b}) = -b_1^2 - b_2^2 + b_3^2 \longrightarrow \phi_2(Y) = -Y_1^2 - Y_2^2 + Y_3^2 \tag{9.17}$$

Notice that the Casimir invariant operator for $\mathfrak{su}(1,1)$ is the analytic continuation of the Casimir invariant operator for $\mathfrak{su}(2)$.

If $\mathfrak{m}$ is some matrix Lie algebra of $n \times n$ matrices, then any operator in $\mathfrak{m}$ can be written as a linear combination of matrices M_{ij}, with entry $+1$ at the intersection of the ith row and jth column and zeroes elsewhere

$$M : \sum a^i{}_j M^j{}_i \tag{9.18}$$

The coefficients of the secular equation for this algebra of $n \times n$ matrices are shown in Fig. 9.1.

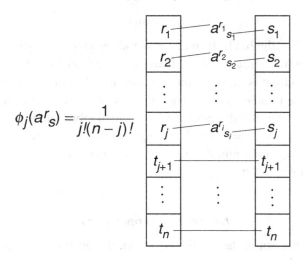

Figure 9.1. Coefficients in the secular equation are expressed in terms of the fully antisymmetric Levi–Civita tensor on n symbols.

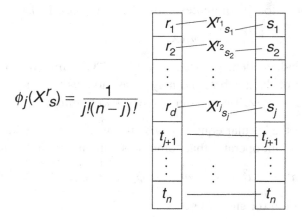

Figure 9.2. Invariant operators $\phi_j(X)$ expressed in terms of the fully antisymmetric Levi–Civita tensor on n symbols. The invariant operators are obtained by replacing the coordinates a^r_s by the operators X^r_s in the coefficients ϕ_j of the secular equation. Here the general element in the Lie algebra is $X = a^r_s X^r_s$.

In this figure the vertical symbol is the Levi–Civita symbol for n dimensions (e.g., in R^3, $= \epsilon_{ijk} = +1$ for (ijk) a cyclic permutation of (123), -1 for a cyclic permutation of (321), and zero otherwise). Contracted dummy indices are connected by lines. The invariant operators for the Lie algebra of $n \times n$ matrices are shown in Fig. 9.2. Contracted dummy indices are connected by lines. The invariance of these operators depends only on the commutation relations of the Lie algebra. Therefore

9.4 Invariant operators

these invariant operators $\phi_j(X^r{}_s)$ remain invariant when the matrices are replaced by any set of operators (see Chapter 6) with isomorphic commutation relations.

Example The orthogonal groups $O(n)$ and their subgroups $SO(n)$ have Lie algebras that consist of $n \times n$ antisymmetric matrices. The secular equation is far easier to compute in the defining representation of $n \times n$ antisymmetric matrices than in the $d \times d$ (the dimension of $\mathfrak{so}(n)$ is $d = n(n-1)/2$) regular matrix representation

$$\| \det(X) - \lambda I_n \| = \sum (-\lambda)^{n-j} \phi_j(X) = 0 \tag{9.19}$$

Further, the secular equation for a matrix and its transpose are equal, but since the Lie algebra consists of antisymmetric matrices, $\det(X)^t = -\det(X)$, and we find

$$\phi_j(X) = \phi_j(-X) = (-)^j \phi_j(X) \tag{9.20}$$

As a result, the only nonzero coefficients in the secular equation for $\mathfrak{so}(n)$ are the even coefficients. Therefore the algebra $\mathfrak{so}(n)$ has rank $[n/2]$.

Example The second order Casimir invariant operator for $\mathfrak{so}(n)$ is obtained by setting $j = 2$ in Fig. 9.2 for the generators X_{ij} of $SO(n)$. Since $X_{ij} = -X_{ji}$, it is possible to "rearrange" the contractions between the operators and the two different antisymmetric tensors, as shown in Fig. 9.3.

As a result, we can write for $\mathfrak{so}(n)$

$$C_2[\mathfrak{so}(n)] = \sum X_{ij}^2 \tag{9.21}$$

Similar "rearrangement" arguments can be used to simplify the expressions for higher order Casimir invariant operators. For example, for $\mathfrak{so}(5)$ the fourth order

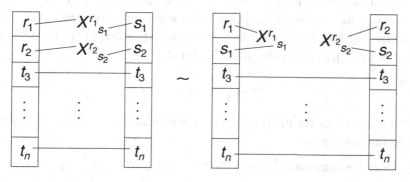

Figure 9.3. If the operators X are antisymmetric, $X^r{}_s = -X^s{}_r$, contractions in the expressions for the Casimir operators can be rearranged as shown.

Casimir operator is

$$C_4[\mathfrak{so}(5)] = \sum_{i=1}^{5} v_i^2 \tag{9.22}$$

where the components of the five-vector $\mathbf{v}$ are $v^m = \epsilon^{ijklm} X_{ij} X_{kl}$, for example

$$v^5 = \epsilon^{ijkl5} X_{ij} X_{kl} \sim X_{12} X_{34} - X_{13} X_{24} + X_{14} X_{23} \tag{9.23}$$

For $\mathfrak{so}(4)$ the fourth order Casimir is a perfect square

$$C_4[\mathfrak{so}(4)] = (\epsilon^{ijkl} X_{ij} X_{kl})^2 \sim (X_{12} X_{34} - X_{13} X_{24} + X_{14} X_{23})^2 \tag{9.24}$$

In general, for n even, the nth order Casimir invariant operator for $\mathfrak{so}(n)$ is a perfect square. Its square root, of order $n/2$, should be taken as an appropriate functionally independent Casimir operator. The existence of two second-order Casimir operators for $\mathfrak{so}(4)$ is another piece of evidence that this algebra is semisimple rather than simple.

9.5 Regular elements

It is useful to choose elements Z in the Lie algebra (Eq. (9.1)) that maximize the amount of information that can be extracted from the secular equation. (At the opposite extreme, the choice $Z = 0$ is not clever since all X obey the same eigenvalue equation $[Z, X] = 0X$.)

We do this by choosing a Z for which we:

1. maximize the number of nonzero roots;
2. minimize the degeneracy of each nonzero root;
3. minimize the degeneracy of the zero root.

Such elements Z in the Lie algebra can always be found. In fact, this is a 'generic' property. 'Almost all' elements Z in the Lie algebra have this property.

As an example of this eigenoperator decomposition we treat again the six-dimensional algebra of two photon operators spanned by $\hat{n} + \frac{1}{2} = \frac{1}{2}\{a, a^\dagger\}$, $a^{\dagger 2}, a^\dagger, I = [a, a^\dagger], a, a^2$. A useful choice for Z is

$$Z = z_1 \left(\hat{n} + \frac{1}{2}\right) + z_2 I \tag{9.25}$$

The secular equations for the 6×6 regular representation and the 4×4 defining matrix representations are

regular representation $\quad (\lambda)^2(\lambda + 2z_1)(\lambda - 2z_1)(\lambda + z_1)(\lambda - z_1) = 0$

$$\tag{9.26}$$

defining representation $\quad (\lambda)^2(\lambda + z_1)(\lambda - z_1) = 0$

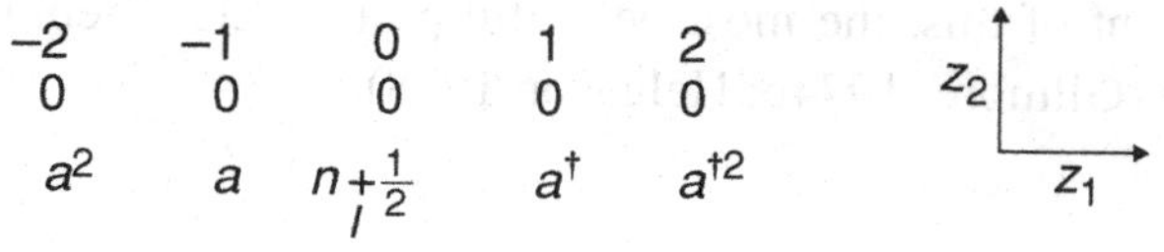

$$\begin{array}{cccccc} -2 & -1 & 0 & 1 & 2 \\ 0 & 0 & 0 & 0 & 0 \\ a^2 & a & n+\tfrac{1}{2} & a^\dagger & a^{\dagger 2} \end{array} \qquad \begin{array}{c} z_2 \\ \\ \longrightarrow z_1 \end{array}$$

Figure 9.4. The six operators in the two-photon algebra can be organized according to their roots, which are eigenvalues of a secular equation. Two operators have zero root.

Each secular equation has only one independent coefficient ϕ. The nontrivial coefficients of the secular equation for the regular representation are

$$\begin{aligned} \phi_2(z_1, z_2) &= -5z_1^2 \\ \phi_4(z_1, z_2) &= 4z_1^4 = 4(-\phi_2(z_1, z_2)/5)^2 \end{aligned} \quad (9.27)$$

For the 4×4 matrix representation the one nontrivial coefficient is

$$\phi_2(z_1, z_2) = -z_1^2 \quad (9.28)$$

This is a rank-one Lie algebra since there is only one functionally independent coefficient in the secular equation. The roots of the secular equation of the regular representation are $\pm 2z_1, \pm z_1, 0, 0$ and the commutation relations can be summarized in the 'root space diagram' shown in Fig. 9.4.

From this diagram we learn

$$\begin{aligned} \left[\hat{n} + \tfrac{1}{2}, X_{(k,0)}\right] &= k X_{(k,0)} \\ \left[I, X_{(k,0)}\right] &= 0 X_{(k,0)} \end{aligned} \quad (9.29)$$

where $X_{(2,0)} = a^{\dagger 2}$, $X_{(1,0)} = a^\dagger$, $X_{(0,0)} = \hat{n} + \tfrac{1}{2} I$, I, $X_{(-1,0)} = a$, $X_{(-2,0)} = a^2$. We also see that if $k, l \in \{-2, -1, 0, +1, +2\}$

$$[X_{(k,0)}, X_{(l,0)}] \sim X_{(k+l,0)} \quad (9.30)$$

if $k + l$ is in the set $\{-2, -1, 0, +1, +2\}$ and zero otherwise. If $k + l = 0$ the commutator is some linear combination of the two operators that span the subspace $(0, 0)$: $\hat{n} + \tfrac{1}{2}$ and I.

9.6 Semisimple Lie algebras

For simple and semisimple Lie algebras the Cartan–Killing inner product is nonsingular. When this inner product is nonsingular, the decomposition of the algebra into its subspaces, one for each root of the secular equation, has additional properties. We list these properties here, providing only an occasional proof. A more complete treatment of this, the most beautiful part of Lie algebra theory, can be found elsewhere (Gilmore, 1974b; Helgason 1978).

9.6.1 Rank

For semisimple Lie algebras the rank l is:

(i) the number of independent coefficients in the secular equation;
(ii) the number of independent roots $\alpha_1, \alpha_2, \ldots, \alpha_l$ of the secular equation; these l independent roots can be collected together as the components of an l-dimensional vector $(\alpha_1, \alpha_2, \ldots, \alpha_l)$ in a **root space**;
(iii) the dimension of the subspace V_0 (which is a subalgebra) of the root space;
(iv) the number of independent invariant operators (Casimir operators).

9.6.2 Properties of roots

Further, the roots have the following properties.

(i) If α and β are roots with subspaces V_α and V_β in the Lie algebra, then

$$[V_\alpha, V_\beta] \subset V_{\alpha+\beta} \tag{9.31}$$

That is, the commutator of any vector in V_α with any vector in V_β is a vector in $V_{\alpha+\beta}$. If $\alpha + \beta$ is not a root, the commutator vanishes.

(ii) The l basis vectors $H_1, H_2, \ldots, H_l$ in the l-dimensional subspace V_0 commute:

$$[H_i, H_j] = 0 \qquad 1 \leq i, j \leq l \tag{9.32}$$

(iii) Each subspace V_α ($\alpha \neq 0$) is one dimensional. Therefore each subspace V_α is spanned by an operator E_α that can be labeled by the root α. As a result (i.e., $[V_0, V_\alpha] \subset V_\alpha$), each H_i maps E_α into a multiple of itself

$$\begin{aligned}[H_i, E_\alpha] &= \alpha_i E_\alpha \\ [\mathbf{H}, E_\alpha] &= \alpha\, E_\alpha\end{aligned} \tag{9.33}$$

(iv) If α is a root, $-\alpha$ is a root. If α is a root and $c\alpha$ is a root, then $|c| = 1$. Thus, nonzero roots occur in pairs of opposite sign. In addition, the only root collinear with 0 and α is $-\alpha$.

(v) The commutator of E_α and $E_{-\alpha}$ is in V_0, so can be expanded as a linear superposition of the H_i:

$$[E_\alpha, E_{-\alpha}] = \alpha^i H_i \tag{9.34}$$

(vi) An inner product relating α^i and α_j by $\alpha^i = h^{ij}\alpha_j$ can be introduced in this root space

$$(\alpha, \beta) = \alpha_i \beta^i = \alpha^j \beta_j = \alpha_i h^{ij} \beta_j \tag{9.35}$$

9.6 Semisimple Lie algebras

This inner product is positive-definite. If the lengths of the roots are normalized so that

$$\sum_{\alpha \neq 0} \alpha_i \alpha_j = \delta_{ij} \quad \text{or} \quad \sum_{\alpha \neq 0} \alpha \cdot \alpha = \text{rank} = l \quad (9.36)$$

then $h^{ij} = \delta^{ij}$ and we can identify α^i with α_i: $\alpha^i = \alpha_i$.

(vii) It remains to compute

$$[E_\alpha, E_\beta] \to \begin{cases} 0 & \alpha + \beta \text{ not a root} \\ N_{\alpha,\beta} E_{\alpha+\beta} & \alpha + \beta \text{ a root} \\ \alpha \cdot \mathbf{H} & \alpha + \beta = 0 \end{cases}$$

Three cases arise, as indicated. The only detail remaining is to determine the coefficient $N_{\alpha,\beta}$ when $\alpha + \beta$ is a nonzero root.

9.6.3 Structure constants

To compute these coefficients we first apply the Jacobi identity to the generators $E_\alpha, E_\beta, E_\gamma$ of three nonzero roots that sum to zero

$$[[E_\alpha, E_\beta], E_\gamma] + [[E_\beta, E_\gamma], E_\alpha] + [[E_\gamma, E_\alpha], E_\beta] = 0 \quad (9.37)$$

From this we derive the symmetry

$$\begin{array}{ll} \text{when} & \alpha + \beta + \gamma = 0 \\ \text{then} & \alpha N_{\beta,\gamma} + \beta N_{\gamma,\alpha} + \gamma N_{\alpha,\beta} = 0 \\ \text{and} & N_{\beta,\gamma} = N_{\gamma,\alpha} = N_{\alpha,\beta} \end{array} \quad (9.38)$$

Next we compute a recursion relation involving these coefficients. This is done by embedding β in a chain of roots involving α additively, as shown in Fig. 9.5. In this chain

$$\beta - m\alpha \quad \beta - (m-1)\alpha \quad \cdots \quad \beta \quad \beta + \alpha \quad \cdots \quad \beta + n\alpha$$

are all roots but

$$\begin{aligned} \beta - (m+1)\alpha \\ \beta + (n+1)\alpha \end{aligned} \quad (9.39)$$

are not roots. By applying the Jacobi identity to roots α, $\beta + k\alpha$, and $-\alpha$ we obtain the recursion relation

$$N^2_{\alpha,\beta+(k-1)\alpha} = N^2_{\alpha,\beta+k\alpha} + \alpha \cdot (\beta + k\alpha) \quad (9.40)$$

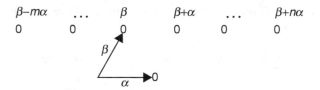

Figure 9.5. α chain containing β. This chain is used to compute coefficients $N_{\alpha,\beta}$ in commutators $[E_\alpha, E_\beta] = N_{\alpha,\beta} E_{\alpha+\beta}$.

This recursion relation satisfies the boundary conditions

$$N^2_{-\alpha,\beta-m\alpha} = 0$$
$$N^2_{+\alpha,\beta+n\alpha} = 0 \tag{9.41}$$

The initial condition $N_{\alpha,\beta+n\alpha} = 0$ leads to

$$N^2_{\alpha,\beta+(k-1)\alpha} = (n-k+1)\left(\alpha \cdot \beta + \frac{1}{2}(n+k)\alpha \cdot \alpha\right) \tag{9.42}$$

The other boundary condition $N^2_{-\alpha,\beta-m\alpha} = N^2_{\alpha,\beta-(m+1)\alpha} = 0$ leads to

$$N^2_{\alpha,\beta-(m+1)\alpha} = (n+m+1)\left(\alpha \cdot \beta + \frac{1}{2}(n-m)\alpha \cdot \alpha\right) = 0 \tag{9.43}$$

9.6.4 Root reflections

From this we extract the following information

(i) $N^2_{\alpha,\beta+k\alpha} = (n-k)(m+k+1)(\alpha \cdot \alpha)/2 \geq 0$. We use this expression because it shows clearly how the boundary conditions are imposed. We note that $\alpha \cdot \beta > 0$ when $m - n > 0$ and $\alpha \cdot \beta < 0$ when $m - n < 0$.

(ii) The inner products obey

$$-n \leq \frac{2\alpha \cdot \beta}{\alpha \cdot \alpha} = -n + m \leq m \tag{9.44}$$

where m and n are nonnegative integers.

(iii) If β is a root, then

$$\beta' = \beta - 2\frac{\beta \cdot \alpha}{\alpha \cdot \alpha}\alpha \tag{9.45}$$

is also a root. This root is obtained by reflecting β in the hyperplane orthogonal to α.

All of the rank-two root space diagrams are shown in Fig. 9.6. There the symmetries of root spaces under reflection and rotation may be seen.

9.7 Canonical commutation relations   151

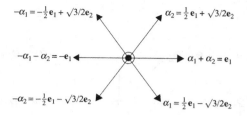

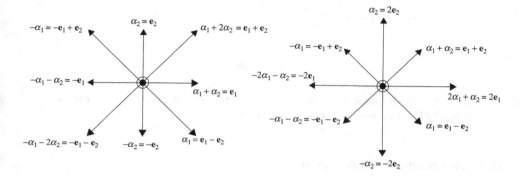

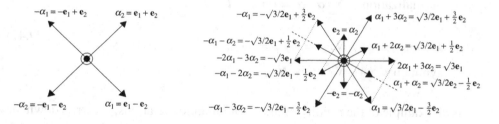

Figure 9.6. Two-dimensional root space diagrams. Top: A_2, B_2, C_2. Bottom: D_2, G_2.

9.7 Canonical commutation relations

The root space diagram encapsulates in a very convenient way all the structure constants of a semisimple Lie algebra. The basis vectors are the l (l is the rank) operators $\mathbf{H} = (H_1, H_2, \ldots, H_l)$ and the "shift" operators E_α, one corresponding to each nonzero root. The root vector $\alpha = (\alpha_1, \alpha_2, \ldots, \alpha_l)$ has l components. The commutation relations are

$$
\begin{aligned}
[H_i, H_j] &= 0 \\
[\mathbf{H}, E_\alpha] &= \alpha E_\alpha \\
[E_\alpha, E_\beta] &= \alpha \cdot \mathbf{H} && \alpha + \beta = 0 \\
&= N_{\alpha,\beta} E_{\alpha+\beta} && \alpha + \beta \neq 0, \text{ a root} \\
&= 0 && \alpha + \beta \text{ not a root}
\end{aligned}
\qquad (9.46)
$$

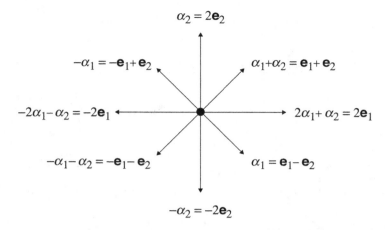

Figure 9.7. Root space C_2. The roots are expressed in terms of a Cartesian (orthogonal) set $\mathbf{e}_1$, $\mathbf{e}_2$ and a fundamental set α_1, α_2.

These commutation relations are subject to:

$$\text{normalization} \quad \sum_{\alpha \neq 0} \alpha \cdot \alpha = \text{rank} = l$$

$$\text{Jacobi} \quad N_{\alpha,\beta} = N_{\beta,\gamma} = N_{\gamma,\alpha} \qquad \alpha + \beta + \gamma = 0 \tag{9.47}$$

$$\text{symmetry} \quad N_{\alpha,\beta} = -N_{-\alpha,-\beta} = -N_{\beta,\alpha}$$

As an example of the rigidity of these commutation relations, we write down the commutation relations described by the rank-two root space C_2 shown in Fig. 9.7. If we choose orthogonal vectors $\mathbf{e}_1$ and $\mathbf{e}_2$ in a two-dimensional Euclidean space, the nonzero roots for C_2 are $\pm 2\mathbf{e}_1, \pm 2\mathbf{e}_2, \pm \mathbf{e}_1 \pm \mathbf{e}_2$. The 10 basis vectors in the Lie algebra are H_i, $i = 1, 2$, and E_α, with α the eight nonzero roots. We normalize these roots by $\sum \alpha \cdot \alpha = 2$ so that

$$(\mathbf{e}_i, \mathbf{e}_j) = \frac{1}{12}\delta_{ij} \tag{9.48}$$

Under this normalization condition the commutation relations are given in Table 9.1. All commutators not explicitly shown in this table vanish. For this rank-two algebra two phases may be set arbitrarily. The two commutators at which the phase choices have been made are indicated by * in Table 9.1. Both choices are $+1$. Other phase choices (-1) give isomorphic commutation relations.

Table 9.1. *Nonzero commutators for Lie algebras with root space* C_2

$[H_1, H_2]$	=	0
$[\mathbf{H}, E_{\pm 2e_1}]$	=	$(\pm 2/\sqrt{12}, 0) E_{\pm 2e_1}$
$[\mathbf{H}, E_{\pm 2e_2}]$	=	$(0, \pm 2/\sqrt{12}) E_{\pm 2e_2}$
$[\mathbf{H}, E_{\pm e_1 \pm e_2}]$	=	$(\pm 1/\sqrt{12}, \pm 1/\sqrt{12}) E_{\pm e_1 \pm e_2}$
$[E_{+2e_1}, E_{-2e_1}]$	=	$(2/\sqrt{12}) H_1$
$[E_{+2e_2}, E_{-2e_2}]$	=	$(2/\sqrt{12}) H_2$
$[E_{\pm e_1 \pm e_2}, E_{-(\pm e_1 \pm e_2)}]$	=	$(1/\sqrt{12})(\pm H_1 \pm H_2)$
$[E_{+2e_1}, E_{-(e_1+e_2)}]$	=	$*(1/\sqrt{6}) E_{e_1-e_2}$
$[E_{-e_1+e_2}, E_{-e_1-e_2}]$	=	$(1/\sqrt{6}) E_{-2e_1}$
$[E_{-e_1-e_2}, E_{+2e_1}]$	=	$(1/\sqrt{6}) E_{+e_1-e_2}$
$[E_{-2e_1}, E_{e_1-e_2}]$	=	$(-1/\sqrt{6}) E_{-e_1-e_2}$
$[E_{+e_1-e_2}, E_{+e_1+e_2}]$	=	$(-1/\sqrt{6}) E_{+2e_1}$
$[E_{+e_1+e_2}, E_{-2e_2}]$	=	$(-1/\sqrt{6}) E_{+e_1-e_2}$
$[E_{+2e_2}, E_{-e_1-e_2}]$	=	$*(1/\sqrt{6}) E_{-e_1+e_2}$
$[E_{-e_1-e_2}, E_{+e_1-e_2}]$	=	$(1/\sqrt{6}) E_{-2e_2}$
$[E_{+e_1-e_2}, E_{+2e_2}]$	=	$(1/\sqrt{6}) E_{+e_1+e_2}$
$[E_{-2e_2}, E_{+e_1+e_2}]$	=	$(-1/\sqrt{6}) E_{+e_1-e_2}$
$[E_{+e_1+e_2}, E_{-e_1+e_2}]$	=	$(-1/\sqrt{6}) E_{+2e_2}$
$[E_{-e_1+e_2}, E_{-2e_2}]$	=	$(-1/\sqrt{6}) E_{-e_1-e_2}$

9.8 Conclusion

The structure constants for a Lie algebra have been reduced to a canonical form by studying the properties of its regular representation. Using the Cartan–Killing inner product it is possible to determine the semisimple part of a Lie algebra and its complement, the maximal solvable invariant subalgebra. An eigenvalue decomposition can be used to put the commutation relations of the semisimple part into a standard form.

When the algebra is simple or semisimple the commutation relations are elegantly summarized by a root space diagram. This is a simple geometric structure in a Euclidean space of dimension l, where l is the rank of the Lie algebra. The rank is:

(i) the number of functionally independent coefficients in the secular equation;
(ii) the number of independent roots of the secular equation;
(iii) the number of Casimir invariant operators;
(iv) the dimension of the root space diagram;
(v) the number of mutually commuting operators in the Lie algebra.

We have illustrated how to extract commutation relations from a root space diagram for C_2.

In classifying simple Lie algebras by their root space diagram, we were forced to extend the field of the Lie algebra from the real to the complex numbers in order to guarantee that the secular equation had as many roots as basis vectors in the Lie algebra. In doing so, we have introduced a situation in which different algebras have the same complex extension (e.g., $\mathfrak{sl}(2;\mathbb{R})$ and $\mathfrak{so}(3)$ have common complex extension $\mathfrak{sl}(2;\mathbb{C})$). Root spaces classify commutation relations of these complex Lie algebras. Root spaces also summarize the commutation relations for the various real subalgebras of these complex algebras – some roots α_i and structure constants will be imaginary. However, determining the real subalgebras of a complex Lie algebra is a not entirely trivial task to which we return in Chapter 10.

9.9 Problems

1. Construct the regular representation for the two-photon operator algebra: $\frac{1}{2}\{a^\dagger, a\}$, $a^{\dagger 2}, a^\dagger, I, a, a^2$. Determine the secular equation for this matrix. Determine the rank of this Lie algebra.

2. Construct the 4×4 defining matrix representation and the 6×6 regular matrix representation of the Lie algebra $\mathfrak{so}(4)$. Construct the secular equation. This equation factors into two independent equations, each with one independent coefficient ϕ. Both are quadratic. Construct these coefficients. Use these to construct the two quadratic invariant operators on this semisimple Lie algebra. Show that in the canonical basis $X_{ij} = x^i \partial_j - x^j \partial_i$ ($1 \le i < j \le 4$) these operators are $C_2 = \sum_{i<j} X_{ij}^2$ and $C_2' = X_{12}X_{34} - X_{13}X_{24} + X_{14}X_{23}$.

3. The Lie algebra $\mathfrak{su}(4)$ has a 4×4 defining matrix representation and a 15×15 regular matrix representation. Show that the secular equation of the regular representation has just three independent coefficients. Do this by showing that there is a relation between the secular equation for the regular representation and the secular equation for the defining matrix representation. What is this relation? The three independent coefficients in the secular equation for the defining representation are of degree 2, 3, 4. Construct the invariant operators on $\mathfrak{su}(4)$ of degree 2, 3, and 4.

4. For $\mathfrak{so}(2n + 1)$ the invariant operators (Casimir operators) are of degree $2, 4, \ldots, 2n$. This is true also for $\mathfrak{so}(2n)$, with one difference: the invariant operator of degree $2n$ is a perfect square. Show that its square root, an invariant operator of degree n, is $C_n' = \epsilon^{i_1 i_2 \cdots i_{2n}} X_{i_1 i_2} X_{i_3 i_4} \cdots X_{i_{2n-1}, i_{2n}}$. Explicitly write out C_2' for $\mathfrak{so}(4)$ and C_3' for $\mathfrak{so}(6)$. Compare your results with Fig. 9.3.

5. In Chapter 11 we will show that $\mathfrak{su}(4) = \mathfrak{so}(6)$. Both Lie algebras have invariant operators of degree 2, 3, 4. Constuct the isomorphism between these Lie algebras and their invariant operators.

6. Summarize the commutation relations satisfied by the algebra of photon operators for two modes. This algebra is ten dimensional. It contains the four operators $a_i^\dagger a_j + \frac{1}{2}\delta_{ij}$

($1 \leq i, j \leq 2$) and the two pairs of three operators $a_i^\dagger a_j^\dagger$ and $a_i a_j$ ($a_i a_j = a_j a_i$). Show that this root space diagram is isomorphic to C_2, shown in Fig. 9.7. The identification is: $a_i^\dagger a_i + \frac{1}{2} \leftrightarrow H_i$, $a_i^\dagger a_j^\dagger \leftrightarrow E_{+e_i+e_j}$ ($i \neq j$), $a_i^\dagger a_j \leftrightarrow E_{+e_i-e_j}$ ($i \neq j$), $a_i a_j \leftrightarrow E_{-e_i-e_j}$ ($i \neq j$), $a_i^\dagger a_i^\dagger \leftrightarrow E_{+2e_i}$, $a_i a_i \leftrightarrow E_{-2e_i}$.

7. Repeat Problem 6 for the algebra of two fermion operators for two modes. This algebra is six dimensional. Show that the resulting root space diagram is D_2 (Fig. 9.6). Why the difference? (Hint: $f_i^\dagger f_i^\dagger = 0$.)

8. The Lie algebras $\mathfrak{su}(2)$ and $\mathfrak{so}(3)$ are isomorphic. In fact, the latter is the regular representation for the former. Choose $X, Y \in \mathfrak{su}(2)$ and compute $(X, Y) = \operatorname{tr}[\mathfrak{Def}(X)\mathfrak{Def}(Y)]$ by taking the trace of the 2×2 matrices in $\mathfrak{su}(2)$ that represent X and Y. Now compute the inner product using instead the Lie algebra $\mathfrak{so}(3)$, that is, the regular matrix representation of $\mathfrak{su}(2)$: $(X, Y) = \operatorname{tr}[\mathfrak{Reg}(X)\mathfrak{Reg}(Y)]$. Show that the two results are proportional. What is the proportionality constant?

9. Choose two vectors X and Y in the Lie algebra $\mathfrak{su}(n)$. Compute their inner product in the $n \times n$ defining matrix representation and in the $(n^2 - 1) \times (n^2 - 1)$ regular matrix representation. The two inner products are proportional. What is the proportionality constant? (Hint: set $Y = X$ and choose a special X, for example $X = H_1$.)

10. Express the Lie algebras spanned by the following ten sets of operators in canonical form (b boson operators; f fermion operators; $1 \leq i, j \leq N$):

$b_i^\dagger b_j$ $b_i^\dagger b_j + \frac{1}{2}\delta_{ij}, b_i^\dagger b_j^\dagger, b_i b_j$ $b^\dagger, b^\dagger b, b^\dagger b b$ $b, b^\dagger b, b^\dagger b^\dagger b$

$f_i^\dagger f_j$ $f_i^\dagger f_j + \frac{1}{2}\delta_{ij}, f_i^\dagger f_j^\dagger, f_i f_j$ $x, x\partial, x\partial^2$ $\partial, x\partial, x^2\partial$

$x^i \partial_j$ $x^i \partial_j + \frac{1}{2}\delta_{ij}, x^i x^j, \partial_i \partial_j$

11. Compute $\mathbf{R} = \frac{1}{2}\sum_{\alpha>0} \alpha$, half the sum over all positive roots, in each of the simple Lie algebras. This vector plays a major role in computing the spectrum of the quadratic Casimir operator for each of the irreducible representations of each of the simple Lie algebras. For example, for B_n, $R_i = \frac{1}{2}(2n+1) - i$ and the spectrum is

$$C^2(\mathbf{M}) = (\mathbf{M} + \mathbf{R}) \cdot (\mathbf{M} + \mathbf{R}) - (\mathbf{R}) \cdot (\mathbf{R}) = \mathbf{M} \cdot \mathbf{M} + \mathbf{M} \cdot 2\mathbf{R}$$

where $\mathbf{M}$ is the highest weight in the representation. For the $(2j+1)$ dimensional representation of $\mathfrak{so}(3)$, $\mathbf{M} = j$, $\mathbf{R} = R_1 = \frac{1}{2}$ and $C^2(j) = (j+\frac{1}{2})^2 - (0+\frac{1}{2})^2 = j(j+1)$.

12. The Weyl group of reflections for a simple Lie algebra is generated by reflections in planes orthogonal to all the nonzero roots.
    a. Show that the Weyl group for A_{n-1} is of order $n!$, the Weyl group for D_n is of order $2^{n-1}n!$, and the Weyl groups for B_n and C_n are of order $2^n n!$.
    b. Show that the product of the degrees of the functionally independent coefficients in the secular equation for each of these algebras is equal to the order of the Weyl group.

c. Show that the product of the degrees of the Casimir operators for each of these algebras is equal to the order of the Weyl group.

13. Compute the dimensions of each of the classical Lie algebras as a function of the rank, and show

	ratio	$n \to \infty$	algebra
$\dfrac{\dim(\mathfrak{g})}{\{\text{rank}(\mathfrak{g})\}^2} =$	$1 + \frac{2}{n}$	1	A_n
	$2 - \frac{1}{n}$	2	D_n
	$2 + \frac{1}{n}$	2	B_n, C_n

14. Multilinear operations can be defined on a matrix Lie algebra by

$$(A_1, A_2, \ldots, A_r)_{\mathfrak{Reg}} = \text{tr}\,\mathfrak{Reg}(A_1)\mathfrak{Reg}(A_2)\cdots\mathfrak{Reg}(A_r)$$

A multilinear operator can be defined similarly in other representations as well: for example, the defining representation.

a. Show

$$\frac{(A_1, A_2, \ldots, A_r)_{\mathfrak{Reg}}}{f_r(\mathfrak{Reg})} = (A_1, A_2, \ldots, A_r) = \frac{(A_1, A_2, \ldots, A_r)_\Gamma}{f_r(\Gamma)}$$

where Γ is some irreducible representation of the Lie algebra. This relation defines the **index** $f_r(\Gamma)$.

b. Show

$$\frac{f_r(\Gamma)}{f_r(\text{def})} = \frac{\text{tr}\,(\Gamma(A))^r}{\text{tr}\,(\text{def}(A))^r} = \frac{\dim(\Gamma)\,C^r(\Gamma)}{\dim(\text{def})\,C^r(\text{def})}$$

In this expression C^r is the value of the rth Casimir invariant in the representation indicated.

c. For $\mathfrak{su}(2)$

$$f_2(j) = \tfrac{1}{6}\{(2j)(2j+1)(2j+2)\}\,f_2(j=\tfrac{1}{2})$$

15. The matrix Lie algebras $\mathfrak{so}(2n)$, $\mathfrak{so}(2n+1)$, $\mathfrak{sp}(2n)$ have the form $\sum_{ij} a^{ij} M_{ij}$, where M_{ij} is a square matrix with $+1$ in the ith row and jth column and zeroes elsewhere, M is $2n \times 2n$ for $\mathfrak{so}(2n)$, $\mathfrak{sp}(2n)$ and $(2n+1) \times (2n+1)$ for $\mathfrak{so}(2n+1)$, and suitable reality restrictions are imposed on the coefficients a^{ij}.
   a. What are the conditions on a^{ij} for each matrix Lie algebra?
   b. Write down the coefficients $\phi_r(a^{ij})$ that occur in the secular equation for each of these matrix Lie algebras.
   c. Show that all odd coefficients $\phi_r(a^{ij})$ vanish for each of these matrix Lie algebras.
   d. Express the even coefficients in terms of the Levi–Civita skew tensors $\epsilon_{i_1 i_2 \cdots i_l}$ ($l = 2n, 2n+1, 2n$).
   e. Show that the even coefficients are all functionally independent.
   f. Conclude that each of these three matrix Lie algebras has rank n.
   g. Show that $\phi_{2n}(a^{ij})$ is a perfect square for $\mathfrak{so}(2n)$; write down its square root; show that it is of degree n.

16. Replace the scalar parameters θ_i in the 3×3 regular representation of $\mathfrak{so}(3)$ or $\mathfrak{su}(2)$ by the corresponding operators:

$$M = \begin{bmatrix} 0 & \theta_3 & -\theta_2 \\ -\theta_3 & 0 & \theta_1 \\ \theta_2 & -\theta_1 & 0 \end{bmatrix} \longrightarrow \mathcal{M} = \begin{bmatrix} 0 & J_3 & -J_2 \\ -J_3 & 0 & J_1 \\ J_2 & -J_1 & 0 \end{bmatrix}$$

  a. Show $\operatorname{tr} M^2 = -2\theta \cdot \theta$.
  b. Show $\operatorname{tr} \mathcal{M}^2 = -2\mathbf{J} \cdot \mathbf{J}$.
  c. Show $[\mathbf{J}, \operatorname{tr} \mathcal{M}^2] = 0$.
  d. Show $\operatorname{tr} \mathcal{M}^{2n+1} = 0$ and $\operatorname{tr} \mathcal{M}^{2n} = (-2)^n (\mathbf{J} \cdot \mathbf{J})^n$.

17. **Casimir covariants** A semisimple Lie algebra has basis vectors X_i that satisfy commutation relations $[X_i, X_j] = C_{ij}{}^k X_k$. There are two linear vector spaces, $V^{(1)}$ and $V^{(2)}$, that carry irreducible representations of this Lie algebra: $X_i \to \Gamma^{(1)}(X_i) = Y_i$ and $X_i \to \Gamma^{(2)}(X_i) = Z_i$. Show that the Casimir covariant $g^{ij} Y_i Z_j$ commutes with $(Y+Z)_k$ (more accurately, with $\Gamma^{(1)}(X_i) \otimes I_{\dim V^{(2)}} + I_{\dim V^{(1)}} \otimes \Gamma^{(2)}(X_i)$).

18. The Cayley–Hamilton theorem guarantees that a polynomial or analytic function of a square $n \times n$ matrix X can be expressed as a finite polynomial in the first n powers of X, starting at $X^0 = I_n$:

$$f(X) = f_0 I_n + f_1 X^1 + f_2 X^2 + \cdots + f_{n-1} X^{n-1} = \sum_{j=0}^{j=n-1} f_j X^j$$

The challenge is to compute the coefficients f_j in this expansion.
  a. Show that each coefficient f_j is a function of the invariants of the matrix X.
  b. Show that the invariants can variously be chosen as either the independent eigenvalues $\lambda_i(X)$ or the independent coefficients $\phi_i(X)$ of the secular equation for X.
  c. Show that the Cayley–Hamilton expansion simplifies considerably if the matrix X is chosen as generic diagonal. In fact it reduces to

$$\begin{bmatrix} 1 & \lambda_1 & \lambda_1^2 & \cdots & \lambda_1^{n-1} \\ 1 & \lambda_2 & \lambda_2^2 & \cdots & \lambda_2^{n-1} \\ \vdots & \vdots & \vdots & \ddots & \vdots \\ 1 & \lambda_n & \lambda_n^2 & \cdots & \lambda_n^{n-1} \end{bmatrix} \begin{bmatrix} f_0 \\ f_1 \\ \vdots \\ f_{n-1} \end{bmatrix} = \begin{bmatrix} f(\lambda_1) \\ f(\lambda_2) \\ \vdots \\ f(\lambda_n) \end{bmatrix}$$

The square matrix on the left is a vanderMonde matrix.
  d. Compute $e^{i\phi J_z}$ for the $(2j+1)$ dimensional matrix representations of $SU(2)$ by computing the vanderMonde matrices. Show that for $j = \frac{1}{2}, 1, \frac{3}{2}, 2$ the resulting

matrices are

$$\begin{bmatrix} 1 & \frac{1}{2} \\ 1 & -\frac{1}{2} \end{bmatrix} \quad \begin{bmatrix} 1 & 1 & 1 \\ 1 & 0 & 0 \\ 1 & -1 & 1 \end{bmatrix} \quad \begin{bmatrix} 1 & \frac{3}{2} & (\frac{3}{2})^2 & (\frac{3}{2})^3 \\ 1 & \frac{1}{2} & (\frac{1}{2})^2 & (\frac{1}{2})^3 \\ 1 & -\frac{1}{2} & (-\frac{1}{2})^2 & (-\frac{1}{2})^3 \\ 1 & -\frac{3}{2} & (-\frac{3}{2})^2 & (-\frac{3}{2})^3 \end{bmatrix}$$

and

$$\begin{bmatrix} 1 & 2 & 4 & 8 & 16 \\ 1 & 1 & 1 & 1 & 1 \\ 1 & 0 & 0 & 0 & 0 \\ 1 & -1 & 1 & -1 & 1 \\ 1 & -2 & 4 & -8 & 16 \end{bmatrix} \begin{bmatrix} f_0 \\ (i\phi)^1 f_1 \\ (i\phi)^2 f_2 \\ (i\phi)^3 f_3 \\ (i\phi)^4 f_4 \end{bmatrix} = \begin{bmatrix} e^{2i\phi} \\ e^{i\phi} \\ 1 \\ e^{-i\phi} \\ e^{-2i\phi} \end{bmatrix}$$

e. Invert each of these van der Monde matrices and determine the functions $f_j(\phi)$ in the expansions of e^X for $X \in \mathfrak{su}(2)$. In particular, show

	Representation			
$(i\phi)^j f_j$	$\frac{1}{2}$	1	$\frac{3}{2}$	2
	2	3	4	5
f_0	$\cos(\phi/2)$	1	$\frac{9}{8}\cos(\frac{\phi}{2}) - \frac{1}{8}\cos(\frac{3\phi}{2})$	1
$(i\phi)^1 f_1$	$2i\sin(\phi/2)$	$i\sin(\phi)$	$\frac{9i}{4}\sin(\frac{\phi}{2}) - \frac{i}{12}\sin(\frac{3\phi}{2})$	$\frac{i}{3}\sin(\phi) - \frac{i}{6}\sin(2\phi)$
$(i\phi)^2 f_2$		$\cos(\phi) - 1$	$-\frac{1}{2}\cos(\frac{\phi}{2}) + \frac{1}{2}\cos(\frac{3\phi}{2})$	$-\frac{5}{4} + \frac{4}{3}\cos(\phi) - \frac{1}{12}\cos(2\phi)$
$(i\phi)^3 f_3$			$-i\sin(\frac{\phi}{2}) + \frac{i}{3}\sin(\frac{3\phi}{2})$	$-\frac{i}{3}\sin(\phi) + \frac{i}{6}\sin(2\phi)$
$(i\phi)^4 f_4$				$\frac{1}{4} - \frac{1}{3}\cos(\phi) + \frac{1}{12}\cos(2\phi)$

f. Recover the two well-known expansions for $j = \frac{1}{2}$ and $l = 1$:

$$j = \frac{1}{2} \quad e^X = \cos\left(\frac{\phi}{2}\right) I_2 + \frac{\sin(\phi/2)}{\phi/2} X$$

$$l = 1 \quad e^X = I_3 + \frac{\sin(\phi)}{\phi} X + \frac{1 - \cos(\phi)}{\phi^2} X^2$$

g. Show that the $(2j+1) \times (2j+1)$ real antisymmetric matrix $X \in \mathfrak{su}(2)$ and its invariant ϕ are related by (cf. Problem 9.14)

$$\operatorname{tr} X^2 = -\frac{j(j+1)(2j+1)}{3} \phi^2$$

10
Root spaces and Dynkin diagrams

In the previous chapter the canonical commutation relations for semisimple Lie algebras were elegantly expressed in terms of roots. Although roots were introduced to simplify the expression of commutation relations, they can be used to classify Lie algebras and to provide a complete list of simple Lie algebras. We achieve both aims in this chapter. However, we use two different methods to accomplish this. We classify Lie algebras by specifying their root space diagrams. This is a relatively simple job using a "building up" approach, adding roots to rank l root space diagrams to construct rank $l+1$ root space diagrams. However, it is not easy to prove the completeness of root space diagrams by this method. Completeness is obtained by introducing Dynkin diagrams. These specify the inner products among a fundamental set of basis roots in the root space diagram. In this approach completeness is relatively simple to prove, while enumeration of the remaining roots within a root space diagram is less so.

10.1 Properties of roots

In an effort to cast the commutation relations of a semisimple Lie algebra into an eigenvalue-eigenvector format, a secular equation was constructed from the regular representation. The rank of an algebra is, among other things:

(i) the number of independent functions in the secular equation;
(ii) the number of independent roots of the secular equation;
(iii) the number of mutually commuting operators in the Lie algebra;
(iv) the number of invariant operators that commute with all elements in the Lie algebra (Casimir operators);
(v) the dimension of the positive-definite root space that summarizes the commutation relations.

In terms of the root space decomposition the commutation relations of the l ($=$ rank) operators H_i and the shift operators E_α are

$$[H_i, H_j] = 0$$
$$[\mathbf{H}, E_\alpha] = \alpha E_\alpha$$
$$\begin{aligned}[E_\alpha, E_\beta] &= \alpha \cdot \mathbf{H} &&\alpha + \beta = 0 \\ &= N_{\alpha,\beta} E_{\alpha+\beta} &&\alpha + \beta \neq 0 \text{ a root} \\ &= 0 &&\alpha + \beta \text{ not a root}\end{aligned} \quad (10.1)$$

The coefficients $N_{\alpha,\beta}$ are defined in terms of the nonnegative integers m, n by

$$N^2_{\alpha,\beta+k\alpha} = (n-k)(m+k+1)(\alpha \cdot \alpha)/2 \quad (10.2)$$

where $\beta + k\alpha$ is a root only for $k = -m, \ldots, +n$. The roots are normalized by

$$\sum_{\alpha \neq 0} \alpha \cdot \alpha = \text{rank} = l \quad (10.3)$$

In deriving the value for the structure constant $N_{\alpha,\beta}$ we observed

$$\frac{2(\alpha \cdot \beta)}{\alpha \cdot \alpha} \quad \text{is an integer}$$
$$\beta' = \beta - \frac{2(\alpha \cdot \beta)}{\alpha \cdot \alpha} \alpha \quad \text{is a root} \quad (10.4)$$

The root β' is obtained by reflecting β in the hyperplane orthogonal to α. These two observations are all that is required to construct root space diagrams of any rank.

If we write $2(\alpha \cdot \beta)/(\alpha \cdot \alpha) = n$ and $2(\alpha \cdot \beta)/(\beta \cdot \beta) = n'$, where n and n' are integers, then by the Schwarz inequality

$$0 \leq \cos^2(\alpha, \beta) = \left(\frac{\alpha \cdot \beta}{\alpha \cdot \alpha}\right)\left(\frac{\alpha \cdot \beta}{\beta \cdot \beta}\right) = \frac{n}{2}\frac{n'}{2} \leq 1 \quad (10.5)$$

These two results severely constrain the possible angles between two roots and their relative length. The results are summarized in Table 10.1.

10.2 Root space diagrams

The procedure for constructing root space diagrams in spaces of any dimension ($=$ rank) is simple. Begin with the rank-one root space. It is unique, with nonzero vectors $\pm \mathbf{e}_1$. To construct rank-two root spaces, add a noncollinear vector to this root space in such a way that the constraints exhibited in Table 10.1 are obeyed, and complete the root space by reflections in hyperplanes orthogonal to all roots. Only a small number of rank-two root spaces can be constructed in this way. These are A_2, $B_2 = C_2$, D_2 and G_2, as shown in Fig. 9.6.

10.2 Root space diagrams

Table 10.1. *Properties of roots in a root space diagram*

$\cos^2(\alpha,\beta)$	$\theta(\alpha,\beta)$	$n = \dfrac{2\alpha\cdot\beta}{\alpha\cdot\alpha}$	$n' = \dfrac{2\alpha\cdot\beta}{\beta\cdot\beta}$	$\dfrac{\alpha\cdot\alpha}{\beta\cdot\beta} = \dfrac{n'}{n}$
1	$\dfrac{\pi}{2}\pm\dfrac{\pi}{2}$	± 2	± 2	1
$\dfrac{3}{4}$	$\dfrac{\pi}{2}\pm\dfrac{\pi}{3}$	± 3, ± 1	± 1, ± 3	3^{-1}, 3^{+1}
$\dfrac{2}{4}$	$\dfrac{\pi}{2}\pm\dfrac{\pi}{4}$	± 2, ± 1	± 1, ± 2	2^{-1}, 2^{+1}
$\dfrac{1}{4}$	$\dfrac{\pi}{2}\pm\dfrac{\pi}{6}$	± 1	± 1	1
0	$\dfrac{\pi}{2}$	0	0	—

Rank-three root spaces are constructed from rank-two root spaces by the same process. A noncoplanar vector is added to a rank-two root space diagram subject to the condition that all the requirements of Table 10.1 are satisfied. The resultant set of roots is completed by reflection in hyperplanes orthogonal to all roots. If any pair of roots in the completed diagram does not satisfy these conditions, the resulting diagram is not an allowed root space diagram. The allowed rank-three root space diagrams are shown in Fig. 10.1.

This procedure is inductive. All rank-l root space diagrams are constructed in this way from rank-$(l-1)$ root space diagrams. We find by this building-up process that there are four infinite series of root spaces with the following sets of roots:

$$\begin{array}{llll}
A_{l-1} & +e_i - e_j & 1 \le i \ne j \le l & l-1 \ge 1 \\
D_l & \pm e_i \pm e_j & 1 \le i \ne j \le l & l > 3 \\
B_l & \pm e_i \pm e_j, \pm e_i & 1 \le i \ne j \le l & l > 2 \\
C_l & \pm e_i \pm e_j, \pm 2e_i & 1 \le i \ne j \le l & l > 1
\end{array} \qquad (10.6)$$

The subscript on the letter indicates the rank of the root space. It is easily seen that D_l is constructed by adding roots $\pm(e_i + e_j)$ to A_{l-1}, and B_l, C_l are constructed by adding roots $\pm e_i$, $\pm 2e_i$ to D_l. The root spaces A_{l-1}, D_l, B_l, C_l are all inequivalent with the following exceptions

$$\begin{aligned}
A_1 &= B_1 = C_1 \\
B_2 &= C_2 \\
A_3 &= D_3
\end{aligned} \qquad (10.7)$$

The root space D_2 is semisimple

$$D_2 = A_1 + A_1 \qquad (10.8)$$

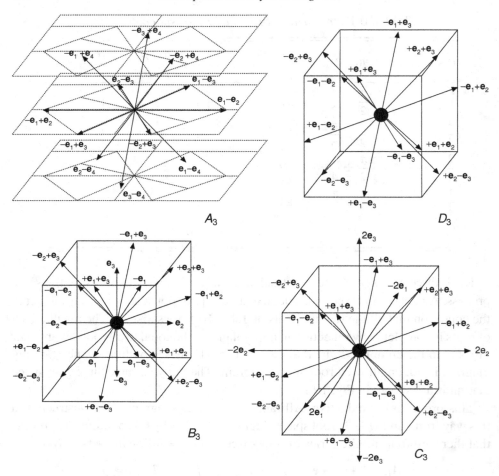

Figure 10.1. Rank-three root space diagrams. Top: A_3, D_3. Bottom: B_3, C_3.

In addition to these four unending series there are five exceptional root spaces:

$$G_2 \quad +\mathbf{e}_i - \mathbf{e}_j$$
$$\pm\left[(\mathbf{e}_i + \mathbf{e}_j) - 2\mathbf{e}_k\right] \qquad 1 \leq i \neq j \neq k \leq 3$$

$$F_4 \quad \pm\mathbf{e}_i \pm \mathbf{e}_j$$
$$\pm 2\mathbf{e}_i$$
$$\pm\mathbf{e}_1 \pm \mathbf{e}_2 \pm \mathbf{e}_3 \pm \mathbf{e}_4 \qquad 1 \leq i \neq j \leq 4$$

$$E_6 \quad \pm\mathbf{e}_i \pm \mathbf{e}_j$$
$$\tfrac{1}{2}\underbrace{(\pm\mathbf{e}_1 \pm \mathbf{e}_2 \pm \mathbf{e}_3 \pm \mathbf{e}_4 \pm \mathbf{e}_5)}_{\text{even number of + signs}} \pm \tfrac{\sqrt{3}}{4}\mathbf{e}_6 \quad 1 \leq i \neq j \leq 5$$

10.2 Root space diagrams

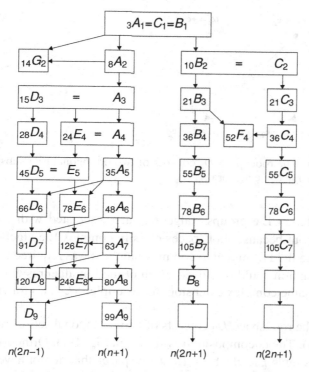

Figure 10.2. Root spaces constructed by the building-up principle. There are four infinite series and five exceptional Lie algebras. The root spaces are organized by rank.

E_7 $\pm e_i \pm e_j$

$\frac{1}{2} \underbrace{(\pm e_1 \pm e_2 \pm e_3 \pm e_4 \pm e_5 \pm e_6)}_{\text{even number of + signs}} \pm \frac{\sqrt{2}}{4} e_7$ $\quad 1 \le i \ne j \le 6$

E_8 $\pm e_i \pm e_j$

$\frac{1}{2} \underbrace{(\pm e_1 \pm e_2 \pm e_3 \pm e_4 \pm e_5 \pm e_6 \pm e_7 \pm e_8)}_{\text{even number of + signs}}$ $\quad 1 \le i \ne j \le 8$ (10.9)

The building-up principle is summarized in Fig. 10.2. In this figure all root spaces are shown by rank. Arrows connect pairs related by the building-up principle.

Remark 1. The following classical groups are associated with these root spaces

$$\begin{array}{lll} A_{l-1} & SU(l), SL(l; \mathbb{R}), SU(p,q) & p+q = l \\ D_l & SO(2l), SO(p,q) & p+q = 2l \\ B_l & SO(2l+1), SO(p,q) & p+q = 2l+1 \\ C_l & Sp(l), Sp(p,q) & p+q = l \end{array} \quad (10.10)$$

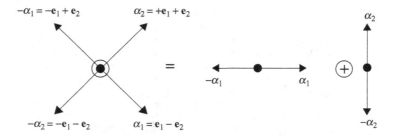

Figure 10.3. The root space D_2 consists of two orthogonal root subspaces. Both describe the rank-one algebra A_1.

Several different Lie groups (algebras) are associated with each root space. This comes about because root spaces classify complex Lie algebras. Recall that extension of the field from real to complex numbers was required to guarantee that the secular equation could be solved. Each of the Lie algebras with the same root space has the same complex extension, for example, $SL(l; C)$ for A_{l-1}.

Remark 2. The root space D_2 consists of two orthogonal sets of roots $\pm(\mathbf{e}_1 - \mathbf{e}_2)$ and $\pm(\mathbf{e}_1 + \mathbf{e}_2)$. The decomposition is shown in Fig. 10.3. Orthogonal root spaces describe semisimple Lie algebras. Root subspaces that do not have an orthogonal decomposition describe simple Lie algebras. Complete reducibility of the regular representation corresponds to decomposition of the root space into disjoint (orthogonal) root spaces and of the semisimple Lie algebras to simple invariant subalgebras.

Remark 3. The root spaces B_2 and C_2 are equivalent, as is easily seen by rotation. The root space B_2 describes $SO(5)$ while C_2 describes $Sp(2) = U(2; \mathbb{Q})$, which has a four-dimensional matrix representation obtained by replacing each quaternion by a complex 2×2 matrix. Therefore we should expect $SO(5)$ to have a four-dimensional "spinor" representation based on $U(2; \mathbb{Q})$ in the same way that $SO(3)$ (B_1) has a two-dimensional spinor representation based on $U(1; Q)$ or $SU(2)$ (A_1).

Remark 4. In the building-up construction the roots in each root space diagram are explicitly constructed. What is not immediately obvious is that there are no more simple root spaces than those listed. How are we sure that there are no more than five exceptional root spaces? This question is not easy to resolve in the context of root space constructions alone. However, it is easily resolved by another algorithmic procedure. This procedure yields a beautiful completeness argument. The price we pay is a somewhat greater difficulty in constructing the complete set of roots for

10.3 Dynkin diagrams

A plane through the origin of a root space diagram that does not contain any nonzero roots divides the roots into two sets, one "positive," the other negative (cf. Fig. 9.6). Among the positive roots the l nearest to this hyperplane in a rank-l root space are linearly independent. They can therefore be chosen as a basis set in this space. These roots are called **fundamental roots**, and denoted $\alpha_1, \alpha_2, \ldots, \alpha_l$. Every positive root can be expressed in terms of this basis as a linear combination of these fundamental roots with integer coefficients. The integers are all positive or zero, because every shift operator defined by a positive root can be written as a multiple commutator of shift operators with fundamental positive roots. By symmetry, every negative root is a linear combination of fundamental roots with nonpositive integer coefficients. The fundamental roots for G_2 are shown in Fig. 10.4. Fundamental roots for the

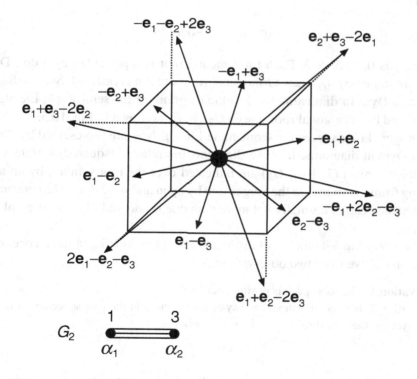

Figure 10.4. Root space for G_2. Fundamental roots are $\alpha_1 = e_1 - e_2$ and $\alpha_2 = -e_1 + 2e_2 - e_3$. All roots are orthogonal to $\mathbf{R} = e_1 + e_2 + e_3$.

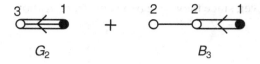

Figure 10.5. Disconnected Dynkin diagrams describe semisimple Lie algebras. Here the disconnected diagram describes $G_2 \oplus B_3$.

root spaces A_{l-1}, D_l, B_l, C_l are

$$
\begin{array}{cccccc}
& \alpha_1 & \alpha_2 & \alpha_3 & \alpha_{l-1} & \alpha_l \\
A_{l-1} & e_1 - e_2 & e_2 - e_3 & e_3 - e_4 & \cdots & e_{l-1} - e_l \\
D_l & " & " & " & \cdots & " & e_{l-1} + e_l \\
B_l & " & " & " & \cdots & " & e_l \\
C_l & " & " & " & \cdots & " & 2e_l
\end{array}
\qquad (10.11)
$$

Inner products among the fundamental roots are summarized conveniently in a diagrammatic form. The inner product between two fundamental roots is negative or zero

$$(\alpha_i, \alpha_j) = -\sqrt{n_{ij}/4} \qquad (10.12)$$

where n_{ij} is 0, 1, 2, or 3. Each fundamental root is represented by a dot. Dots i and j are joined by n_{ij} lines. Orthogonal roots are not connected. Such a diagram is called a **Dynkin diagram**. The Dynkin diagram for the semisimple Lie algebra represented by orthogonal root spaces $G_2 + B_3$ is shown in Fig. 10.5.

Orthogonal root spaces for semisimple Lie algebras are represented by disconnected Dynkin diagrams. In these diagrams the relative (squared) lengths of the fundamental roots (3, 1 for G_2) are indicated over the root symbol, by an arrow pointing from the shorter to the longer, and by open and solid dots. The conventions are interchangeable: normally not more than one is adopted. We will use only one at a time.

Only a very limited number of distinct kinds of Dynkin diagrams can occur. The limitations derive from two observations.

Observation 1 The root space is positive-definite.
Observation 2 If v_i is an orthonormal system of vectors in the root space and $\mathbf{u}$ is a unit vector, then the direction cosines $\mathbf{u} \cdot v_i$ obey

$$\sum (\mathbf{u} \cdot v_i)^2 \leq 1 \qquad (10.13)$$

These two observations are now used to list a set of properties that constrain the allowed Dynkin diagrams ever more tightly.

10.3 Dynkin diagrams

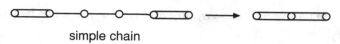

Figure 10.6. A simple linear chain can be removed. If the original is an allowed Dynkin diagram, the shortened diagram is also an allowed Dynkin diagram. In this case the original diagram is not an allowed Dynkin diagram.

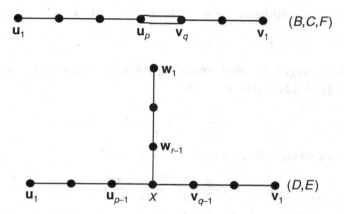

Figure 10.7. General forms of allowed root space diagrams after the process of contraction has been performed.

Property 1 There are no loops. A diagram containing a loop has at least as many lines as vertices. With $\mathbf{u}_i = \alpha_i/|\alpha_i|$ the inner product

$$\left(\sum \mathbf{u}_i, \sum \mathbf{u}_j\right) = n + 2\sum\sum \mathbf{u}_i \cdot \mathbf{u}_j \geq 0 \qquad (10.14)$$

cannot be positive since $2\mathbf{u}_i \cdot \mathbf{u}_j \leq -1$ if α_i and α_j are connected.

Property 2 The number of lines connected to any node is less than four. This results from Observation 2. If $\mathbf{v}_i$ are connected to $\mathbf{u}$, then

$$\sum (\mathbf{u} \cdot \mathbf{v}_i)^2 = \sum n_i/4 < 1 \qquad (10.15)$$

Property 3 A simple chain connecting any two dots can be shrunk. An allowed diagram is transformed to an allowed diagram. This allows the construction shown in Fig. 10.6. Since the constructed diagram violates Property 2, so also does the original diagram.

The only possibilities remaining are shown in Fig. 10.7.

For the diagrams (B, C, F) with a single double link, the Schwarz inequality applied to the vectors

$$\mathbf{u} = \sum_{i=1}^{p} i\mathbf{u}_i \qquad \mathbf{v} = \sum_{j=1}^{q} j\mathbf{v}_j \qquad (10.16)$$

where $\mathbf{u}_i$, $\mathbf{v}_j$ are unit vectors $\mathbf{u}_i = \alpha_i/|\alpha_i|$ and $\mathbf{v}_i = \alpha_j/|\alpha_j|$, can be transformed to the inequality

$$\left(1 + \frac{1}{p}\right)\left(1 + \frac{1}{q}\right) > 2 \tag{10.17}$$

This has the following solutions with $p \geq q$

$$\begin{aligned} p \text{ arbitrary}, \quad q = 1, \quad B_l, C_l \quad & l = p + 1 \\ p = 2, \quad q = 2, \quad F_4 & \end{aligned} \tag{10.18}$$

For the diagrams (D, E) Observation 2 applied to the vectors $\mathbf{u}$, $\mathbf{v}$, and $\mathbf{w}$ defined as in Eq. (10.16) leads to the inequality

$$\frac{1}{p} + \frac{1}{q} + \frac{1}{r} > 1 \tag{10.19}$$

This has the following solutions with $p \geq q \geq r \geq 2$

$$\begin{array}{cccc} p & q & r & \text{Root space} \\ \hline p & 2 & 2 & D_{p+2} \\ 3 & 3 & 2 & E_6 \\ 4 & 3 & 2 & E_7 \\ 5 & 3 & 2 & E_8 \end{array} \tag{10.20}$$

The allowed Dynkin diagrams are summarized in Table 10.2. This table provides a complete list of simple root spaces. Each root space was constructed in Section 10.2. The complete set of roots in each of the root spaces is listed in that section.

10.4 Conclusion

The canonical commutation relations for a semisimple Lie algebra have been expressed in terms of root space diagrams. These diagrams have been used to classify all simple root space diagrams of rank l by constructing a complete set of roots inductively from each root space diagram of rank $l - 1$. The completeness of this construction is guaranteed by the 1:1 correspondence between the root space diagrams constructed in Section 10.2 and the allowed connected Dynkin diagrams constructed in Section 10.3.

10.5 Problems

1. Show that the following three statements for a semisimple Lie algebra are equivalent:
   a. the Lie algebra has two simple invariant subalgebras;
   b. the nonzero roots in its root space diagram fall into two mutually orthogonal subsets;

10.5 Problems

Table 10.2. *Allowed root spaces*

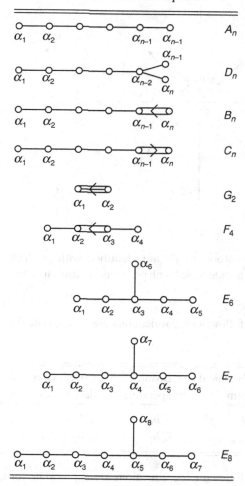

c. its Dynkin diagram is disconnected, with two connected components.
Do these statements extend to semisimple Lie algebras with three or more simple invariant subalgebras?

2. Show that bilinear combinations of two boson creation and/or annihilation operators can be identified with the roots in the ten-dimensional Lie algebra C_2 as shown in Fig. 10.8(a). Identify H_1 and H_2.

3. Show that bilinear combinations of two fermion creation and/or annihilation operators can be identified with the roots in the six-dimensional Lie algebra D_2 as shown in Fig. 10.8(b). Identify H_1 and H_2.

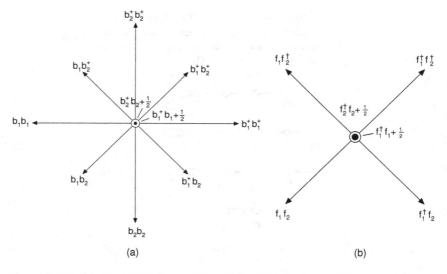

Figure 10.8. (a) Roots of C_2 are identified with products of boson operators. (b) Roots of D_2 are identified with products of fermion operators. Note that $f_1^\dagger f_1^\dagger = 0$, etc.

4. Show that the following identifications are appropriate for the generators of the Lie group $U(l)$:

Canonical form	Boson operators	Coordinates and derivatives	Fermion operators
H_i	$b_i^\dagger b_i$	$x^i \partial_i$	$f_i^\dagger f_i$
$E_{+e_i - e_j}$	$b_i^\dagger b_j$	$x^i \partial_j$	$f_i^\dagger f_j$

5. Show that the following identifications are appropriate for the eigenoperators for the root spaces C_l and D_l:

	C_l		D_l	
Canonical form	Boson operators	Coordinates and derivatives	Fermion operators	Coordinates and derivatives
H_i	$b_i^\dagger b_i + \frac{1}{2}$	$x^i \partial_i$	$f_i^\dagger f_i + \frac{1}{2}$	$x^i \partial_i + \frac{1}{2}$
$E_{+e_i - e_j}$	$b_i^\dagger b_j$	$x^i \partial_j$	$f_i^\dagger f_j$	$x^i \partial_j$
$E_{+e_i + e_j}$	$b_i^\dagger b_j^\dagger$	$x^i x^j$	$f_i^\dagger f_j^\dagger$	$x^i x^j$
$E_{-e_i - e_j}$	$b_i b_j$	$\partial_i \partial_j$	$f_i f_j$	$\partial_i \partial_j$
E_{+2e_i}	$b_i^\dagger b_i^\dagger$	$x^i x^i$		
E_{-2e_i}	$b_i b_i$	$\partial_i \partial_i$		

10.5 Problems

6. Apply the Schwartz inequality to the two vectors in Eq. (10.16) and show that the result can be expressed in the form of the inequality given in Eq. (10.17).

7. Use the projection inequality of Eq. (10.13) with the three vectors constucted for the Dynkin diagrams of type (D, E) to obtain the inequality of Eq. (10.19).

8. A Lie algebra is spanned by n^2 operators of the form $a_i^\dagger a_j$, with $1 \le i, j \le n$. Show that the linear vector space for this algebra can be written as the direct sum of two subspaces: **L**, **Q** spanned by the operators

$$\begin{array}{cc} \mathbf{L} & \mathbf{Q} \\ L_{ij} = a_i^\dagger a_j - a_j^\dagger a_i = -L_{ji} & Q_{ij} = a_i^\dagger a_j + a_j^\dagger a_i = +Q_{ji} \end{array}$$

For $n = 3$ the subspaces transform like an angular momentum vector and a quadrupole tensor. Show that the commutation relations are

$$\begin{array}{rl} [\mathbf{L}, \mathbf{L}] = \mathbf{L} & [L_{ij}, L_{rs}] = +\delta_{jr} L_{is} + \delta_{is} L_{jr} - \delta_{ir} L_{js} - \delta_{js} L_{ir} \\ [\mathbf{L}, \mathbf{Q}] = \mathbf{Q} & [L_{ij}, Q_{rs}] = +\delta_{jr} Q_{is} - \delta_{is} Q_{jr} - \delta_{ir} Q_{js} + \delta_{js} Q_{ir} \\ [\mathbf{Q}, \mathbf{Q}] = \mathbf{L} & [Q_{ij}, Q_{rs}] = +\delta_{jr} L_{is} + \delta_{is} L_{jr} + \delta_{ir} L_{js} + \delta_{js} L_{ir} \end{array}$$

The quadrupole tensor, in turn, with six components, can be written as the sum of a traceless tensor $\hat{\mathbf{Q}}$ and a scalar N:

$$\hat{N} = \sum_{i=1}^{3} a_i^\dagger a_i \qquad \hat{Q}_{ij} = Q_{ij} - \frac{2}{3} \hat{N} \delta_{ij}$$

The operator $\hat{N}$ commutes with all operators $a_i^\dagger a_j$. Interpret these commutation relations in physical terms (scalars, vectors, and traceless quadrupole tensors) and in mathematical terms (commutative invariant subalgebra $\hat{N}$, Cartan decomposition of a simple Lie algebra $\mathbf{L} + \hat{\mathbf{Q}}$).

9. Carry out a similar decomposition for any value of n. Show that the only changes in the discussion of Problem 8 are the dimensions of the spaces **L** $(3 \to n(n-1)/2)$, **Q** $(6 \to n(n+1)/2)$, and the definition of $\hat{N}$ $(3 \to n)$.

11
Real forms

Root space diagrams classify all the simple Lie algebras and summarize their commutation relations. The Lie algebras so classified exist over the field of complex numbers. Each simple Lie algebra over $\mathbb{C}$ of complex dimension n has a number of inequivalent real subalgebras over $\mathbb{R}$ of real dimension n. These are obtained by putting reality restrictions on the coordinates in the complex Lie algebra. The different real forms of a complex simple Lie algebra are obtained systematically by a simple eigenvalue decomposition. For the classical (matrix) Lie algebras, three different procedures suffice to construct all real forms. These are: block submatrix decomposition; subfield restriction; and field embedding.

11.1 Preliminaries

In our attempt to find a canonical form for the commutation relations of a real simple Lie algebra with elements $Z = r^i X_i$ (r^i are real numbers, X_i the generators of the Lie group, or basis vectors in the Lie algebra), we were led to an eigenvalue equation of the form $\sum \sum r^i [R_i{}^j(Z) - \lambda \delta_i{}^j] X_j = 0$. This equation cannot be solved in general unless the field is extended from the real to the complex numbers. Allowing that extension, we were able to find a canonical form for the operators in semisimple Lie algebras. The general operator in such algebras has the form

$$\sum_{i=1}^{\text{rank}} h^i H_i + \sum_{\alpha \neq 0} e^\alpha E_\alpha \tag{11.1}$$

where h^i and e^α are complex numbers and the "diagonal" and "shift" operators were defined in Section 9.7. The commutation relations were classified in terms of a root space diagram. These diagrams were used to enumerate all the simple Lie algebras over the complex field.

11.1 Preliminaries

We return now to the question of determining the real forms associated with each of the root space diagrams or, more accurately, the complex Lie algebra associated with each root space diagram. We do this by first presenting Cartan's method of decomposing a Lie algebra into two subspaces with very special commutation relations and orthogonality properties. Three simple decompositions of this type are applied to the compact matrix Lie algebra to generate all the real forms of the classical simple Lie algebras A_{n-1}, D_n, B_n, C_n. These decompositions are: block submatrix decomposition; subfield restriction; and field embeddings.

Example The noncompact Lie algebras $\mathfrak{sl}(2;\mathbb{R})$ and $\mathfrak{su}(1,1)$ have commutation relations described by the root space A_1. The nonisomorphic Lie algebra $\mathfrak{su}(2)$ has the same root space. To see why, we compute the regular representation of $\mathfrak{sl}(2;\mathbb{R})$ and $\mathfrak{su}(2)$ and their secular equations

Algebra	Defining representation	Regular representation
$\mathfrak{sl}(2;\mathbb{R})$	$\dfrac{1}{2}\begin{bmatrix} a_1 & a_2+a_3 \\ a_2-a_3 & -a_1 \end{bmatrix} \longrightarrow$	$\begin{bmatrix} 0 & -a_3 & -a_2 \\ a_3 & 0 & a_1 \\ -a_2 & a_1 & 0 \end{bmatrix}$
		$-\lambda[\lambda^2 + (-a_1^2 - a_2^2 + a_3^2)] = 0$
$\mathfrak{su}(2)$	$\dfrac{i}{2}\begin{bmatrix} b_3 & b_1-ib_2 \\ b_1+ib_2 & -b_3 \end{bmatrix} \longrightarrow$	$\begin{bmatrix} 0 & -b_3 & b_2 \\ b_3 & 0 & -b_1 \\ -b_2 & b_1 & 0 \end{bmatrix}$
		$-\lambda[\lambda^2 + (b_1^2 + b_2^2 + b_3^2)] = 0$

(11.2)

In the case of $\mathfrak{sl}(2;\mathbb{R})$ it is possible to find three real roots of the secular equation for certain choices of the real parameters a_1, a_2, a_3 while in the compact case this is not possible. If the real parameters (a_1, a_2, a_3) and (b_1, b_2, b_3) are allowed to become complex the two Lie algebras become algebras of 2×2 complex traceless matrices – the Lie algebra for $SL(2;\mathbb{C})$. This relation is shown in Fig. 11.1.

The complex extension Lie algebra has root space A_1 describing canonical commutation relations for the diagonal and shift operators shown in Fig. 11.1. The most general element in this Lie algebra is a complex linear combination of the three matrices shown. The algebras $\mathfrak{sl}(2;\mathbb{R})$ and $\mathfrak{su}(2)$ have real dimension 3 while their common complex extension has complex dimension 3 (real dimension 6). In the following sections we present a systematic way for determining how to restrict the complex parameters to real parameters in order to construct all inequivalent real Lie algebras with the same dimension as the complex Lie algebra whose commutation relations are described by a root space diagram.

174   Real forms

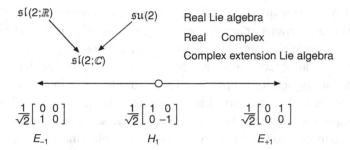

Figure 11.1. Lie groups $SL(2; \mathbb{R})$ and $SU(2)$ are related by analytic continuation. The canonical form for the diagonal and shift operators in their Lie algebras is also shown.

11.2 Compact and least compact real forms

The Cartan–Killing inner product for the basis vectors H_i, E_α is

$$\begin{bmatrix} 1 & & & & & & & & \\ & 1 & & & & & & & \\ & & 1 & & & & & & \\ & & & \ddots & & & & & \\ & & & & 1 & & & & \\ & & & & & 0 & 1 & & \\ & & & & & 1 & 0 & & \\ & & & & & & & 0 & 1 \\ & & & & & & & 1 & 0 \\ & & & & & & & & & \ddots \end{bmatrix} \begin{array}{l} H_1 \\ H_2 \\ H_3 \\ \vdots \\ H_n \\ E_\alpha \\ E_{-\alpha} \\ E_\beta \\ E_{-\beta} \\ \vdots \end{array} \quad (11.3)$$

The inner product can be brought to diagonal form by choosing linear combinations of basis vectors $\frac{1}{\sqrt{2}}(E_\alpha \pm E_{-\alpha})$:

$$\begin{bmatrix} 1 & & & & & & \\ & 1 & & & & & \\ & & 1 & & & & \\ & & & \ddots & & & \\ & & & & 1 & & \\ & & & & & 1 & \\ & & & & & & -1 \\ & & & & & & & 1 \\ & & & & & & & & -1 \\ & & & & & & & & & \ddots \end{bmatrix} \begin{array}{l} H_1 \\ H_2 \\ H_3 \\ \vdots \\ H_n \\ \frac{1}{\sqrt{2}}(E_\alpha + E_{-\alpha}) \\ \frac{1}{\sqrt{2}}(E_\alpha - E_{-\alpha}) \\ \frac{1}{\sqrt{2}}(E_\beta + E_{-\beta}) \\ \frac{1}{\sqrt{2}}(E_\beta - E_{-\beta}) \\ \vdots \end{array} \quad (11.4)$$

11.2 Compact and least compact real forms

If we restrict the coefficients of H_i, $\frac{1}{\sqrt{2}}(E_\alpha + E_{-\alpha})$, and $\frac{1}{\sqrt{2}}(E_\alpha - E_{-\alpha})$ (all $\alpha \neq 0$) to be real, then the generators $\frac{1}{\sqrt{2}}(E_\alpha - E_{-\alpha})$ span the maximal compact subalgebra (closure under commutation must be verified; this is left an an exercise) while the generators H_i and $\frac{1}{\sqrt{2}}(E_\alpha + E_{-\alpha})$ span a noncompact subspace.

On the other hand, if we restrict the coefficients of H_i and $\frac{1}{\sqrt{2}}(E_\alpha + E_{-\alpha})$ to be imaginary and those of $\frac{1}{\sqrt{2}}(E_\alpha - E_{-\alpha})$ to be real

$$ih^i H_i + ie^\alpha \frac{1}{\sqrt{2}}(E_\alpha + E_{-\alpha}) + e^{-\alpha}\frac{1}{\sqrt{2}}(E_\alpha - E_{-\alpha}) \tag{11.5}$$

then the factors i can be absorbed within the generators. With respect to these redefined generators the Cartan–Killing inner product is negative-definite and the algebra constructed is compact

$$\begin{array}{c} h^1 \\ h^2 \\ h^3 \\ \vdots \\ h^n \\ e^{+\alpha} \\ e^{-\alpha} \\ e^{+\beta} \\ e^{-\beta} \\ \vdots \end{array} \left[\begin{array}{ccccccccc} -1 & & & & & & & & \\ & -1 & & & & & & & \\ & & -1 & & & & & & \\ & & & \ddots & & & & & \\ & & & & -1 & & & & \\ & & & & & -1 & & & \\ & & & & & & -1 & & \\ & & & & & & & -1 & \\ & & & & & & & & -1 \\ & & & & & & & & & \ddots \end{array}\right] \begin{array}{l} iH_1 \\ iH_2 \\ iH_3 \\ \vdots \\ iH_n \\ i\frac{1}{\sqrt{2}}(E_\alpha + E_{-\alpha}) \\ \frac{1}{\sqrt{2}}(E_\alpha - E_{-\alpha}) \\ i\frac{1}{\sqrt{2}}(E_\beta + E_{-\beta}) \\ \frac{1}{\sqrt{2}}(E_\beta - E_{-\beta}) \\ \vdots \end{array} \tag{11.6}$$

real coefficients | Cartan–Killing inner product | basis vectors in Lie algebra

Two real forms of A_1, $\mathfrak{sl}(2; \mathbb{R})$ and $\mathfrak{su}(1, 1)$, are obtained as follows

$$\begin{array}{ccc} H_1 & \frac{1}{\sqrt{2}}(E_{+1} + E_{-1}) & \frac{1}{\sqrt{2}}(E_{+1} - E_{-1}) \end{array}$$

$$(h_r + ih_i)\begin{bmatrix}1 & 0 \\ 0 & -1\end{bmatrix} + (a_r + ia_i)\begin{bmatrix}0 & 1 \\ 1 & 0\end{bmatrix} + (b_r + ib_i)\begin{bmatrix}0 & 1 \\ -1 & 0\end{bmatrix} \tag{11.7}$$

$$(h_r, a_r, b_r) \longrightarrow \begin{bmatrix} h_r & a_r + b_r \\ a_r - b_r & -h_r \end{bmatrix} \quad \mathfrak{sl}(2; \mathbb{R})$$

$$(ih_i, ia_i, b_r) \longrightarrow i\begin{bmatrix} h_i & a_i - ib_r \\ a_i + ib_r & -h_i \end{bmatrix} \quad \mathfrak{su}(2) \tag{11.8}$$

Here the six parameters $h_r, h_i; a_r, a_i; b_r, b_i$ are real.

It is useful to specify how compact a real form is by specifying its **index** (n_+, n_-), where n_+ is the dimension of the subspace on which the nonsingular Cartan–Killing inner product is positive-definite and n_- is the dimension of the subspace (subalgebra) on which it is negative-definite. These two pieces of information may be abbreviated to a single integer, the **character** $\chi = n_+ - n_-$, to describe a real form. This is the trace of the normalized Cartan–Killing form. Inspection of (11.3) and (11.6) shows that the character is $+$(rank) of the root space for the real Lie algebra spanned by H_i, $\frac{1}{\sqrt{2}}(E_\alpha + E_{-\alpha})$, and $\frac{1}{\sqrt{2}}(E_\alpha - E_{-\alpha})$ and is $-$(dimension) for the compact real form spanned by real linear combinations of iH_i, $i\frac{1}{\sqrt{2}}(E_\alpha + E_{-\alpha})$, and $\frac{1}{\sqrt{2}}(E_\alpha - E_{-\alpha})$. In general, for all real forms the character satisfies the bounds

$$- \text{dimension} \leq \chi = \text{character} \leq +\text{rank} \tag{11.9}$$

11.3 Cartan's procedure for constructing real forms

Cartan has proposed a simple and elegant procedure for constructing all the real forms of a (complex) simple Lie algebra. This procedure constructs one real form from another by "analytic continuation." It is modeled on Minkowski's transformation of space-time (x, y, z, ct) with indefinite metric $g_{\mu,\nu} = \text{diag}(1, 1, 1, -1)$ to space-time with imaginary time (x, y, z, ict) and positive-definite metric $g_{\mu,\nu} = \text{diag}(1, 1, 1, 1)$.

Since the compact real form can always be constructed easily for a simple Lie algebra (see Eq. (11.6)) it is useful to begin with that form. The compact Lie algebra $\mathfrak{g}$ is divided into two pieces with the following commutation relations and orthogonality properties

$$\mathfrak{g} = \mathfrak{h} + \mathfrak{p}$$

$$\begin{array}{ll} [\mathfrak{h}, \mathfrak{h}] \subseteq \mathfrak{h} & (\mathfrak{h}, \mathfrak{h}) < 0 \\ [\mathfrak{h}, \mathfrak{p}] \subseteq \mathfrak{p} & (\mathfrak{h}, \mathfrak{p}) = 0 \\ [\mathfrak{p}, \mathfrak{p}] \subseteq \mathfrak{h} & (\mathfrak{p}, \mathfrak{p}) < 0 \end{array} \tag{11.10}$$

In short, the subspace $\mathfrak{h}$ is a subalgebra and $\mathfrak{p}$ is its orthogonal complement. A concrete example of this decomposition is

$$\begin{array}{ccccc} \mathfrak{su}(2) & = & \mathfrak{u}(1) & + & \mathfrak{su}(2) - \mathfrak{u}(1) \\ \frac{i}{2}\begin{bmatrix} a_3 & a_1 - ia_2 \\ a_1 + ia_2 & -a_3 \end{bmatrix} & = & \frac{i}{2}\begin{bmatrix} a_3 & 0 \\ 0 & -a_3 \end{bmatrix} & + & \frac{i}{2}\begin{bmatrix} 0 & a_1 - ia_2 \\ a_1 + ia_2 & 0 \end{bmatrix} \end{array} \tag{11.11}$$

The Lie algebra $\mathfrak{g}$ is mapped into a noncompact Lie algebra $\mathfrak{g}'$ by means of "Minkowski's trick": $\mathfrak{p} \to \mathfrak{p}' = i\mathfrak{p}$. The mapping, commutation relations, and orthogonality relations are

$$\begin{array}{rcl} \mathfrak{g} & = & \mathfrak{h} + \mathfrak{p} \\ [\mathfrak{h}, \mathfrak{h}] & \subseteq & \mathfrak{h} \\ [\mathfrak{h}, \mathfrak{p}'] & \subseteq & \mathfrak{p}' \\ [\mathfrak{p}', \mathfrak{p}'] & \subseteq & \mathfrak{h} \end{array} \longrightarrow \begin{array}{rcl} \mathfrak{g}' & = & \mathfrak{h} + i\mathfrak{p} = \mathfrak{h} + \mathfrak{p}' \\ (\mathfrak{h}, \mathfrak{h}) & < & 0 \\ (\mathfrak{h}, \mathfrak{p}') & = & 0 \\ (\mathfrak{p}', \mathfrak{p}') & > & 0 \end{array} \quad (11.12)$$

In $\mathfrak{g}'$, $\mathfrak{h}$ is the maximal compact subalgebra and $\mathfrak{p}'$ consists of all the noncompact generators. The character of this algebra is

$$\chi(\mathfrak{g}') = \dim(\mathfrak{p}') - \dim(\mathfrak{h}) = \dim(\mathfrak{g}) - 2\dim(\mathfrak{h}) = 2\dim(\mathfrak{p}') - \dim(\mathfrak{g}) \quad (11.13)$$

As a concrete example of this mapping, we have from (11.11)

$$\mathfrak{su}(2) \to \mathfrak{su}(1, 1) : \quad \frac{i}{2} \begin{bmatrix} a_3 & 0 \\ 0 & -a_3 \end{bmatrix} - \frac{1}{2} \begin{bmatrix} 0 & a_1 - ia_2 \\ a_1 + ia_2 & 0 \end{bmatrix} \quad (11.14)$$

The mapping is reversible: noncompact $\mathfrak{g}'$ can be mapped back to compact $\mathfrak{g}$.

A systematic method exists for finding Cartan decompositions (11.12). Assume T is a linear mapping of the Lie algebra $\mathfrak{g}$ onto itself that preserves inner products, and that also obeys

$$T^2 = I \quad (11.15)$$

("involutive automorphism"). Then T has two eigenvalues: ± 1. Under T, one eigenspace of T is mapped into itself while the other (its orthogonal complement) is mapped into its negative. The map T splits $\mathfrak{g}$ into eigenspaces $\mathfrak{h}$ and $\mathfrak{p}$

$$\begin{array}{rcl} \mathfrak{g} & = & \mathfrak{h} + \mathfrak{p} \\ T(\mathfrak{g}) & = & T(\mathfrak{h}) + T(\mathfrak{p}) \\ \mathfrak{g} & = & (+1)\mathfrak{h} + (-1)\mathfrak{p} \end{array} \quad (11.16)$$

The two subspaces are orthogonal

$$(\mathfrak{h}, \mathfrak{p}) = (T^2\mathfrak{h}, \mathfrak{p}) = (T\mathfrak{h}, T\mathfrak{p}) = (\mathfrak{h}, -\mathfrak{p}) = -(\mathfrak{h}, \mathfrak{p}) = 0 \quad (11.17)$$

and satisfy commutation relations (11.12).

As a consequence of this result, a search for all real forms of a complex semisimple Lie algebra reduces to a hunt for all metric-preserving mappings T of the compact real form of that Lie algebra to itself that obey $T^2 = I$.

11.4 Real forms of simple matrix Lie algebras

All of the real forms of all of the simple classical (matrix) Lie algebras can be constructed from one of three types of mappings T of matrices into themselves that obey $T^2 = I$. These three mapping types are derived from block matrix decomposition, subfield restriction, and field embeddings. We discuss each in the next three subsections, indicating the real forms that are produced. In all instances we begin with the compact Lie algebras.

11.4.1 Block matrix decomposition

In a block matrix decomposition the compact Lie algebras $\mathfrak{u}(n, \mathbb{F})$ have the form

$$\mathfrak{u}(n; \mathbb{F}) \qquad \begin{bmatrix} A_p & 0 \\ 0 & A_q \end{bmatrix} + \begin{bmatrix} 0 & B \\ -B^\dagger & 0 \end{bmatrix} \qquad (11.18)$$

where $A_p = -A_p^\dagger$, $A_q = -A_q^\dagger$, and B is an arbitrary $p \times q$ matrix. Under the procedure described in the previous section the off-diagonal block is multiplied by i. This is equivalent to changing the metric I_{p+q} that is preserved by $\mathfrak{u}(n; \mathbb{F})$ to the metric $I_{p,q}$ that is preserved by $\mathfrak{u}(p, q; \mathbb{F})$, where $p + q = n$. The factor i can be absorbed into the $p \times q$ off-diagonal blocks, so that the noncompact algebra has matrix form

$$\mathfrak{u}(p, q; \mathbb{F}) \qquad \begin{bmatrix} A_p & 0 \\ 0 & A_q \end{bmatrix} + \begin{bmatrix} 0 & B \\ +B^\dagger & 0 \end{bmatrix} \qquad (11.19)$$

For the fields $\mathbb{F} = \mathbb{R}, \mathbb{C}, \mathbb{Q}$ related to the root spaces $(D, B), A, C$ the real forms are

$$\begin{array}{lll} \mathbb{R} & \mathfrak{so}(p, q) & D, B \\ \mathbb{C} & \mathfrak{su}(p, q) & A \\ \mathbb{Q} & \mathfrak{sp}(p, q) & C \end{array} \qquad (11.20)$$

11.4.2 Subfield restriction

The real numbers form a subset (subfield) of the complex numbers; the complex numbers form a subset (subfield) of the quaternions. A Lie algebra over the complex numbers can be divided into two subsets: real matrices and the remainder, imaginary matrices. Similarly, a matrix algebra over the quaternions can be divided into two subsets: complex matrices and the remainder

$$\begin{array}{rcccl} \mathfrak{g} & = & \mathfrak{h} + & \mathfrak{p} & \longrightarrow & \mathfrak{g}' \\ \mathfrak{su}(n) & = & \mathfrak{so}(n) + & [\mathfrak{su}(n) - \mathfrak{so}(n)] & \longrightarrow & \mathfrak{sl}(n; \mathbb{R}) \\ \mathfrak{sp}(n) & = & \mathfrak{u}(n) + & [\mathfrak{sp}(n) - \mathfrak{u}(n)] & \longrightarrow & \mathfrak{sp}(2n; \mathbb{R}) \end{array} \qquad (11.21)$$

Under the Cartan procedure, $\mathfrak{su}(n)$ is mapped to $\mathfrak{sl}(n; \mathbb{R})$ and $\mathfrak{sp}(n)$ is mapped to $\mathfrak{sp}(2n; \mathbb{R})$.

11.4 Real forms of simple matrix Lie algebras

We illustrate this for $\mathfrak{su}(2)$:

$$\mathfrak{su}(2) = \frac{1}{2}\begin{bmatrix} ia_3 & ia_1+a_2 \\ ia_1-a_2 & -ia_3 \end{bmatrix} \rightarrow \frac{1}{2}\begin{bmatrix} 0 & +a_2 \\ -a_2 & 0 \end{bmatrix} + \frac{i}{2}\begin{bmatrix} a_3 & a_1 \\ a_1 & -a_3 \end{bmatrix}$$

$$\downarrow \qquad\qquad\qquad\qquad\qquad\qquad\qquad p \rightarrow ip \downarrow$$

$$\mathfrak{sl}(2;\mathbb{R}) = \frac{1}{2}\begin{bmatrix} a_3 & a_1+a_2 \\ a_1-a_2 & -a_3 \end{bmatrix} \leftarrow \frac{1}{2}\begin{bmatrix} 0 & +a_2 \\ -a_2 & 0 \end{bmatrix} + \frac{1}{2}\begin{bmatrix} a_3 & a_1 \\ a_1 & -a_3 \end{bmatrix}$$

(11.22)

The transformation from $\mathfrak{sp}(n) = \mathfrak{u}(n;\mathbb{Q})$ to $\mathfrak{sp}(2n;\mathbb{R})$ is somewhat less familiar. To make the mapping more comprehensible, it is useful to recall the mappings of complex numbers into real 2×2 matrices and of quaternions into complex 2×2 matrices (cf. Eqs. (3.3) and (3.4))

$$\alpha + i\beta \longrightarrow \begin{bmatrix} \alpha & \beta \\ -\beta & \alpha \end{bmatrix}$$

$$\alpha + \mathcal{I}\beta + \mathcal{J}\gamma + \mathcal{K}\delta \longrightarrow \begin{bmatrix} \alpha+i\delta & i\beta+\gamma \\ i\beta-\gamma & \alpha-i\delta \end{bmatrix}$$

(11.23)

where α, β, γ and δ are real. With these replacements the Lie algebra of $n \times n$ complex matrices for $\mathfrak{u}(n)$ is replaced by a set of $2n \times 2n$ real matrices. We call these matrices $\mathfrak{ou}(2n)$, since they form an orthogonal representation of the unitary algebra in terms of $\underline{2n} \times 2n$ matrices. Similarly, the Lie algebra of $n \times n$ quaternion matrices for $\mathfrak{sp}(n)$ is replaced by a set of $2n \times 2n$ complex matrices $\mathfrak{usp}(2n)$, the unitary representation of the symplectic algebra of $\underline{2n} \times 2n$ matrices:

(11.23)

$$\mathfrak{u}(n) \longrightarrow \mathfrak{ou}(2n)$$
$$\mathfrak{sp}(n) \longrightarrow \mathfrak{usp}(2n)$$

(11.24)

Since $\mathfrak{usp}(2n)$ consists of complex matrices, the algebra can be decomposed into the subalgebra of real matrices, which is $\mathfrak{ou}(2n)$, and the complementary subspace of imaginary matrices

$$\mathfrak{sp}(n) \quad=\quad \mathfrak{u}(n) \quad+\quad [\mathfrak{sp}(n) - \mathfrak{u}(n)]$$
$$\downarrow \qquad\qquad \downarrow \qquad\qquad\qquad \downarrow$$
$$\mathfrak{usp}(2n) \;=\; \underbrace{\mathfrak{ou}(2n)}_{\text{real}} + \underbrace{[\mathfrak{usp}(2n) - \mathfrak{ou}(2n)]}_{\text{imaginary}}$$

(11.25)

$$\downarrow \qquad\qquad \downarrow \qquad\qquad \downarrow\, p \rightarrow ip$$
$$\mathfrak{sp}(2n;R) = \underbrace{\mathfrak{ou}(2n)}_{\text{real}} + i\,\underbrace{[\mathfrak{usp}(2n) - \mathfrak{ou}(2n)]}_{\text{real}}$$

Both $\mathfrak{sl}(n;\mathbb{R})$ and $\mathfrak{sp}(2n;\mathbb{R})$ are the least compact real forms associated with their respective root spaces A_{n-1} and C_n.

Remark The matrix Lie group $Sp(2n;\mathbb{R})$ leaves invariant a nonsingular antisymmetric metric in R^{2n}. It is possible to choose coordinates $p_1, q_1, p_2, q_2, \ldots, p_n, q_n$ in this space so that the inner product between two vectors $v'_i G_{ij} v_j$ is

$$v'_i G_{ij} v_j = (p_1, q_1, p_2, q_2, \ldots, p_n, q_n)' \begin{bmatrix} \begin{array}{cc}0 & 1 \\ -1 & 0\end{array} & & & \\ & \begin{array}{cc}0 & 1 \\ -1 & 0\end{array} & & \\ & & \ddots & \\ & & & \begin{array}{cc}0 & 1 \\ -1 & 0\end{array} \end{bmatrix} \begin{bmatrix} p_1 \\ q_1 \\ p_2 \\ q_2 \\ \vdots \\ \vdots \\ p_n \\ q_n \end{bmatrix}$$

$$= \sum_{i=1}^{n} (p'_i q_i - q'_i p_i) \tag{11.26}$$

Then symplectic transformations $M \in Sp(2n;\mathbb{R})$ leave this metric matrix G invariant: $M^t G M = G$. Symplectic transformations leave invariant the canonical form of the hamiltonian equations of motion in classical mechanics.

11.4.3 Field embeddings

The algebras for the orthogonal and unitary groups of even dimension have the following decompositions:

$$\begin{array}{c} \mathfrak{so}(2n) = \mathfrak{ou}(2n) + [\mathfrak{so}(2n) - \mathfrak{ou}(2n)] \\ \downarrow \qquad\qquad \downarrow \\ \mathfrak{ou}(2n) + i\,[\mathfrak{so}(2n) - \mathfrak{ou}(2n)] = \mathfrak{so}^*(2n) \end{array} \tag{11.27}$$

$$\begin{array}{c} \mathfrak{su}(2n) = \mathfrak{usp}(2n) + [\mathfrak{su}(2n) - \mathfrak{usp}(2n)] \\ \downarrow \qquad\qquad \downarrow \\ \mathfrak{usp}(2n) + i\,[\mathfrak{su}(2n) - \mathfrak{usp}(2n)] = \mathfrak{su}^*(2n) \end{array}$$

Application of the map $\mathfrak{p} \to i\mathfrak{p}$ produces the real forms $\mathfrak{so}^*(2n)$ and $\mathfrak{su}^*(2n)$.

Remark The real forms $\mathfrak{so}^*(2n)$ of D_n and $\mathfrak{su}^*(2n)$ of A_{2n-1} do not occur explicitly in the list of matrix Lie algebras given in Chapter 5.

11.5 Results

We summarize in Table 11.1 the real forms of the simple classical Lie algebras. This table indicates the root space associated with each real form.

Some of the low-dimensional root spaces are equivalent. For example, A_1 (where the compact real form is $\mathfrak{su}(2)$), B_1 ($\mathfrak{so}(3)$), and C_1 ($\mathfrak{sp}(1)$) are equivalent, as are B_2 ($\mathfrak{so}(5)$) and C_2 ($\mathfrak{sp}(2)$). So also are A_3 ($\mathfrak{su}(4)$) and D_3 ($\mathfrak{so}(6)$). As a result, there are equivalences between the real forms of these Lie algebras. These equivalences are summarized in Table 11.2.

Table 11.1. *Real forms of the simple classical Lie algebras*

Mapping	Real form	Root space	Condition
Block submatrix	$\mathfrak{so}(p,q)$	D_n	$p+q = 2n$
	$\mathfrak{so}(p,q)$	B_n	$p+q = 2n+1$
	$\mathfrak{su}(p,q)$	A_{n-1}	$p+q = n$
	$\mathfrak{sp}(p,q)$	C_n	$p+q = n$
Subfield restriction	$\mathfrak{sl}(n;\mathbb{R})$	A_{n-1}	
	$\mathfrak{sp}(2n;\mathbb{R})$	C_n	
Field embedding	$\mathfrak{so}^*(2n)$	D_n	
	$\mathfrak{su}^*(2n)$	A_{2n-1}	

Table 11.2. *Equivalence among real forms of the simple classical Lie algebras*

A_1		B_1		C_1
$\mathfrak{su}(2)$	$\sim$	$\mathfrak{so}(3)$	$\sim$	$\mathfrak{sp}(1) = \mathfrak{usp}(2)$
$\mathfrak{su}(1,1) = \mathfrak{sl}(2;\mathbb{R})$	$\sim$	$\mathfrak{so}(2,1)$	$\sim$	$\mathfrak{sp}(2;\mathbb{R})$
D_2	$=$	A_1	$+$	A_1
$\mathfrak{so}(4)$	$=$	$\mathfrak{so}(3)$	$+$	$\mathfrak{so}(3)$
$\mathfrak{so}^*(4)$	$\sim$	$\mathfrak{so}(3)$	$+$	$\mathfrak{so}(2,1)$
$\mathfrak{so}(3,1)$	$\sim$	$\mathfrak{sl}(2;\mathbb{C})$		
$\mathfrak{so}(2,2)$	$\sim$	$\mathfrak{so}(2,1)$	$+$	$\mathfrak{so}(2,1)$
B_2	$=$	C_2		
$\mathfrak{so}(5)$	$\sim$	$\mathfrak{sp}(2) = \mathfrak{usp}(4)$		
$\mathfrak{so}(4,1)$	$\sim$	$\mathfrak{sp}(1,1) = \mathfrak{usp}(2,2)$		
$\mathfrak{so}(3,2)$	$\sim$	$\mathfrak{sp}(4;R)$		
D_3	$=$	A_3		
$\mathfrak{so}(6)$	$\sim$	$\mathfrak{su}(4)$		
$\mathfrak{so}(5,1)$	$\sim$	$\mathfrak{su}^*(4)$		
$\mathfrak{so}^*(6)$	$\sim$	$\mathfrak{su}(3,1)$		
$\mathfrak{so}(4,2)$	$\sim$	$\mathfrak{su}(2,2)$		
$\mathfrak{so}(3,3)$	$\sim$	$\mathfrak{sl}(4;\mathbb{R})$		

Table 11.3. *Real forms of the exceptional Lie algebras*

Root space	Class$_{\text{rank(character)}}$	Maximal compact subgroup	
		Root space	Dimension
G_2	$G_{2(-14)}$	G_2	14
	$G_{2(+2)}$	$A_1 + A_1$	6
F_4	$F_{4(-52)}$	F_4	52
	$F_{4(-20)}$	B_4	36
	$F_{4(+4)}$	$C_3 + A_1$	24
E_6	$E_{6(-78)}$	E_6	78
	$E_{6(-26)}$	F_4	52
	$E_{6(-14)}$	$D_5 + D_1$	46
	$E_{6(+2)}$	$A_5 + A_1$	38
	$E_{6(+6)}$	C_4	36
E_7	$E_{7(-133)}$	E_7	133
	$E_{7(-25)}$	$E_6 + D_1$	79
	$E_{7(-5)}$	$D_6 + A_1$	69
	$E_{7(+7)}$	A_7	63
E_8	$E_{8(-248)}$	E_8	248
	$E_{8(-24)}$	$E_7 + A_1$	136
	$E_{8(+8)}$	D_8	120

For completeness, we list the real forms for the exceptional Lie algebras in Table 11.3. The subscript in parentheses after the rank is the character of the real form.

11.6 Conclusion

Connected root space diagrams summarize the commutation relations of simple Lie algebras over the field of complex numbers. By placing various reality restrictions on the coefficients of the complex algebra, a spectrum of real subalgebras is obtained, each of which has the same complex extension. To each root space there corresponds a unique real form that is compact. All other real forms are obtained from this compact real form by "analytic continuation." The analytic continuation is carried out by determining all linear mappings T on the compact algebra $\mathfrak{g}$ that preserve the inner product and obey $T^2 = I$. The subspace $\mathfrak{p}$ of $\mathfrak{g}$ that obeys $T(\mathfrak{p}) = -\mathfrak{p}$ is analytically continued by $\mathfrak{p} \to \mathfrak{p}' = i\mathfrak{p}$; the subspace $\mathfrak{h}$ of $\mathfrak{g}$ that obeys $T(\mathfrak{h}) = +\mathfrak{h}$ is the maximal compact subalgebra of the noncompact real form $\mathfrak{g}'$: $\mathfrak{g} = \mathfrak{h} + \mathfrak{p} \to \mathfrak{g}' = \mathfrak{h} + i\mathfrak{p}'$. For the simple classical (matrix) Lie algebras three types of mappings T suffice to construct all real forms: block submatrix decomposition; subfield restriction; and field embedding.

11.7 Problems

1. Four operators $a_i^\dagger a_j$ can be constructed from boson operators for two modes $1 \leq i$, $j \leq 2$. These operators close under commutation.

   a. Show that the regular representation of this Lie algebra is

   $$\mathfrak{Reg}(wa_1^\dagger a_1 + xa_1^\dagger a_2 + ya_2^\dagger a_1 + za_2^\dagger a_2) = \begin{bmatrix} 0 & -x & y & 0 \\ -y & w-z & 0 & y \\ x & 0 & -w+z & -x \\ 0 & x & -y & 0 \end{bmatrix}$$

   b. Show that the Cartan–Killing inner product is

   $$\operatorname{tr} \mathfrak{Reg}^2(wa_1^\dagger a_1 + xa_1^\dagger a_2 + ya_2^\dagger a_1 + za_2^\dagger a_2) = 2(w-z)^2 + 8xy$$

   c. Set

   $$\begin{aligned} w &= \alpha + \beta & x &= \gamma + \delta \\ z &= \alpha - \beta & y &= \gamma - \delta \end{aligned} \quad \text{inner product} \to 8(\beta^2 + \gamma^2 - \delta^2)$$

   $$wa_1^\dagger a_1 + xa_1^\dagger a_2 + ya_2^\dagger a_1 + za_2^\dagger a_2 = \alpha(a_1^\dagger a_1 + a_2^\dagger a_2)$$
   $$+ \beta(a_1^\dagger a_1 - a_2^\dagger a_2) + \gamma(a_1^\dagger a_2 + a_2^\dagger a_1) + \delta(a_1^\dagger a_2 - a_2^\dagger a_1)$$

   Conclude that $a_1^\dagger a_1 + a_2^\dagger a_2$ spans the maximum commutative subalgebra, $a_1^\dagger a_2 - a_2^\dagger a_1$ spans the maximal compact subalgebra, and the two generators $a_1^\dagger a_1 - a_2^\dagger a_2$ and $a_1^\dagger a_2 + a_2^\dagger a_1$ are noncompact.

   d. Identify the simple three-dimensional subalgebra as $\mathfrak{sl}(2;\mathbb{R})$ or $\mathfrak{su}(1,1)$. Show that the compact real form is obtained by multiplying the two noncompact generators by i.

   e. Construct a 2×2 matrix representation of the three operators that span $\mathfrak{su}(1,1)$ using the methods of Chapter 6. Multiply the two noncompact operators by i. Show that the three matrices that result are exactly $i\sigma_j$, where σ_j are the Pauli spin matrices.

2. The classical matrix groups $SO(n)$ are not simply connected, so they are $k \to 1$ images of their universal covering groups $\overline{SO(n)} = Spin(n)$, for some integer k. Show that the covering groups $Spin(n)$ are classical matrix groups for $n = 3, 4, 5, 6$, and make these identifications:

   $$\begin{aligned} Spin(3) &= SU(2) \\ Spin(4) &= SU(2) \otimes SU(2) \\ Spin(5) &= USp(4) \\ Spin(6) &= SU(4) \end{aligned}$$

   Show that for $n > 6$ the groups $Spin(n)$ are not equal to any classical matrix Lie groups.

3. **Spectrum of quadratic Casimir**
   a. Use the metric (11.3) for a simple Lie algebra to show that the quadratic Casimir operator is

   $$C^2 = \sum H_i^2 + \sum E_\alpha E_{-\alpha} + E_{-\alpha} E_\alpha$$

   b. Since the H_i are mutually commuting, in a hermitian/unitary representation they are simultaneously diagonalizable. Identify basis states in a Hilbert space by their eigenvalues under the operators H_i: $|n_1, n_2, \ldots, n_r\rangle$,

   $$H_i |n_1, n_2, \ldots, n_r\rangle = n_i |n_1, n_2, \ldots, n_r\rangle$$

   c. For the orthogonal groups $SO(n)$, impose suitable reality conditions (i.e., $H_j \to iH_j$, etc.), choose a Hilbert space containing the state $|l, 0, \ldots, 0\rangle$ and show that the value of C^2 on every vector (i.e., apply shift operators E_α until no new states are created) is

   $$C^2 |\text{state}\rangle = -l(l + n - 2)|\text{state}\rangle$$

   The $-$ sign indicates that $SO(n)$ is compact. This spectrum reduces to the well-known spectrum $-m^2$ for $SO(2)$ (on $e^{im\phi}$) and $-l(l+1)$ for $SO(3)$ (on $Y_m^l(\theta, \phi)$).

4. **Master analytic representation for A_1** The complex Lie algebra with root space diagram A_1 has two real forms

   $$\mathfrak{su}(2) \quad J_3, J_\pm \quad [J_3, J_\pm] = \pm J_\pm \quad [J_+, J_-] = +2J_3$$
   $$\mathfrak{su}(1,1) \quad K_3, K_\pm \quad [K_3, K_\pm] = \pm K_\pm \quad [K_+, K_-] = -2K_3$$

   In Problem 2 of Chapter 6 we exploited the isomorphism between the Lie algebra $\mathfrak{su}(2)$ and bilinear combinations of creation and annihilation operators for two modes in order to construct matrix elements of the angular momentum operators. These are matrix elements of a *hermitian* representation of J_i, $i = 1, 2, 3$. Exponentials of the form $EXP(ir^k J_k)$, with r^k real, provide *unitary* representations of the compact Lie group $SU(2)$. *All* unitary irreducible representations of $SU(2)$ are finite dimensional $(2j + 1)$ and are obtained in this way. In this problem we will review the construction of the UIR (unitary irreducible representations) of the compact group $SU(2)$ and will use similar methods to construct all the UIR of its analytic continuation, the non-compact Lie group $SU(1, 1)$. Since the algebras are related by analytic continuation, so also are the UIR. We will begin with the analytic hermitian matrix elements for $\mathfrak{su}(2)$ and continue to hermitian matrix elements for the analytically continued algebra $\mathfrak{su}(1, 1)$.

   a. Make the identifications

   $$\begin{aligned} K_3 &= \tfrac{1}{2}(a_1^\dagger a_1 - a_2^\dagger a_2) = J_3 \\ K_+ &= i a_1^\dagger a_2 = i J_+ \\ K_- &= i a_2^\dagger a_1 = i J_- \end{aligned} \quad (11.28)$$

   Verify all commutation relations are satisfied.

b. Recall that in both $SU(2)$ and $SU(1, 1)$, rotation about the z-axis by 4π radians returns to the same group operation. Show that in any matrix representation of $SU(2)$ with J_3 diagonal, or in any matrix representation of $SU(1, 1)$ with K_3 diagonal, the matrix is diagonal with matrix elements $e^{im\phi}\delta_{m'm}$. Show that the single-valuedness condition under $\phi \to \phi + 4\pi$ requires that $m = \frac{1}{2}(n_1 - n_2)$ is integer or half-integer. Show that the shift operators $J_\pm$ and $K_\pm$ require that all m values in a UIR with J_3 or K_3 diagonal are either integer or half-integer.

c. Relax the assumption that all indices n_1, n_2 in the basis states $|n_1, n_2\rangle = |n_1\rangle \otimes |n_2\rangle$ must be nonnegative integers. Construct the matrix elements of the operators $J_3, J_\pm, K_3, K_\pm$ under this relaxed assumption. Show that all commutation relations are satisfied in the representation afforded by this set of basis states.

d. Show that the matrices for J_x and J_y are also hermitian provided that

$$\langle n_1 + 1, n_2 - 1 | J_+ | n_1, n_2 \rangle = \sqrt{(n_1 + 1)n_2}$$
$$\|\qquad\qquad\qquad\qquad\qquad\qquad\qquad\|$$
$$\langle n_1, n_2 | J_- | n_1 + 1, n_2 - 1\rangle^* = \sqrt{(n_1 + 1)n_2}^*$$

Show that these conditions are satisfied for

$$n_1 \geq 0 \text{ and } n_2 \geq 0 \qquad \text{or} \qquad n_1 \leq -1 \text{ and } n_2 \leq -1$$

Show that the lattice sites in quadrants I and III of Fig. 11.2, with vertices at $(0, 0)$ (QI) and $(-1, -1)$ (QIII), satisfy these conditions.

e. With the identification $\left|{j \atop m}\right\rangle = |n_1, n_2\rangle$, $j = \frac{1}{2}(n_1 + n_2)$, $m = \frac{1}{2}(n_1 - n_2)$, show that

$$\left\langle {j \atop m \pm 1} \middle| J_\pm \middle| {j \atop m} \right\rangle = \sqrt{(j \mp m)(j \pm m + 1)} = \sqrt{\left(j + \frac{1}{2}\right)^2 - \left(m \pm \frac{1}{2}\right)^2} \quad (11.29)$$

f. Show that the operators $J_\pm$ act diagonally. In order for all states connected by successive application of these operators to remain in QI or QIII, n_1 and n_2 must be integers so that in the various quadrants the shift operators vanish on the edges as shown:

Quadrant	Operator	Edge
I	J_+	$n_2 = 0$
I	J_-	$n_1 = 0$
III	J_+	$n_1 = -1$
III	J_-	$n_2 = -1$

g. Show that the matrix elements for the shift operators $K_\pm$ in $\mathfrak{su}(1, 1)$ are

$$\left\langle {j \atop m \pm 1} \middle| K_\pm \middle| {j \atop m} \right\rangle = i\sqrt{(j \mp m)(j \pm m + 1)} = \sqrt{\left(m \pm \frac{1}{2}\right)^2 - \left(j + \frac{1}{2}\right)^2}$$

$$(11.30)$$

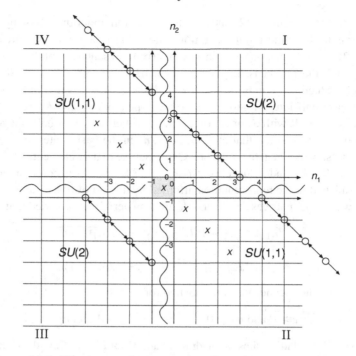

Figure 11.2. The integer lattice in two dimensions carries representations of the algebras $\mathfrak{su}(2)$ and $\mathfrak{su}(1,1)$ that exponentiate to unitary irreducible representations with careful choice of the basis set. All points in this plane are mapped to other points along diagonals of the form $n_1 + n_2 = $ constant. The subspaces of basis vectors for the unitary irreducible representations of $SU(2)$ and $SU(1,1)$ are separated by a "no man's land" defined by $-1 < n_1 < 0$ and $-1 < n_2 < 0$ (wavy lines). All points labeled x belong to the principal series of representations of $SU(1,1)$ with $n_1 + n_2 = -\frac{1}{2} + i\beta$.

h. For $\mathfrak{su}(1,1)$ show the hermiticity condition is satisfied for all real numbers except those in QI and QIII.

i. In order to ensure that a set of states $|n_1 + k, n_2 - k\rangle$ (k integer) mapped into each other by the shift operators $K_\pm$ do not enter QI from QIV, show that the edge (lowest m) state must be $|n_1, n_2 = -1\rangle$ for $n_1 = 0, 1, 2, \ldots$. The basis states for this bounded discrete series of representations are $|{}^{\,j}_{m}\rangle = |n_1, n_2\rangle$ with $j + \frac{1}{2} = 0, \frac{1}{2}, 1, \frac{3}{2}, \ldots$ or $2j + 1 = 0, 1, 2, 3, \ldots$ and $m = j+1, j+2, \ldots$. This is the discrete series of representations that is bounded below: $\mathcal{D}_+^j$.

j. In order to ensure that a set of states $|n_1 + k, n_2 - k\rangle$ (k integer) mapped into each other by the shift operators $K_\pm$ do not enter QI from QII, show that the edge (highest m) state must be $|n_1 = -1, n_2\rangle$ for $n_2 = 0, 1, 2, \ldots$. The basis states for this bounded discrete series of representations are $|{}^{\,j}_{m}\rangle = |n_1, n_2\rangle$ with $j + \frac{1}{2} = 0, \frac{1}{2}, 1, \frac{3}{2}, \ldots$ or $2j + 1 = 0, 1, 2, 3, \ldots$ and $m = -j-1, -j-2, \ldots$. This is the discrete series of representations that is bounded above: $\mathcal{D}_-^j$.

k. Advance similar arguments to guarantee that states do not enter QIII from QIV ($\mathcal{D}_+^j$) or from QII ($\mathcal{D}_-^j$).

l. Now relax the condition that n_1 and n_2 are integers. The set of states $|n_1 + k, n_2 - k\rangle$ ($k = \ldots, -2, -1, 0, +1, +2, \ldots$) connected by $K_\pm$ carries a hermitian representation of $\mathfrak{su}(1, 1)$ if one of the states falls in the square with corners on the vertices of the four quadrants. If this state is $|p, q\rangle$, with $-1 \leq p, q \leq 0$ then the single-valuedness condition requires $\frac{1}{2}(p - q)$ = integer or half-integer. In the latter case, it is not possible for the matrix element in Eq. (11.30) to be real for all values of the $U(1)$ index m. Therefore $p - q = 0$ and $-1 \leq p = q \leq 0$. The states $|{}_m^j\rangle$ with m integer, j real and $-\frac{1}{2} \leq j + \frac{1}{2} \leq +\frac{1}{2}$ carry representations $\mathcal{D}^p$ of the complementary series of UIR for $SU(1, 1)$.

m. By setting $j + \frac{1}{2} = i\beta$ (β real) the matrix elements in Eq. (11.30) become

$$\left\langle {}_{m\pm 1}^{\ j} \middle| K_\pm \middle| {}_m^j \right\rangle = i\sqrt{(j \mp m)(j \pm m + 1)} = \sqrt{\left(m \pm \frac{1}{2}\right)^2 + \beta^2} \qquad (11.31)$$

These matrix elements are always positive, for both representations with m integer and those with m half-integer. These states carry UIR belonging to the principal series of representations of $SU(1, 1)$.

n. The four series of UIR for $SU(1, 1)$ are

principal	$j + \frac{1}{2} = i\beta$	β real	$m = 0, \pm 1, \pm 2, \ldots$
			$m = \pm\frac{1}{2}, \pm\frac{3}{2}, \ldots$
complementary	$j + \frac{1}{2} = p$	$-\frac{1}{2} \leq p \leq +\frac{1}{2}$	$m = 0, \pm 1, \pm 2, \ldots$
discrete, +	$2j + 1 = 0, \pm 1, \pm 2, \ldots$		$m = +|j| + 1, +|j| + 2, \ldots$
discrete, −	$2j + 1 = 0, \pm 1, \pm 2, \ldots$		$m = -|j| - 1, -|j| - 2, \ldots$

Show that states with $j' < -\frac{1}{2}$ obtained by reflection through the diagonal containing the central point in the shaded square with coordinates $(n_1, n_2) = (-\frac{1}{2}, -\frac{1}{2})$ support representations equivalent to those with index $j > -\frac{1}{2}$. The relation among indices is $j + \frac{1}{2} = -(j' + \frac{1}{2})$ and the relation among states is $|{}_{m'}^{j'}\rangle \simeq |{}_m^j\rangle_{m=m'}$.

o. Show that the following equivalences occur among representations of these four series:

principal series	$j' = -\frac{1}{2} - i|\beta| \leftrightarrow j = j'^* = -\frac{1}{2} + i|\beta|$
complementary series	$-\frac{1}{2} \leq j' + \frac{1}{2} \leftrightarrow (j + \frac{1}{2}) = -(j' + \frac{1}{2}) < 0$
discrete series, +	$j' + \frac{1}{2} < 0 \leftrightarrow j + \frac{1}{2} = -(j' + \frac{1}{2})$
discrete series, −	$j' + \frac{1}{2} < 0 \leftrightarrow j + \frac{1}{2} = -(j' + \frac{1}{2})$

5. A real simple Lie algebra of rank l and dimension n has basis vectors H_i and E_α. An element X in the Lie algebra is a real linear combination of these generators: $X = h^i H_i + e^\alpha E_\alpha$, with h^i, e^α real. Show that the real subalgebra spanned by the $\frac{1}{2}(n - l)$ linear combinations of the form $(E_\alpha - E_{-\alpha})/\sqrt{2}$ is the maximal compact subalgebra of this simple Lie algebra.

6. The noncompact real form $\mathfrak{sp}(p, q)$ of the symplectic algebra was constructed from the compact real form $\mathfrak{sp}(p + q)$ by "Minkowski's trick," or analytic continuation. This procedure is delicate: one must be careful of the complex unit i with quaternions. Show by more careful arguments that the result stated is correct.

12
Riemannian symmetric spaces

In the classification of the real forms of the simple Lie algebras we encountered subspaces $\mathfrak{p}, i\mathfrak{p}$ on which the Cartan–Killing inner product was negative-definite (on $\mathfrak{p}$) or positive-definite (on $i\mathfrak{p}$). In both cases these subspaces exponentiate onto algebraic manifolds on which the invariant metric g_{ij} is definite, either negative or positive. Manifolds with a definite metric are Riemannian spaces. These spaces are also globally symmetric in the sense that every point looks like every other point – because each point in the space EXP($\mathfrak{p}$) or EXP($i\mathfrak{p}$) is the image of the origin under some group operation. We briefly discuss the properties of these Riemannian globally symmetric spaces in this chapter.

12.1 Brief review

In the discussion of the group $SL(2; \mathbb{R})$ we encountered three symmetric spaces. These were $S^2 \sim SU(2)/U(1)$, which is compact, and its dual $H_{2+}^2 = SL(2; \mathbb{R})/SO(2) = SU(1, 1)/U(1)$, which is the upper sheet of the two-sheeted hyperboloid. "Between" these two spaces occurs $H_1^2 = SL(2; \mathbb{R})/SO(1, 1)$, which is the single-sheeted hyperboloid. These spaces are shown in Fig. 12.1.

The Cartan–Killing inner product in the linear vector subspace $\mathfrak{su}(2) - \mathfrak{u}(1)$ is negative definite. This is mapped, under the EXPonential function, to the Cartan–Killing metric on the space $SU(2)/U(1) \sim S^2$, the sphere. On S^2 the Cartan–Killing metric is negative-definite. We may just as well take it as positive-definite. Under this metric the sphere becomes a Riemannian manifold since there is a metric on it with which to measure distances.

The Cartan–Killing inner product on $\mathfrak{su}(1, 1) - \mathfrak{u}(1) \simeq \mathfrak{sl}(2; \mathbb{R}) - \mathfrak{so}(2)$ is positive-definite. It maps to a positive-definite metric on $H_{2+}^2 = SU(1, 1)/SO(2)$. The upper sheet of the two-sheeted hyperboloid is topologically equivalent to the flat space R^2 but geometrically it is not: it has intrinsic curvature that can be computed, via its Cartan–Killing metric and the curvature tensor derived from it.

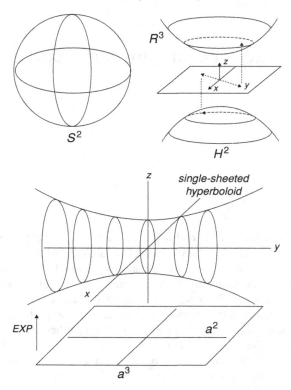

Figure 12.1. $S^2 = SO(3)/SO(2) = SU(2)/U(1)$, $H^2_{2+} = SO(2,1)/SO(2) = SU(1,1)/U(1)$, $H^2_1 = SO(2,1)/SO(1,1) = SL(2;\mathbb{R})/SO(1,1)$. The first two are Riemannian symmetric spaces, the third is a pseudo-Riemannian symmetric space.

The most interesting of these spaces is the single-sheeted hyperboloid H^2_1. It is obtained by exponentiating $\mathfrak{su}(1,1) - \mathfrak{so}(1,1)$. The Cartan–Killing inner product in this linear vector space is indefinite. Therefore the Cartan–Killing metric on the topological space $\text{EXP}[\mathfrak{su}(1,1) - \mathfrak{so}(1,1)] = SU(1,1)/SO(1,1)$ is indefinite. The space is a pseudo-Riemannian manifold. In addition it is multiply connected.

12.2 Globally symmetric spaces

The three cases for A_1 reviewed in the previous section serve as a model for the description of all other Riemannian symmetric spaces. For a compact simple Lie algebra $\mathfrak{g}$ (i.e., $\mathfrak{so}(n)$, $\mathfrak{su}(n)$, $\mathfrak{sp}(n)$) the Cartan decompositions have the form (11.10)

$$\begin{aligned} \mathfrak{g} &= \mathfrak{h} + \mathfrak{p} \quad (\mathfrak{p}, \mathfrak{p}) < 0 \\ \mathfrak{g}' &= \mathfrak{h} + i\mathfrak{p} \quad (i\mathfrak{p}, i\mathfrak{p}) > 0 \end{aligned} \qquad (12.1)$$

On the linear vector space $\mathfrak{p}$ ($i\mathfrak{p}$) the Cartan–Killing inner product is negative (positive) definite. On the topological spaces EXP($\mathfrak{p}$) (EXP($i\mathfrak{p}$)) the Cartan–Killing metric is negative- (positive-) definite also:

$$\begin{aligned} G/H &= \mathrm{EXP}(\mathfrak{p}) & ds^2 &= g_{\mu,\nu}\,dx^\mu dx^\nu < 0 \\ G'/H &= \mathrm{EXP}(i\mathfrak{p}) & ds^2 &= g_{\mu,\nu}\,dx^\mu dx^\nu > 0 \end{aligned} \quad (12.2)$$

In both cases, the metric is definite and defines a Riemannian space. This space is globally symmetric. That is, every point "looks like" every other point. This is because they all look like the identity EXP(0), since the identity and its neighborhood can be shifted to any other point in the space by multiplication by the appropriate group operation (for example, by EXP($\mathfrak{p}$) or EXP($i\mathfrak{p}$)).

The space $P = G/H = \mathrm{EXP}(\mathfrak{p})$ (e.g., S^2) is compact. The exponential of a straight line through the origin in $\mathfrak{p}$ returns periodically to the neighborhood of the identity. The space P is not topologically equivalent to any Euclidean space, in which a straight line (geodesic) through the origin never returns to the origin. The space P may be simply connected or multiply connected.

The space $P' = G'/H = \mathrm{EXP}(i\mathfrak{p})$ (i.e., H^2_{2+}) is noncompact. The exponential of a straight line through the origin in $i\mathfrak{p}$ (a geodesic through the identity in EXP($i\mathfrak{p}$)) simply goes away from this point without ever returning. The space $P' = \mathrm{EXP}(i\mathfrak{p})$ is topologically equivalent to a Euclidean space R^n, where $n = \dim i\mathfrak{p}$. Geometrically it is not Euclidean since it has nonzero curvature. This space is simply connected.

The Riemannian spaces $P = \mathrm{EXP}(\mathfrak{p})$ and $P' = \mathrm{EXP}(i\mathfrak{p})$ are symmetric but not isotropic unless the rank of the space is 1, as it is for S^2 and H^2_{2+}.

If $\mathfrak{g}$ is simple with a Cartan decomposition of the form $\mathfrak{g} = \mathfrak{k} + \mathfrak{p}$, with standard commutation relations $[\mathfrak{k}, \mathfrak{k}] \subseteq \mathfrak{k}$, $[\mathfrak{k}, \mathfrak{p}] \subseteq \mathfrak{p}$, and $[\mathfrak{p}, \mathfrak{p}] \subseteq \mathfrak{k}$, the quotient coset $P = G/K$ is a globally symmetric space as every point "looks like" every other point.

12.3 Rank

Rank for a symmetric space can be defined in exactly the same way as rank for a Lie group or a Lie algebra. This should not be surprising, as a symmetric space consists of points (coset representatives $P = G/H$ or $P' = G'/H$) in the Lie group.

To compute the rank of a symmetric space one starts from the secular equation for the associated algebra $\mathfrak{g} = \mathfrak{h} + \mathfrak{p}$

$$\| \mathfrak{Reg}(\mathfrak{h} + \mathfrak{p}) - \lambda I_n \| = \sum_{j=0}^{n} (-\lambda)^{n-j} \phi_j(\mathfrak{h}, \mathfrak{p}) \qquad (12.3)$$

and restricts to the subspace $\mathfrak{p}$. Calculation of the rank can be carried out in any faithful matrix representation, for example the defining $n \times n$ matrix representation. The secular equations for the spaces $SO(p,q)/SO(p) \times SO(q)$,

$SU(p,q)/S[U(p) \times U(q)]$, $Sp(p,q)/Sp(p) \times Sp(q)$ are

$$\left\| \begin{bmatrix} 0 & B \\ B^\dagger & 0 \end{bmatrix} - \lambda I_{p+q} \right\| = \sum_{j=0}^{n=p+q} (-\lambda)^{n-j} \phi_j(B, B^\dagger) \tag{12.4}$$

It is easy to check that the function ϕ_j depends on the $q \times q$ matrix $B^\dagger B$ or the $p \times p$ matrix $BB^\dagger$, whichever is smaller. The rank of these spaces is $\min(p,q)$.

For Riemannian globally symmetric spaces the rank is (cf. Section 10.1):

(i) the number of independent functions in the secular equation;
(ii) the number of independent roots of the secular equation;
(iii) the maximal number of mutually commuting operators in the subspace $\mathfrak{p}$ or $\mathfrak{p}'$;
(iv) the number of invariant (Laplace–Beltrami) operators defined over the space P (P');
(v) the dimension of a positive-definite root space that can be used to define diagrammatically the properties of these spaces (Araki–Satake root diagrams);
(vi) the number of distinct, nonisotropic directions;
(vii) the dimension of the largest Euclidean submanifold in P.

We will not elaborate on these points here. We mention briefly that the Laplace–Beltrami operators on $P = G/H$ are the Casimir operators of its parent group G, restricted to the subspace P. The number of nonisotropic directions is determined by computing the number of distinct eigenvalues of the Cartan–Killing metric on P, or equivalently and more easily, of the Cartan–Killing inner product on $\mathfrak{p}$ (same as the metric at the identity). In each of the spaces P there is a Euclidean subspace (submanifold). For S^2, any great circle is Euclidean.

12.4 Riemannian symmetric spaces

Table 12.1 lists all the classical noncompact Riemannian symmetric spaces of the form G'/H, where G' is simple and noncompact and H is the maximal compact

Table 12.1. *All classical noncompact Riemannian symmetric spaces*

Root space	Quotient G'/H	Dimension P	Rank P
A_{p+q-1}	$SU(p,q)/S[U(p) \otimes U(q)]$	$2pq$	$\min(p,q)$
A_{n-1}	$SL(n;\mathbb{R})/SO(n)$	$\frac{1}{2}(n+2)(n-1)$	$n-1$
A_{2n-1}	$SU^*(2n)/USp(2n)$	$(2n+1)(n-1)$	$n-1$
B_{p+q}	$SO(p,q)/SO(p) \otimes SO(q)$	pq	$\min(p,q)$
D_{p+q}	$SO(p,q)/SO(p) \otimes SO(q)$	pq	$\min(p,q)$
D_n	$SO^*(2n)/U(n)$	$n(n-1)$	$[n/2]$
C_{p+q}	$USp(2p,2q)/USp(2p) \otimes USp(2q)$	$4pq$	$\min(p,q)$
C_n	$Sp(2n;\mathbb{R})/U(n)$	$n(n+1)$	n

Table 12.2. *All exceptional noncompact Riemannian symmetric spaces*

Root space	G'/H	Dim G'	Dim H	Dim P	Rank P
G_2	$G_{2(+2)}/(A_1 \oplus A_1)$	14	6	8	2
F_4	$F_{4(-20)}/B_4$	52	36	16	1
	$F_{4(+4)}/(C_3 \oplus A_1)$	52	24	28	4
E_6	$E_{6(-26)}/F_4$	78	52	26	2
	$E_{6(-14)}/(D_5 \oplus D_1)$	78	46	32	2
	$E_{6(+2)}/(A_5 \oplus A_1)$	78	38	40	4
	$E_{6(+6)}/C_4$	78	36	42	6
E_7	$E_{7(-25)}/(E_6 \oplus D_1)$	133	79	54	3
	$E_{7(-5)}/(D_6 \oplus A_1)$	133	69	64	4
	$E_{7(+7)}/A_7$	133	63	70	7
E_8	$E_{8(-24)}/(E_7 \oplus A_1)$	248	136	112	4
	$E_{8(+8)}/D_8$	248	120	128	8

subgroup in G'. To each there is a compact real form under $G'/H \to G/H$. For example, $SO(p,q)/SO(p) \otimes SO(q)$ and $SO(p+q)/SO(p) \otimes SO(q)$ are dual. These spaces are classical because they involve the classical series of Lie groups: the orthogonal, the unitary, and the symplectic.

Table 12.2 lists all the exceptional noncompact Riemannian symmetric spaces. As before, to each there is a dual compact real form.

12.5 Metric and measure

The metric tensor on the spaces P, P' is computed by defining a metric at the identity and then moving it elsewhere by group multiplication. The metric at the identity is chosen as the Cartan–Killing inner product on $i\mathfrak{p}$, or its negative on $\mathfrak{p}$.

If $dx(\text{Id})$ are infinitesimal displacements at the identity that are translated to infinitesimal displacements $dx(p)$ at point p, then these two sets of infinitesimals are linearly related by a nonsingular linear transformation (cf. Eq. (4.44))

$$dx^i(\text{Id}) = M^i{}_\mu \, dx^\mu(p) \tag{12.5}$$

The metrics and invariant volume elements are related by (cf. Eqs. (4.47) and (4.49))

$$\begin{aligned} ds^2 &= g_{ij}(\text{Id}) dx^i(\text{Id}) dx^j(\text{Id}) \\ &= g_{\mu\nu}(p) dx^\mu(p) dx^\nu(p) \\ &\Rightarrow g_{\mu\nu}(p) = g_{ij}(\text{Id}) M^i{}_\mu M^j{}_\nu \end{aligned} \tag{12.6}$$

$$\begin{aligned} dV &= \rho(\text{Id}) dx^1(\text{Id}) \wedge dx^2(\text{Id}) \wedge \cdots \wedge dx^n(\text{Id}) \\ &= \rho(p) dx^1(p) \wedge dx^2(p) \wedge \cdots \wedge dx^n(p) \\ &\Rightarrow \rho(p) = \| M(p) \| \, \rho(\text{Id}) \sim \sqrt{\det g(p)} \end{aligned}$$

The matrix $M^i{}_\mu(p)$ is not easy to compute in general. For the rank-one spaces $SO(n, 1)/SO(n)$, $SU(n, 1)/U(n)$, $Sp(n, 1)/Sp(n) \times Sp(1)$ defined by

$$P' = \begin{bmatrix} W & X \\ X^\dagger & Y \end{bmatrix} \qquad X = \begin{bmatrix} x^1 \\ x^2 \\ \vdots \\ x^n \end{bmatrix} \qquad (12.7)$$

$$W^2 = I_n + XX^\dagger$$
$$Y^2 = 1 + X^\dagger X$$

the matrix $M^i{}_\mu(X)$ is determined from

$$dx(X) = W dx(\text{Id}) \qquad (12.8)$$

The matrix $M^i{}_\mu(X)$ is given by W^{-1}. Since the Cartan–Killing inner product is I_n at the identity, we find

$$g_{\mu\nu}(X) = W^{-1} I_n W^{-1} = \{I_n + XX^\dagger\}^{-1}_{\mu\nu}$$
$$\rho(X) = \| W \|^{-1} = 1/\sqrt{1 + X^\dagger X} = Y^{-1} \qquad (12.9)$$

12.6 Applications and examples

The coset representatives for the Riemannian symmetric spaces $SO(2, 1)/SO(2)$ and $SO(3)/SO(2)$ are

$$\begin{array}{cc} SO(2, 1)/SO(2) & SO(3)/SO(2) \\[4pt] \begin{bmatrix} W & X \\ +X^t & Y \end{bmatrix} & \begin{bmatrix} W & X \\ -X^t & Y \end{bmatrix} \end{array} \qquad (12.10)$$

$$W^2 = I_2 + \begin{pmatrix} x \\ y \end{pmatrix} (x \; y) \qquad W^2 = I_2 - \begin{pmatrix} x \\ y \end{pmatrix} (x \; y)$$

$$Y^2 = I_1 + (x \; y) \begin{pmatrix} x \\ y \end{pmatrix} \qquad Y^2 = I_1 - (x \; y) \begin{pmatrix} x \\ y \end{pmatrix}$$

From these coset representatives we can compute the metric tensors on the noncompact hyperboloid $H_2^2 = SO(2, 1)/SO(2)$ and the compact sphere

$S^2 = SO(3)/SO(2)$. The metric tensors in the two cases are the 2×2 matrices

$$SO(2,1)/SO(2) \qquad SO(3)/SO(2)$$

$$g_{*,*} = W^{-2} = \left[I_2 + \begin{pmatrix} x \\ y \end{pmatrix} (x \; y) \right]^{-1} \qquad g_{*,*} = W^{-2} = \left[I_2 - \begin{pmatrix} x \\ y \end{pmatrix} (x \; y) \right]^{-1}$$

$$g^{*,*} = W^{+2} = \begin{bmatrix} 1+x^2 & +xy \\ +yx & 1+y^2 \end{bmatrix} \qquad g^{*,*} = W^{+2} = \begin{bmatrix} 1-x^2 & -xy \\ -yx & 1-y^2 \end{bmatrix}$$

(12.11)

The noncompact Riemannian symmetric space $H_2^2 = SO(2,1)/SO(2)$ is parameterized by the entire x–y plane while its dual compact Riemannian symmetric space $SO(2+1)/SO(2)$ is parameterized by the interior of the unit circle $Y^2 = 1 - (x^2 + y^2) \geq 0$.

Since the (intrinsic) properties of the Riemannian symmetric space are entirely encoded in its metric tensor, we can begin to compute its important properties, for example, the curvature tensor. It is first useful to compute the Christoffel symbols as a way-station on the road to computing the full Riemannian curvature tensor. The Christoffel symbols (not a tensor!), the Riemannian curvature tensor, the Ricci tensor, and the curvature scalars are constructed in terms of the metric tensor as follows:

Christoffel $\qquad \Gamma^\sigma_{\mu\nu} = \frac{1}{2} g^{\sigma\alpha} \left(\frac{\partial g_{\mu\alpha}}{\partial x^\nu} + \frac{\partial g_{\nu\alpha}}{\partial x^\mu} - \frac{\partial g_{\mu\nu}}{\partial x^\alpha} \right)$

Riemann curvature tensor $\qquad R^\mu_{\sigma,\alpha\beta} = \frac{\partial \Gamma^\mu_{\sigma\beta}}{\partial x^\alpha} - \frac{\partial \Gamma^\mu_{\sigma\alpha}}{\partial x^\beta} + \Gamma^\mu_{\rho\alpha} \Gamma^\rho_{\sigma\beta} - \Gamma^\mu_{\rho\beta} \Gamma^\rho_{\sigma\alpha}$

Ricci tensor $\qquad R_{\sigma\beta} = R^\mu_{\sigma,\mu\beta}$

curvature scalar $\qquad R = g^{\sigma\beta} R_{\sigma\beta}$

(12.12)

In general, computing these objects is not easy. This task is greatly simplified in a symmetric space, for all points look the same and we can compute the tensors wherever the computation is easiest. This turns out to be at the origin. We illustrate by carrying out the computations in the neighborhood of the identity for the compact case, the sphere. Instead of using the pair x, y as coordinates, we use indexed coordinates x^i, $i = 1, 2, \ldots, N$, and set $N = 2$ at the end of this computation.

We first note that it is sufficient to estimate the behavior of the metric tensor in the neighborhood of the origin (identity in the coset) only up to quadratic terms, so that

$$g_{ij} = W^{-2} = [I_N - XX^t]^{-1}_{ij} \simeq [I_N + XX^t]_{ij} \to \delta_{ij} + x^i x^j \qquad (12.13)$$

The inverse (contravariant metric) is $g^{ij} \simeq \delta^{ij} - x^i x^j$, but we will not need this result. In the neighborhood of the identity ($g^{ij} \to \delta^{ij}$)

$$\Gamma^\sigma_{\mu\nu} \to \frac{1}{2}\left(\frac{\partial g_{\mu\sigma}}{\partial x^\nu} + \frac{\partial g_{\nu\sigma}}{\partial x^\mu} - \frac{\partial g_{\mu\nu}}{\partial x^\sigma}\right)$$

$$= \frac{1}{2}\left\{\begin{matrix}\delta_{\nu\mu}x^\sigma + \delta_{\mu\nu}x^\sigma - \delta_{\sigma\mu}x^\nu \\ \delta_{\nu\sigma}x^\mu + \delta_{\mu\sigma}x^\nu - \delta_{\sigma\nu}x^\mu\end{matrix}\right\} \qquad (12.14)$$

$$\to \delta_{\mu\nu}x^\sigma \quad (\to 0 \text{ at origin})$$

Computation of the components of the Riemann curvature tensor at the orign is even simpler. At the origin the components of the Christoffel symbols all vanish, so it is sufficient to retain only the first two terms in the expression for the curvature tensor. We find

$$R^\mu_{\sigma,\alpha\beta} \to \frac{\partial}{\partial x^\alpha}(\delta_{\sigma\beta}x^\mu) - \frac{\partial}{\partial x^\beta}(\delta_{\sigma\alpha}x^\mu) = \delta_{\sigma\beta}\delta_\alpha{}^\mu - \delta_{\sigma\alpha}\delta_\beta{}^\mu \qquad (12.15)$$

The contravariant index μ can be lowered with the metric tensor, which is the delta function at the origin, and the resulting fully covariant metric tensor $R_{\mu\sigma,\alpha\beta} = \delta_{\alpha\mu}\delta_{\beta\sigma} - \delta_{\alpha\sigma}\delta_{\beta\mu}$ exhibits the full spectrum of expected symmetries.

The Ricci tensor is obtained by contraction

$$R_{\sigma\beta} = R^\mu_{\sigma,\mu\beta} = \delta_{\sigma\beta}\delta_{\mu\mu} - \delta_{\sigma\mu}\delta_{\beta\mu} = N\delta_{\sigma\beta} - \delta_{\sigma\beta} \qquad (12.16)$$

The curvature scalar is obtained from the Ricci tensor by saturating its covariant indices by the contravariant components of the metric tensor, which is simply a delta function at the origin:

$$R = g^{\sigma\beta}R_{\sigma\beta} \to \delta^{\sigma\beta}(N-1)\delta_{\sigma\beta} = N(N-1) \qquad (12.17)$$

For $N = 2$ (sphere S^2), $R = 2$.

The computation can be carried out just as easily for the noncompact space H_2^2. The major change occurs in the first step, where the metric in the neighborhood of the origin undergoes the change

$$SO(2+1)/SO(2) \quad SO(2,1)/SO(2)$$

$$g_{ij} \to \delta_{ij} + x^i x^j \to g_{ij} \to \delta_{ij} - x^i x^j \qquad (12.18)$$

The net result is that a negative sign attaches itself at each step in the computation: for example $\Gamma^\sigma_{\mu\nu} \to -\delta_{\mu\nu}x^\sigma$. The end result for H_2^2 is that $R = -2$.

12.7 Pseudo-Riemannian symmetric spaces

Topological spaces on which a "metric tensor" can be defined that is neither positive-definite ($ds^2 = g_{\mu\nu} dx^\mu dx^\nu > 0$, equality $\Rightarrow dx = 0$) nor negative-definite ($ds^2 < 0$), but which is nonsingular ($\| g \| \neq 0$) are called **pseudo-Riemannian spaces**. Pseudo-Riemannian spaces that are globally symmetric can be constructed following the procedures described in Sections 12.1 and 12.2. As the example of the single-sheeted hyperboloid H_1^2 shows, these spaces are even more interesting than the Riemannian globally symmetric spaces.

To make these statements more explicit, assume a Lie algebra $\mathfrak{g}''$ (noncompact) has a decomposition

$$\mathfrak{g}'' = \mathfrak{h}'' + \mathfrak{p}'' \tag{12.19}$$

with commutation relations of the form (11.10)

$$\begin{aligned} [\mathfrak{h}'', \mathfrak{h}''] &\subseteq \mathfrak{h}'' \\ [\mathfrak{h}'', \mathfrak{p}''] &\subseteq \mathfrak{p}'' \\ [\mathfrak{p}'', \mathfrak{p}''] &\subseteq \mathfrak{h}'' \end{aligned} \tag{12.20}$$

Then $\mathfrak{h}''$ and $\mathfrak{p}''$ are orthogonal subspaces in $\mathfrak{g}''$ under the Cartan–Killing inner product. Assume also that the inner product is indefinite on $\mathfrak{p}''$ (also $\mathfrak{h}''$). Then

$$P'' = \mathrm{EXP}(\mathfrak{p}'') = G''/H'' \tag{12.21}$$

is a pseudo-Riemannian globally symmetric space. The metric on this space is indefinite. The space is curved and typically multiply connected. The space $H'' = \mathrm{EXP}(\mathfrak{h}'')$ is also an interesting pseudo-Riemannian symmetric space.

All of the algebraic properties associated with a Riemannian symmetric space hold also for pseudo-Riemannian symmetric spaces. That is, rank can be defined, and carries most of the implications listed in Section 12.3.

There is a systematic method for constructing pseudo-Riemannian symmetric spaces. Begin with a compact simple Lie algebra $\mathfrak{g}$ and suppose T_1, T_2 are two metric-preserving mappings of the Lie algebra onto itself that obey $T_1^2 = I$, $T_2^2 = I$ (cf. Section 11.3) and $T_1 \neq T_2$. Define the eigenspaces of $\mathfrak{g}$ under T_1, T_2 as $\mathfrak{g}_{\pm,\pm}$:

$$\begin{aligned} T_1\, \mathfrak{g}_{\pm,*} &= \pm \mathfrak{g}_{\pm,*} \\ T_2\, \mathfrak{g}_{*,\pm} &= \pm \mathfrak{g}_{*,\pm} \end{aligned} \tag{12.22}$$

Then T_1 can be used to construct a noncompact algebra

$$\mathfrak{g}' = (\mathfrak{g}_{+,+} + \mathfrak{g}_{+,-}) + i(\mathfrak{g}_{-,+} + \mathfrak{g}_{-,-}) \tag{12.23}$$

and T_2 can be used to split $\mathfrak{g}'$ in a different way

$$\mathfrak{g}'' = (\mathfrak{g}_{+,+} + i\mathfrak{g}_{+,-}) + (i\mathfrak{g}_{-,+} + \mathfrak{g}_{-,-}) \qquad (12.24)$$

$$= \mathfrak{h}'' \quad + \quad \mathfrak{p}''$$

The subspaces $\mathfrak{h}''$, $\mathfrak{p}''$ obey commutation relations (12.20). The Cartan–Killing inner product is indefinite on both $\mathfrak{h}''$ and $\mathfrak{p}''$ as long as $T_1 \neq T_2$.

For $\mathfrak{su}(2)$ the only two mappings are T_1 = block diagonal decomposition and T_2 = complex conjugation. The eigenspace decomposition is

Operation	$i\sigma_1$	$i\sigma_2$	$i\sigma_3$
T_1 = block matrix decomposition	-1	-1	$+1$
T_2 = complex conjugation	-1	$+1$	-1
$T_3 = T_1 T_2$	$+1$	-1	-1

(12.25)

This gives $\mathfrak{g}_{+,+} = 0$, $\mathfrak{g}_{+,-} = i\sigma_3$, $\mathfrak{g}_{-,+} = i\sigma_2$, $\mathfrak{g}_{-,-} = i\sigma_1$. Note that each mapping T_i has one positive and two negative eigenvalues, and chooses a different generator for the maximal compact subalgebra $\mathfrak{h}'$ of the noncompact real form $\mathfrak{g}'$.

12.8 Conclusion

Globally symmetric spaces have the form $P = G/K$, where $\mathfrak{g}$ is a real form of a simple Lie algebra, $\mathfrak{g} = \mathfrak{k} + \mathfrak{p}$, with $[\mathfrak{k}, \mathfrak{k}] \subseteq \mathfrak{k}$, $[\mathfrak{k}, \mathfrak{p}] \subseteq \mathfrak{p}$, and $[\mathfrak{p}, \mathfrak{p}] \subseteq \mathfrak{k}$. All Riemannian globally symmetric spaces are constructed as quotients of a simple Lie group G by a maximal compact subgroup K. More specifically, they are exponentials of a subalgebra $\mathfrak{p}$ of a Lie algebra $\mathfrak{g}$ for which commutation relations and inner products are given by (11.10). Pseudo-Riemannian globally symmetric spaces are similarly constructed. For these spaces the rank can be defined. This determines a number of algebraic properties (maximal number of independent mutually commuting generators and Laplace–Beltrami operators) as well as geometric properties (number of nonisotropic directions, dimension of maximal Euclidean subspaces). Metric and measure are determined on these spaces in an invariant way.

12.9 Problems

1. Show that the invariant polynomials $\phi_j(B, B^\dagger)$ in (12.4) actually depend on the invariants of $BB^\dagger$ or $B^\dagger B$. These are the eigenvalues of these square, hermitian matrices. Both the $p \times p$ and $q \times q$ matrix have the same spectrum of nonzero eigenvalues. The remaining $(p - q)$ or $(q - p)$ (whichever is positive) eigenvalues of the larger matrix are zero (singular value decomposition theorem).

12.9 Problems

2. The second order Laplace–Beltrami operator Δ^2 is constructed from the second order Casimir invariant C^2 by restricting the action of the latter to the Riemannian manifold $G/H = P$.

   a. Show that this operator can be expressed in terms of the Cartan–Killing metric tensor on P as $\Delta^2 = g^{ij}(\partial_i \partial_j - \Gamma_{ij}{}^k \partial_k)$.

   b. Show that there is one Laplace–Beltrami on the sphere S^2 and compute it in the standard parameterization in terms of the coordinates (x, y) in the interior of the unit disk $x^2 + y^2 \leq 1$.

   c. Show that there is one Laplace–Beltrami on the two-sheeted hyperboloid H_2^2 and compute it in the standard parameterization in terms of the coordinates on the plane R^2.

   d. Show that these two Laplace–Beltrami operators are dual in some sense. What sense?

   e. Extend these results to the sphere S^n and its dual, H^n, $n > 2$.

3. Show that the two metric-preserving mappings T_1 and T_2 that satisfy $T_1^2 = T_2^2 = I$ generate a third, $T_3 = T_1 T_2$ and that $T_1 T_2 = T_2 T_1$. Show that $T_3 \neq I$ if $T_1 \neq T_2$. Show that these three operators, together with the identity, form a group isomorphic with the "four-group" ("vierergruppe") V_4. Describe the variety of decompositions of a compact Lie algebra $\mathfrak{g} = \mathfrak{g}_{+,+} + \mathfrak{g}_{+,-} + \mathfrak{g}_{-,+} + \mathfrak{g}_{-,-}$ that is available by choosing first, one of these three involutions, and then a second (there are $3!/1! = 6$ choices). Discuss dualities.

4. Show that the secular equation for the symmetric space $SO(3)/SO(2)$ can be obtained from (11.2) by setting $b_3 = 0$:

$$\det |\mathfrak{Reg}(p) - \lambda I_3| = -\lambda \left[\lambda^2 + (b_1^2 + b_2^2)\right] = 0$$

   There is one independent function in this secular equation. There is one independent root. What else can be said about this Riemannian symmetric space?

5. Show that the coefficients $\phi_j(p)$ in the secular equation for a symmetric space are obtained from the coefficients $\phi_j(\mathfrak{h}, p)$ in the secular equation for the parent Lie algebra (Eq. (12.3)) by setting $\mathfrak{h} = 0$.

6. The hyperbolic plane H_2^2 is the Riemannian symmetric space $SO(2,1)/SO(2)$ obtained by exponentiating a real symmetric matrix in the three-dimensional Lie algebra

$$\mathrm{EXP} \begin{bmatrix} 0 & t_1 & t_2 \\ t_1 & 0 & 0 \\ t_2 & 0 & 0 \end{bmatrix} = \begin{bmatrix} x_0 & x_1 & x_2 \\ x_1 & * & * \\ x_2 & * & * \end{bmatrix} \qquad x_0^2 - x_1^2 - x_2^2 = 1$$

   a. Show that the hyperbolic plane is the two-dimensional algebraic manifold defined by the condition $x_0^2 - x_1^2 - x_2^2 = 1$ in the Lorentz 3-space with signature $(1, 2)$.

   b. Show that the invariant metric is induced from the metric $-ds^2 = dx_0^2 - dx_1^2 - dx_2^2$ in this Lorentz 3-space.

c. Use coordinates x_1, x_2 to parameterize the points in H_2^2, and show

   $$ds^2 = \frac{(dx_1 \; dx_2) \begin{bmatrix} 1+x_2^2 & -x_1 x_2 \\ -x_1 x_2 & 1+x_1^2 \end{bmatrix} \begin{pmatrix} dx_1 \\ dx_2 \end{pmatrix}}{1 + x_1^2 + x_2^2}$$

   d. Show that the invariant measure is

   $$d\mu = \frac{dx_1 dx_2}{\sqrt{1 + x_1^2 + x_2^2}}$$

   e. Introduce polar coordinates (r, θ), $x_1 = r\cos(\theta)$, $x_2 = r\sin(\theta)$. Show that

   $$ds^2 = \frac{(dr \; d\theta) \begin{bmatrix} \frac{1}{1+r^2} & 0 \\ 0 & r^2 \end{bmatrix} \begin{pmatrix} dr \\ d\theta \end{pmatrix}}{1 + r^2}$$

   $$d\mu = \frac{r \, dr \, d\theta}{\sqrt{1 + r^2}}$$

   f. Determine the action of a group operation in $SO(1, 2)$ on the point $(x_1, x_2) \in H_2^2$.

7. The metric on a pseudo-Riemannian symmetric space is $g_{ij}(x)$.
   a. Show that the generators of infinitesimal rotations at a point are $X_{rs} = g_{rt} x^t \partial_s - g_{st} x^t \partial_r$.
   b. Show $[X_{ab}, \Delta] = 0$, where $\Delta = G^{ab;rs} X_{ab} X_{rs}$ is the Laplace–Beltrami operator on this space, $G_{ab;rs} = \text{tr}\{\mathfrak{def}(X_{ab})\mathfrak{def}(X_{rs})\}$, and $G^{ab;rs}$ is the inverse of $G_{ab;rs}$.
   c. Show that Δ consists of terms that are both quadratic and linear in the operators ∂_r, and that

   $$\Delta = g^{rs} \partial_r \partial_s - g^{rs} \Gamma_{rs}{}^t \partial_t$$

   The function $\Gamma_{rs}{}^t$ is *not* a tensor. The components of the Christoffel symbol are given by

   $$\Gamma_{rs}{}^t = \frac{1}{2} g^{tu} (\partial_s g_{ru} + \partial_r g_{su} - \partial_u g_{rs})$$

8. Use radial coordinates $(r, \phi_2, \phi_3, \ldots, \phi_n)$ on the sphere $S^n \subset R^{n+1}$.
   a. Show the invariant volume element is

   $$dV = \sqrt{\|g\|} r^{n-1} \sin^{n-2} \phi_2 \sin^{n-3} \phi_3 \cdots \sin^1 \phi_{n-1} \sin^0 \phi_n$$
   $$dr \wedge d\phi_2 \wedge d\phi_3 \wedge \cdots \wedge d\phi_n$$

   b. Show that the second order Laplace–Beltrami operator is

   $$\Delta = \frac{1}{\sqrt{\|g\|}} \partial_\mu \sqrt{\|g\|} g^{\mu\nu} \partial_\nu \quad \text{where} \quad \partial_\nu = \partial/\partial \phi_\mu$$

12.9 Problems

c. Compare this with the second order Casimir operator for $SO(n+1)$:

$$C_2[SO(n+1)/SO(n)] = \sum_{1 \leq r < s}^{n+1} X_{r,s}^2(\phi)$$

d. Show that the Laplace–Beltrami operators on a sphere can be written recursively:

$$\Delta(S^n) = \partial_n(f_1(\phi)\partial_n) + f_2(\phi)\Delta(S^{n-1})$$

Compute $f_1(\phi)$ and $f_2(\phi)$.

9. A quantum system with n degrees of freedom is described by a hamiltonian that is a linear superposition of the bilinear products $a_i^\dagger a_j$ ($\mathcal{H} = h_{ij}(t)a_i^\dagger a_j$, $1 \leq i, j \leq n$), so that $i\mathcal{H}$ is a time-dependent element in the Lie algebra $u(n)$. Assume the system is initially in its ground state. Show that it evolves into a coherent state whose trajectory exists in the rank one symmetric space $SU(n)/U(n-1)$. Write down the coherent state parameters explicitly for a two-level system, and relate the coherent state parameters to the forcing terms in the hamiltonian.

10. Conformal group The inner product on an n-dimensional linear vector space $V^{(n)}$ is defined by $(x,x)_m = m_{ij}x^i x^j$. Define coordinates y in an $n+2$ dimensional linear vector space $W^{(n+2)}$ as follows

$$y^i = sx^i \quad (1 \leq i \leq n)$$
$$y^{n+1} = s$$
$$y^{n+2} = s(x,x)_m$$

and define an inner product M in this space by

$$M = \begin{bmatrix} m_{ij} & & \\ \hline & 0 & -\frac{1}{2} \\ & -\frac{1}{2} & 0 \end{bmatrix}$$

a. Show $(y,y)_M = M_{\mu\nu}y^\mu y^\nu = (sx, sx)_m - \frac{2}{3}s[s(x,x)_m] = 0$.

b. If m is positive definite and Lie group G preserves inner products in $V^{(n)}$, then $G = O(n)$.

c. Show that the Lie group H that preserves inner products in $W^{(n+2)}$ is $O(n+1, 1)$.

d. If the metric m has signature n_1, n_2 ($n_1 + n_2 = n$), show that $G = O(n_1, n_2)$ and $H = O(n_1 + 1, n_2 + 1)$.

e. H is called a conformal group because it preserves angles. Show this.

f. Construct the quotient space $SO(n_1 + 1, n_2 + 1)/SO(n_1, n_2)$.

g. Under a conformal transformation $y \to y'$ and $x \to x'$. Show $x'^i = y'^i/y'^{n+1}$.

h. The Lorentz metric $(+1, -1, -1, -1)$ leaves the four-momentum invariant:

$$E^2 - (pc)^2 = (mc^2)^2$$

Show that the conformal group on space-time is $SO(4,2)$.

i. Show that the infinitesimal generators of the conformal group are

$$L_{\mu\nu} = x_\mu \partial_\nu - x_\nu \partial_\mu$$
$$P_\mu = \partial_\mu$$
$$K_\mu = 2x_\mu(x^\nu \partial_\nu) - (x^\nu x_\nu)\partial_\mu = 2x_\mu(x, \partial) - (x, x)\partial_\mu$$
$$S = x^\nu \partial_\nu$$

The operators $L_{\mu\nu}$ are the infinitesimal generators of the Lorentz group $SO(3, 1)$ and P_μ generate translations. Taken together $L_{\mu\nu}$ and P_μ generate the Poincaré group. The operator S generates dilations and the four operators K_μ generate conformal transformations. Above $x_\mu = g_{\mu\nu} x^\nu$.

j. Show that the additional operators satisfy the commutation relations

$$[L_{\mu\nu}, K_\lambda] = g_{\nu\lambda} K_\mu - g_{\mu\lambda} K_\nu$$
$$[L_{\mu\nu}, S] = 0$$
$$[P_\mu, K_\nu] = 2(g_{\mu\nu} S - L_{\mu\nu})$$
$$[S, P_\mu] = -P_\mu \qquad [P_\mu, P_\nu] = 0$$
$$[S, K_\mu] = +K_\mu \qquad [K_\mu, K_\nu] = 0$$

k. Show that $e^{c^\mu K_\mu}(x^\nu) = x'^\nu = \frac{x^\nu + c^\nu(x,x)}{1 + 2(c,x) + (c,c)(x,x)}$.

l. Show that the conformal group $SO(4, 2)$ is:
- the largest group that leaves the free space (no sources) Maxwell equations form invariant;
- the largest group that maps the (bound, scattering, parabolic) states of the hydrogen atom to themselves.

m. Discuss the duality created by $P_\mu \to P'_\mu = x_\mu$ and $K_\mu \to K'_\mu = 2(x, \partial)\partial_\mu - (\partial, \partial)x_\mu$.

11. The upper half of the complex plane has coordinates $z = x + iy$. This upper half-plane provides a well studied model for the hyperbolic plane when a suitable metric is placed on it. The half-plane is mapped onto itself by linear fractional transformations

$$z \to z' = \frac{az + b}{cz + d} \qquad \begin{bmatrix} a & b \\ c & d \end{bmatrix} \in SL(2; \mathbb{R}), \quad ad - bc = 1$$

This transformation group is called the projective special linear transformation group and denoted $PSL(2, \mathbb{R})$.

a. Show that $M, -M \in SL(2; \mathbb{R})$ generate identical transformations. The group $SL(2; \mathbb{R})$ is a two-fold covering group of $PSL(2, \mathbb{R})$.

b. Show

$$z' = \frac{ac(x^2 + y^2) + (ad + bc)x + bd + iy}{|cz + d|^2}$$

In particular, show that $y' > 0$ if $y > 0$ and $y' = 0$ if $y = 0$. The transformation maps the upper half-plane onto the upper half-plane and its boundary, the real axis ($y = 0$), onto itself.

c. Show that the metric

$$ds^2 = \begin{bmatrix} dx & dy \end{bmatrix} \begin{bmatrix} \frac{1}{y} & 0 \\ 0 & \frac{1}{y} \end{bmatrix} \begin{bmatrix} dx \\ dy \end{bmatrix} = \frac{d\bar{z}\, dz}{y^2}$$

is invariant under these transformations.

d. Show $dz' = dz/|cz+d|^2$

e. Show that the invariant measure is $d\mu = dx\, dy/y^2$

f. Show that the distance between two points z_1 and z_2 is

$$s(z_1, z_2) = 2 \tanh^{-1} \frac{|z_1 - z_2|}{|z_1 - \bar{z}_2|} = \log \left\{ \frac{|z_1 - \bar{z}_2| + |z_1 - z_2|}{|z_1 - \bar{z}_2| - |z_1 - z_2|} \right\}$$

12. The unit disk in the complex plane $w = x + iy$ consists of those points that satisfy $\bar{w}w = x^2 + y^2 \leq 1$. The unit disk, with a suitable metric, provides a second representation of the hyperbolic plane. The unit disk is mapped onto itself by linear fractional transformations

$$w \to w' = \frac{\alpha w + \beta}{\bar{\beta} w + \bar{\alpha}} \qquad \begin{bmatrix} \alpha & \beta \\ \bar{\beta} & \bar{\alpha} \end{bmatrix} \in SU(1,1), \quad \bar{\alpha}\alpha - \bar{\beta}\beta = 1$$

a. Show that $M, -M \in SU(1,1)$ generate identical mappings of the unit disk into itself.

b. Show that $w = e^{i\phi} \to w' = e^{i\psi}$. Compute $\psi(\phi)$.

c. Show that the metric

$$ds^2 = \begin{pmatrix} dx & dy \end{pmatrix} \begin{bmatrix} \frac{1}{(1-\bar{w}w)^2} & 0 \\ 0 & \frac{1}{(1-\bar{w}w)^2} \end{bmatrix} \begin{pmatrix} dx \\ dy \end{pmatrix} = \frac{d\bar{w}\, dw}{(1-\bar{w}w)^2}$$

is invariant under this group.

d. Show that the invariant volume element is

$$d\mu = \frac{dx\, dy}{(1-\bar{w}w)^2} = \frac{d\bar{w}\, dw}{(1-\bar{w}w)^2}$$

e. Show that the distance between two points w_1 and w_2 in this unit disk is

$$s(w_1, w_2) = \tanh^{-1} \left\{ \frac{|w_1 - w_2|}{|1 - w_1 \bar{w}_2|} \right\}$$

13. Show that the mapping from z in the upper half-plane to w in the unit disk given by

$$w = e^{i\phi} \frac{z - z_0}{z - \bar{z}_0}$$

is conformal, that is, it preserves angles. Here z_0 is any point in the upper half-plane.

a. Compute the inverse of this mapping, and show that it maps the interior of the unit disk unto the upper half of the complex plane and the boundary of the unit disk onto the real axis (boundary of the upper half-plane).

b. Choose $z_0 = i$ and $e^{i\phi} = i$ to give the canonical map

$$w = \frac{iz + 1}{z + i}$$

c. Show that the matrices that generate the Möbius transformations of the upper half-plane and the unit disk are related by

$$S \begin{bmatrix} a & b \\ c & d \end{bmatrix} S^{-1} = \begin{bmatrix} \alpha & \beta \\ \bar{\beta} & \bar{\alpha} \end{bmatrix} \qquad S = \frac{1}{\sqrt{2}} \begin{bmatrix} 1 & -i \\ -i & 1 \end{bmatrix}$$

d. Show that this transformation maps the invariant metric and measure on the upper half-plane onto the invariant metric and measure on the unit disk.

13

Contraction

New Lie groups can be constructed from old by a process called group contraction. Contraction involves reparameterization of the Lie group's parameter space in such a way that the group multiplication properties, or commutation relations in the Lie algebra, remain well defined even in a singular limit. In general, the properties of the original Lie group have well-defined limits in the contracted Lie group. For example, the parameter space for the contracted group is well defined and noncompact. Other properties with well-defined limits include: Casimir operators; basis states of representations; matrix elements of operators; and Baker–Campbell–Hausdorff formulas. Contraction provides limiting relations among the special functions of mathematical physics. We describe a particularly simple class of contractions, the Inönü–Wigner contractions, and treat one example of a contraction not in this class.

13.1 Preliminaries

It is possible to construct new Lie algebras from old by a certain limiting process called **contraction**. In this process a new set of basis vectors Y_r is related to the initial set of basis vectors X_i through a parameter-dependent change of basis: $Y_r = M_r^i(\epsilon) X_i$. The structure constants have the transformation properties of a tensor: $C_{rs}^{\ t}(\epsilon) = M_r^i(\epsilon) M_s^j(\epsilon) C_{ij}^{\ k} (M(\epsilon)^{-1})_k^t$ (cf. Eq. (4.22)). As long as the change of basis transformation is nonsingular the Lie algebra is unchanged.

If the transformation becomes singular, the structure constants $C_{rs}^{\ t}(\epsilon)$ may still converge to a well-defined limit. It is often the case that the structure constants

$$C_{rs}^{\ t}(0) = \lim_{\epsilon \to 0} C_{rs}^{\ t}(\epsilon) \tag{13.1}$$

exist and define a Lie algebra that is different from the original Lie algebra.

13.2 Inönü–Wigner contractions

If a Lie algebra $\mathfrak{g}$ has a subalgebra $\mathfrak{h}$ and a complementary subspace $\mathfrak{p}$ with commutation relations of the form

$$\begin{aligned} \mathfrak{g} &= \mathfrak{h} + \mathfrak{p} \\ [\mathfrak{h}, \mathfrak{h}] &\subseteq \mathfrak{h} \quad &\text{subalgebra} \\ [\mathfrak{h}, \mathfrak{p}] &\subseteq \mathfrak{p} \quad &\text{this is important} \\ [\mathfrak{p}, \mathfrak{p}] &\subseteq \mathfrak{h} + \mathfrak{p} \quad &\text{this is always true} \end{aligned} \tag{13.2}$$

then the Inönü–Wigner contraction of $\mathfrak{g} \to \mathfrak{g}'$ involves the following change of basis transformation

$$\begin{bmatrix} \mathfrak{h}' \\ \mathfrak{p}' \end{bmatrix} = \begin{bmatrix} I_{\dim(\mathfrak{h})} & 0 \\ 0 & \epsilon\, I_{\dim(\mathfrak{p})} \end{bmatrix} \begin{bmatrix} \mathfrak{h} \\ \mathfrak{p} \end{bmatrix} \tag{13.3}$$

where $\dim(\mathfrak{h})$ is the dimension of the subalgebra $\mathfrak{h}$. The commutation relations of $\mathfrak{g}'$ are well defined for all values of ϵ, including the singular limit $\epsilon \to 0$:

$$\begin{aligned} \left[\mathfrak{h}', \mathfrak{h}' \right] &= [\mathfrak{h}, \mathfrak{h}] \subseteq \mathfrak{h} \\ \left[\mathfrak{h}', \mathfrak{p}' \right] &= [\mathfrak{h}, \epsilon\mathfrak{p}] = \epsilon [\mathfrak{h}, \mathfrak{p}] \; \lim_{\epsilon \to 0} \; \epsilon\mathfrak{p} \to \mathfrak{p}' \\ \left[\mathfrak{p}', \mathfrak{p}' \right] &= [\epsilon\mathfrak{p}, \epsilon\mathfrak{p}] = \epsilon^2 [\mathfrak{p}, \mathfrak{p}] \; \lim_{\epsilon \to 0} \; \epsilon^2(\mathfrak{h} + \mathfrak{p}) \to 0 \end{aligned} \tag{13.4}$$

In the limit $\epsilon \to 0$ the contracted algebra $\mathfrak{g}'$ is the semidirect sum of the original subalgebra $\mathfrak{h}$ and the subalgebra $\mathfrak{p}'$, where $\mathfrak{p}'$ is commutative and $[\mathfrak{h}, \mathfrak{p}'] \subseteq \mathfrak{p}'$:

$$\begin{array}{lll} \mathfrak{g} = \mathfrak{h} + \mathfrak{p} & \mathfrak{p} \to \mathfrak{p}' = \epsilon\mathfrak{p} & \mathfrak{g}' = \mathfrak{h} + \mathfrak{p}' \\ [\mathfrak{h}, \mathfrak{h}] \subseteq \mathfrak{h} & & [\mathfrak{h}, \mathfrak{h}] \subseteq \mathfrak{h} \\ [\mathfrak{h}, \mathfrak{p}] \subseteq \mathfrak{p} & \longrightarrow & [\mathfrak{h}, \mathfrak{p}'] \subseteq \mathfrak{p}' \\ [\mathfrak{p}, \mathfrak{p}] \subseteq \mathfrak{h} + \mathfrak{p} & & [\mathfrak{p}', \mathfrak{p}'] = 0 \end{array} \tag{13.5}$$

13.3 Simple examples of Inönü–Wigner contractions

In this section we illustrate several facets of Inönü–Wigner contractions by contracting three different orthogonal groups.

13.3.1 The contraction $SO(3) \to ISO(2)$

The infinitesimal generators of the Lie group $SO(3)$ may be chosen as $L_1 = X_{23} = x_2 \partial_3 - x_3 \partial_2 = \epsilon_{1jk} x_j \partial_k$, with L_2 and L_3 defined by cyclic permutation. The commutation relations are

$$\begin{aligned} [L_1, L_2] &= -L_3 \\ [L_2, L_3] &= -L_1 \\ [L_3, L_1] &= -L_2 \end{aligned} \tag{13.6}$$

13.3 Simple examples of Inönü–Wigner contractions

Under contraction with respect to the subalgebra of rotations about the z-axis (infinitesimal generator L_3) the operators L_1 and L_2 go to

$$\begin{bmatrix} \epsilon L_1 \\ \epsilon L_2 \end{bmatrix} \xrightarrow{\epsilon=1/R} \begin{bmatrix} (1/R)L_1 \\ (1/R)L_2 \end{bmatrix} \to \begin{bmatrix} -P_2 \\ +P_1 \end{bmatrix} \tag{13.7}$$

The commutation relations of the contracted algebra, $ISO(2) = E(2)$, are

$$\begin{aligned} [L_3, P_1] &= -P_2 \\ [L_3, P_2] &= +P_1 \\ [P_1, P_2] &= 0 \end{aligned} \tag{13.8}$$

The three operators L_3, P_1, P_2 generate the group of Euclidean motions of the plane, $E(2)$, or inhomogeneous orthogonal transformations in the plane R^2, $ISO(2)$. This group consists of rotations about the z-axis, generated by L_3, and displacements of the origin in the x- and y-directions, generated by $P_1 = \partial_1$ and $P_2 = \partial_2$.

To verify this interpretation we can imagine the group $SO(3)$ acting on the sphere $x^2 + y^2 + z^2 = R^2$ in the neighborhood of the north pole $(0, 0, R)$, as shown in Fig. 13.1. An element in the Lie algebra $\mathfrak{so}(3)$ can be written in the form

$$\theta_1 L_1 + \theta_2 L_2 + \theta_3 L_3 \longrightarrow (-d_2)\left(\frac{L_1}{R}\right) + (+d_1)\left(\frac{L_2}{R}\right) + \theta_3 L_3 \tag{13.9}$$

In the limit $R \to \infty$ we find

$$\frac{1}{R}L_1 = \frac{1}{R}(x^2\partial_3 - x^3\partial_2) = \frac{1}{R}(y\partial/\partial z - R\partial/\partial y) \to -\partial/\partial y = -\partial_2 = -P_2$$

$$\frac{1}{R}L_2 = \frac{1}{R}(x^3\partial_1 - x^1\partial_3) = \frac{1}{R}(R\partial/\partial x - x\partial/\partial z) \to +\partial/\partial x = +\partial_1 = +P_1$$
$$\tag{13.10}$$

The contracted limits of the operators L_1 and L_2 in the limit of a sphere of very large radius are operators $-P_2, +P_1$ describing displacements in the $-y$ and $+x$ directions. In addition, the parameters θ_1, θ_2 and d_1, d_2 are related by

$$\begin{aligned} d_1 &= +R\theta_2 \\ d_2 &= -R\theta_1 \end{aligned} \tag{13.11}$$

As the radius of the sphere becomes very large, the two angles θ_1, θ_2 become small with the product $R\theta_i$ ($i = 1, 2$) approaching a well-defined limit. This corresponds to a rotation through an angle $\theta_2 = d_1/R$ about the y-axis producing a displacement of d_1 in the x-direction, and a rotation through an angle $\theta_1 = d_2/R$ about the x-axis producing a displacement of $-d_2$ in the y-direction.

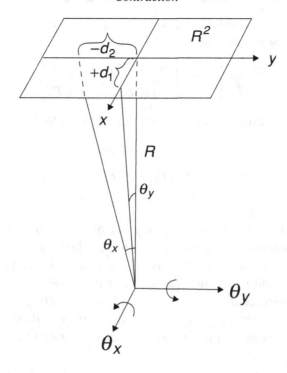

Figure 13.1. Rotations on the surface of a sphere of radius R approach displacements in the plane as $R \to \infty$.

The Casimir operator for the group $SO(3)$ contracts to an invariant operator as follows:

$$\begin{aligned} \mathcal{C}^2[SO(3)] &= L_1^2 + L_2^2 + L_3^2 \\ \mathcal{C}^2[ISO(2)] &= \lim(1/R^2)\,\mathcal{C}^2[SO(3)] \\ &= \lim[(L_1/R)^2 + (L_2/R)^2 + (L_3/R)^2] \\ &= (-P_2)^2 + (+P_1)^2 + 0 = \frac{\partial^2}{\partial y^2} + \frac{\partial^2}{\partial x^2} \end{aligned} \qquad (13.12)$$

This is just the Laplacian operator on the plane R^2.

13.3.2 The contraction SO(4) $\to$ ISO(3)

This group is similar to $SO(3)$ and can be treated similarly. The six generators are

$$\begin{aligned} L_i &= \epsilon_{ijk} x_j \partial_k \\ V_i &= x_i \partial_4 - x_4 \partial_i \end{aligned} \qquad 1 \le i \ne j \ne k \le 3 \qquad (13.13)$$

13.3 Simple examples of Inönü–Wigner contractions

The commutation relations are

$$[L_i, L_j] = -\epsilon_{ijk} L_k$$
$$[L_i, V_j] = -\epsilon_{ijk} V_k \qquad (13.14)$$
$$[V_i, V_j] = -\epsilon_{ijk} L_k$$

We contract with respect to the subgroup $SO(3)$ generated by the angular momentum operators L_i, defining

$$-P_i = \lim_{R \to \infty} \frac{1}{R} V_i = \lim_{R \to \infty} \frac{1}{R}(x_i \partial_4 - x_4 \partial_i) = -\partial_i \qquad (13.15)$$

The commutation relations of the contracted algebra are

$$[L_i, L_j] = -\epsilon_{ijk} L_k$$
$$[L_i, P_j] = -\epsilon_{ijk} P_k \qquad (13.16)$$
$$[P_i, P_j] = 0$$

The operators P_i describe displacements in the x-, y-, and z-directions ($i = 1, 2, 3$). The contracted group is $ISO(3)$, the Euclidean, or inhomogeneous orthogonal group, on R^3.

As in the case $SO(3) \to ISO(2)$, we can contract the second order Casimir operator of $SO(4)$ to that of $ISO(3)$

$$\begin{aligned} \mathcal{C}_1^2[ISO(3)] &= \lim_{R \to \infty} (1/R^2)(\mathbf{L} \cdot \mathbf{L} + \mathbf{V} \cdot \mathbf{V}) \\ &= \lim_{R \to \infty} [(\mathbf{L}/R) \cdot (\mathbf{L}/R) + (\mathbf{V}/R) \cdot (\mathbf{V}/R)] \\ &= 0 + \mathbf{P} \cdot \mathbf{P} = \nabla^2 = \frac{\partial^2}{\partial x^2} + \frac{\partial^2}{\partial y^2} + \frac{\partial^2}{\partial z^2} \end{aligned} \qquad (13.17)$$

As before, this is no surprise. The contracted operator is the Laplacian on R^3. What is a surprise is that there is a second nontrivial invariant operator. For $SO(4)$ this is (cf. Eq. (9.24))

$$\mathcal{C}_2^2[SO(4)] = \epsilon^{ijkl} X_{ij} X_{kl} \to 8 \mathbf{L} \cdot \mathbf{V} \qquad (13.18)$$

The contracted limit of this operator is

$$\begin{aligned} \mathcal{C}_2^2[ISO(3)]/8 &= \lim_{R \to \infty} (1/R)(\mathbf{L} \cdot \mathbf{V}) \\ &= \lim_{R \to \infty} [\mathbf{L} \cdot (\mathbf{V}/R)] = -\mathbf{L} \cdot \mathbf{P} \end{aligned} \qquad (13.19)$$

The two invariant operators $\mathbf{P} \cdot \mathbf{P} = \nabla^2$ and $\mathbf{L} \cdot \mathbf{P} = -\mathbf{L} \cdot \nabla$ form a complete set of invariant operators for the group $ISO(3)$.

13.3.3 The contraction SO(4, 1) → ISO(3, 1)

The group $ISO(3, 1)$ consists of proper Lorentz transformations $[SO(3, 1)]$ that leave invariant the quadratic form

$$x^2 + y^2 + z^2 - (ct)^2 \tag{13.20}$$

as well as displacements of the origin in the three space-like directions and one time-like direction. The inhomogeneous Lorentz group, or Poincaré group, leaves invariant space-time intervals

$$(x - x')^2 + (y - y')^2 + (z - z')^2 - (ct - ct')^2 \tag{13.21}$$

This group can be contracted from either $SO(4, 1)$ or $SO(3, 2)$.

We choose as infinitesimal generators for the group $SO(4, 1)$ the operators

$$
\begin{array}{llll}
X_{ij} = x_i \partial_j - x_j \partial_i = \epsilon_{ijk} L_k & 1 \leq i, j, k \leq 3 & & \text{rotations} \\
B_{i4} = x_i \partial_4 + x_4 \partial_i & 1 \leq i \leq 3 & & \text{boosts} \\
T_{i5} = x_i \partial_5 \pm x_5 \partial_i & i = 1, 2, 3 & -\text{sign} & \text{space displacements} \\
 & i = 4 & +\text{sign} & \text{time displacements}
\end{array}
\tag{13.22}
$$

This set of generators is contracted with respect to the subgroup $SO(3, 1)$ generated by rotations and boosts.

The second order Casimir invariant for $SO(4, 1)$ and its contraction to the second order Casimir invariant for the Poincaré group are

$$C^2[SO(4, 1)] = \mathbf{L} \cdot \mathbf{L} - \mathbf{B} \cdot \mathbf{B} + \mathbf{T} \cdot \mathbf{T} - T_{45}^2$$

$$C^2[ISO(3, 1)] = \quad 0 \quad - \quad 0 \quad + \nabla \cdot \nabla - \frac{1}{c^2} \frac{\partial^2}{\partial t^2}$$

However, $SO(4, 1)$ has a second Casimir operator, since it is a real form for the rank-two root space B_2. This is a fourth-degree operator that is derived by analytic continuation from the fourth order Casimir operator of $SO(5)$ (cf. Eqs. (9.22) and (9.23))

$$
\begin{aligned}
C^4[SO(5)] &= W^\alpha W_\alpha \\
W^\alpha &= \epsilon^{\alpha\beta\gamma\mu\nu} X_{\beta\gamma} X_{\mu\nu}
\end{aligned}
\tag{13.23}
$$

where $\epsilon^{\alpha\beta\gamma\mu\nu}$ is the Levi–Civita symbol (antisymmetric tensor) on five symbols, and W_α is similarly defined. The contracted limit of W^α is nonzero only if one of the four remaining symbols (e.g., ν) is 5:

$$
\lim_{R \to \infty} (1/R) W^\alpha \to \lim_{R \to \infty} \epsilon^{\alpha\beta\gamma\mu 5} X_{\beta\gamma} [(1/R) X_{\mu 5}]
$$

$$
= \epsilon^{\alpha\beta\gamma\mu 5} X_{\beta\gamma} (\partial/\partial x^\mu) \tag{13.24}
$$

The four vector $\epsilon^{\alpha\beta\gamma\mu 5} X_{\beta\gamma}(\partial/\partial x^\mu)$ is fairly complicated. Since $W^\alpha W_\alpha$ is invariant, it is convenient to compute it for a particle of mass m in a frame in which the particle is at rest

$$P_\mu = (0, 0, 0, mc) \tag{13.25}$$

In this frame

$$W^\alpha = \epsilon^{\alpha\beta\gamma\mu} X_{\beta\gamma} mc = L_\alpha mc \tag{13.26}$$

Therefore the invariant is

$$W^\alpha W_\alpha = (\mathbf{L} \cdot \mathbf{L})(\mathbf{P} \cdot \mathbf{P}) \tag{13.27}$$

with $\mathbf{P} \cdot \mathbf{P} = \sum P_\mu P^\mu = -(mc)^2$.

It should be emphasized that if an operator is an invariant and its spectrum or interpretation is desired, the operator should be viewed from the coordinate system which most simplifies its determination (principle of maximum laziness).

13.4 The contraction $U(2) \to H_4$

In this section we consider a group contraction that is not of Inönü–Wigner type. This is the contraction of the compact unitary group $U(2)$ to the solvable group H_4. This contraction relates the angular momentum operators to the single-mode photon operators. These are the infinitesimal generators of the groups $U(2)$ and H_4, respectively. This contraction leads to a number of useful relations that are explored in successive sections.

13.4.1 Contraction of the algebra

The Lie algebra u(2) is spanned by infinitesimal generators $J_3, J_\pm, J_0$ with commutation relations

$$\begin{aligned}{}[J_3, J_\pm] &= \pm J_\pm \\ [J_+, J_-] &= 2J_3 \\ [J_0, \mathbf{J}] &= 0 \end{aligned} \tag{13.28}$$

The operators $h_3, h_\pm, h_0$ are related to $J_3, J_\pm, J_0$ by the following change of basis

$$\begin{bmatrix} h_+ \\ h_- \\ h_3 \\ h_0 \end{bmatrix} = \begin{bmatrix} c & & & \\ & c & & \\ & & 1 & \frac{1}{2c^2} \\ & & & 1 \end{bmatrix} \begin{bmatrix} J_+ \\ J_- \\ J_3 \\ J_0 \end{bmatrix} \tag{13.29}$$

These operators satisfy the following commutation relations

$$[h_3, h_\pm] = \pm h_\pm$$
$$[h_+, h_-] = 2c^2 h_3 - h_0 \qquad (13.30)$$
$$[h_0, \mathbf{h}] = 0$$

In the limit $c \to 0$ the change of basis transformation becomes singular but the commutation relations (Eq. (13.30)) converge to a well-defined limit satisfied by the single-mode photon operators

$$\begin{bmatrix} h_3 \\ h_+ \\ h_- \\ h_0 \end{bmatrix} \xrightarrow{c \to 0} \begin{bmatrix} \hat{n} + \frac{1}{2}I = \frac{1}{2}\{a, a^\dagger\} \\ a^\dagger \\ a \\ I \end{bmatrix} \qquad (13.31)$$

13.4.2 Contraction of the Casimir operators

The group $U(2)$ has rank two. Its two Casimir operators are of first and second order

$$C^1 = J_0$$
$$C^2 = J_3^2 + \tfrac{1}{2}(J_+ J_- + J_- J_+) \qquad (13.32)$$

Under contraction $J_0 \to h_0$ but the second Casimir operator has a more interesting limit

$$\lim_{c \to 0} c^2 C^2 = \lim_{c \to 0} c^2 \left(h_3 - \frac{1}{2c^2} h_0 \right)^2 + \frac{1}{2}[(cJ_+)(cJ_-) + (cJ_-)(cJ_+)]$$

$$= \lim_{c \to 0} c^2 h_3^2 - \frac{1}{2}(h_3 h_0 + h_0 h_3) + c^2 \left(-\frac{1}{2c^2} h_0 \right)^2$$

$$+ \frac{1}{2}[(h_+)(h_-) + (h_-)(h_+)] \qquad (13.33)$$

The operator $(h_0/2c)^2$ is proportional to the square of the first Casimir operator. It therefore commutes with all elements in the Lie algebra. Therefore the remaining set of operators on the right-hand side of (13.34) must also commute with all operators in the Lie algebra. In the limit $c \to 0$, $(ch_3)^2 \to 0$ and the remaining operators go to a well-defined limit

$$\lim_{c \to 0} c^2 C^2[U(2)] - (h_0/2c)^2 \to C^2[H_4] = -\frac{1}{2}\left[\left(\hat{n} + \frac{1}{2}I\right)I + I\left(\hat{n} + \frac{1}{2}I\right)\right]$$

$$+ \frac{1}{2}(aa^\dagger + a^\dagger a) \qquad (13.34)$$

This is a quadratic operator in the generators $\hat{n} + \frac{1}{2}I$, $a^\dagger$, a, and I of H_4. The value of this operator in the standard Fock space spanned by the photon number states $|0\rangle, |1\rangle, |2\rangle, \ldots$ is zero. It is the other "invisible invariant" for H_4.

13.4.3 Contraction of the parameter space

An arbitrary element in the Lie algebra $\mathfrak{u}(2)$ and its counterpart in the algebra $\mathfrak{h}_4$ with basis $h_3, h_\pm, h_0$ is (Arecchi et al., 1972; Gilmore, 1974b)

$$i\theta_\mu J_\mu = \frac{1}{2}\theta e^{-i\phi} J_+ - \frac{1}{2}\theta e^{+i\phi} J_- + i\theta_3 J_3 + i\theta_0 J_0$$

$$= \frac{\theta}{2c} e^{-i\phi} h_+ - \frac{\theta}{2c} e^{+i\phi} h_- + i\theta_3 h_3 + i\left(\theta_0 - \frac{\theta_3}{2c^2}\right) h_0 \quad (13.35)$$

In the limit $c \to 0$ the parameter θ must approach zero so that the limits

$$\begin{aligned}\lim_{c \to 0} +\frac{\theta}{2c} e^{-i\phi} &\to +\alpha \\ \lim_{c \to 0} -\frac{\theta}{2c} e^{+i\phi} &\to -\alpha^*\end{aligned} \quad (13.36)$$

exist. In addition, θ_0 should diverge so that $\theta_0 - \theta_3/2c^2$ remains well defined.

13.4.4 Contraction of representations

The action of the operators h_3 on the angular momentum state $|J, M\rangle$ is

$$h_3 \left|\begin{matrix}J\\M\end{matrix}\right\rangle = \left(J_3 + \frac{1}{2c^2} J_0\right) \left|\begin{matrix}J\\M\end{matrix}\right\rangle = \left(M + \frac{1}{2c^2}\right) \left|\begin{matrix}J\\M\end{matrix}\right\rangle \quad (13.37)$$

It is useful to measure states from the "lowest" state $|J, -J\rangle$ in the angular momentum multiplet. The state with the quantum number M is the ground state if $M = -J$, and the nth state when

$$n = J + M \quad (13.38)$$

In order for the action of h_3 on $|J, M\rangle$ to have a well-defined limit, we insist that

$$\lim_{c \to 0} \left(M + \frac{1}{2c^2}\right) = \lim_{c \to 0} \left(n - J + \frac{1}{2c^2}\right) \quad (13.39)$$

be well defined. This is the case when we go through a sequence of larger and larger representations J of dimension $(2J + 1)$ as c becomes smaller and smaller. Specifically, we require c and J to be related by (Arecchi et al., 1972; Gilmore, 1974b)

$$\lim_{c \to 0} \left(-J + \frac{1}{2c^2}\right) = 0 \quad \text{implies} \quad 2Jc^2 = 1 \quad (13.40)$$

In this case

$$\lim_{\substack{c \to 0 \\ J \to \infty}} h_3 \left| \begin{matrix} J \\ M \end{matrix} \right\rangle = n \left| \begin{matrix} \infty \\ n \end{matrix} \right\rangle \qquad (13.41)$$

13.4.5 Contraction of basis states

The basis states $|J, M\rangle$ for an angular momentum multiplet are constructed by applying the angular momentum shift up operator $n = J + M$ times to the ground state $|J, -J\rangle$. These states are contracted to the harmonic oscillator states as follows

$$\left| \begin{matrix} J \\ M = -J + n \end{matrix} \right\rangle = \frac{(J_+)^n}{[(2J)!n!/(2J-n)!]^{1/2}} \left| \begin{matrix} J \\ -J \end{matrix} \right\rangle$$

$$\left| \begin{matrix} \infty \\ n \end{matrix} \right\rangle = \lim_{J \to \infty} \frac{(cJ_+)^n}{[(2Jc^2)^n n!]^{1/2}} \left| \begin{matrix} \infty \\ 0 \end{matrix} \right\rangle \qquad (13.42)$$

$$= \frac{(a^\dagger)^n}{\sqrt{n!}} \left| \begin{matrix} \infty \\ 0 \end{matrix} \right\rangle$$

13.4.6 Contraction of matrix elements

The matrix elements of the angular momentum operators on the angular momentum basis states contract readily to the matrix elements of the photon operators on the Fock states

$$a^\dagger a \left| \begin{matrix} \infty \\ n \end{matrix} \right\rangle = \lim_{c \to 0} \left(J_3 + \frac{1}{2c^2} \right) \left| \begin{matrix} J \\ M \end{matrix} \right\rangle$$

$$= \lim_{c \to 0} \left[J + M + \left(\frac{1}{2c^2} - J \right) \right] \left| \begin{matrix} J \\ M = n - J \end{matrix} \right\rangle \to (n + 0) \left| \begin{matrix} \infty \\ n \end{matrix} \right\rangle \qquad (13.43)$$

$$a^\dagger \left| \begin{matrix} \infty \\ n \end{matrix} \right\rangle = \lim_{c \to 0} cJ_+ \left| \begin{matrix} J \\ M \end{matrix} \right\rangle$$

$$= \lim_{c \to 0} \left| \begin{matrix} J \\ M+1 \end{matrix} \right\rangle \sqrt{(J-M)(J+M+1)c^2} \to \sqrt{n+1} \left| \begin{matrix} \infty \\ n+1 \end{matrix} \right\rangle \qquad (13.44)$$

$$a \left| \begin{matrix} \infty \\ n \end{matrix} \right\rangle = \lim_{c \to 0} cJ_- \left| \begin{matrix} J \\ M \end{matrix} \right\rangle$$

$$= \lim_{c \to 0} \left| \begin{matrix} J \\ M-1 \end{matrix} \right\rangle \sqrt{(J+M)(J-M+1)c^2} \to \sqrt{n} \left| \begin{matrix} \infty \\ n-1 \end{matrix} \right\rangle \qquad (13.45)$$

13.4.7 Contraction of BCH formulas

Baker–Campbell–Hausdorff formulas, which can easily be derived for $U(2)$ in its faithful 2×2 matrix representation, can readily be contracted to BCH formulas for H_4, which can be derived with only a little more difficulty in its faithful 3×3 matrix representation (cf. Eq. (7.36)). For example, the following BCH formula for $U(2)$

$$e^{(\zeta J_+ - \zeta^* J_-)} = e^{\tau J_+} e^{\ln(1+\tau^*\tau) J_3} e^{-\tau^* J_-} \qquad \frac{\zeta}{|\zeta|} \tan|\zeta| = \tau \qquad (13.46)$$

contracts under $\lim_{c \to 0} \zeta/c \to \alpha$ to the BCH formula for H_4

$$e^{(\alpha a^\dagger - \alpha^* a)} = e^{\alpha a^\dagger} e^{-\frac{1}{2}\alpha^* \alpha I} e^{-\alpha^* a} \qquad \alpha = \lim_{c \to 0} \zeta/c \qquad (13.47)$$

13.4.8 Contraction of special functions

Special functions that are associated with the group $SU(2)$ include Jacobi polynomials, the associated Legendre polynomials and spherical harmonics, and the Legendre polynomials. The special functions associated with the "harmonic oscillator" group H_4 are the Hermite polynomials and the harmonic oscillator wavefunctions. One might reasonably expect that the Hermite polynomials and harmonic oscillator wavefunctions are related to the Jacobi or associated Legendre polynomials in some contraction limit. This is so.

The spherical harmonics $Y_m^l(\theta, \phi)$ and associated Legendre polynomials $P_m^l(\cos\theta)$ are related by (Arecchi et al., 1972; Gilmore, 1974b)

$$Y_m^l(\theta, \phi) = \frac{e^{im\phi}}{\sqrt{2\pi}} P_m^l(\cos\theta) \qquad Y_{-m}^l(\theta, \phi) = (-)^m Y_{+m}^l(\theta, \phi)^* \qquad (13.48)$$

The associated legendre polynomials are defined by

$$P_m^l(u) = (-)^{l+m} \frac{1}{2^l l!} \sqrt{\frac{2l+1}{2}} \sqrt{\frac{(l-m)!}{(l+m)!}} (1-u^2)^{+m/2} \frac{d^{l+m}}{du^{l+m}} (1-u^2)^l \qquad (13.49)$$

These polynomials are contracted to harmonic oscillator wavefunctions under $u \to x/\sqrt{l}$ and $l + m = n$:

$$\lim_{c \to 0} l^{-1/4} P_m^l(u = x/\sqrt{l})$$

$$= \lim_{c \to 0} (-)^n \sqrt{\frac{(2l)! l^{1/2}}{2^{(2l)} l! l!}} \sqrt{\frac{1}{2^n n! (2lc^2)^n}}$$

$$\times [1 - 2c^2 x^2]^{(-1/2c^2)/2} \frac{d^n}{dx^n} [1 - 2c^2 x^2]^{1/2c^2} \qquad (13.50)$$

The limit is taken as $c \to 0$, $l \to \infty$, $l + m = n$, $2lc^2 = 1$. The limit inside the first square root is $1/\sqrt{\pi}$, that within the second is $(2^n n!)^{-1}$. The result of this contraction is

$$\lim_{c \to 0} l^{-1/4} P_m^l(u = x/\sqrt{l}) = \frac{1}{\sqrt{2^n n! \sqrt{\pi}}} e^{x^2/2} \left(-\frac{d}{dx}\right)^n e^{-x^2} = \psi_n(x) \qquad (13.51)$$

where $\psi_n(x)$ is the appropriately normalized harmonic oscillator eigenfunction

$$\psi_n(x) = \frac{1}{\sqrt{2^n n! \sqrt{\pi}}} H_n(x) e^{-x^2/2} \qquad (13.52)$$

and $H_n(x)$ is the nth Hermite polynomial.

Under contraction the orthogonality relations obeyed by the associated Legendre functions go over to the orthogonality relations for the harmonic oscillator eigenfunctions

$$\delta_{mm'} = \int_{-1}^{+1} P_m^l(u) P_{m'}^l(u) du$$

$$\to \lim_{l \to \infty} \int_{-\sqrt{l}}^{+\sqrt{l}} \left(\frac{1}{l^{1/4}} P_m^l(x/\sqrt{l})\right) \left(\frac{1}{l^{1/4}} P_{m'}^l(x/\sqrt{l})\right) d(u\sqrt{l})$$

$$\to \int_{-\infty}^{+\infty} \psi_n(x) \psi_{n'}(x) dx = \delta_{nn'} \qquad (13.53)$$

Unfortunately, it is not possible to derive the completeness relations for the harmonic oscillator eigenfunctions from the completeness relations for the Jacobi or associated Legendre polynomials. However, there is a very simple and beautiful proof of the completeness relations for all special functions associated with compact Lie groups. It is due to Wigner and Stone.

13.5 Conclusion

Contraction of groups to form inequivalent groups can be carried out whenever a singular change of basis can be constructed under which the structure constants have a well-defined limit. Contraction is a particularly useful way to construct non-semisimple Lie groups from simple and semisimple Lie groups. The contracted group is always noncompact. Contraction of groups provides many useful relations between the original group and its contracted limit. These involve the commutation relations in the Lie algebra, the range of values in the parameter spaces that map onto the groups, the Casimir operators, the basis states of representations, operator matrix elements, Baker–Campbell–Hausdorff formulas, and limiting relations among special functions. These relations have all been illustrated by example.

13.6 Problems

1. Under the contraction $SO(3) \to ISO(2)$ the representations of $SO(3)$ contract to representations of $ISO(2)$. Since $ISO(2)$ is a noncompact group it has no faithful finite-dimensional unitary representations. We therefore consider the following limit

$$\lim a \downarrow 0 \quad aJ_\pm \to P_\pm \quad a^2 l(l+1) \to p^2 \text{ finite}$$

$$l \uparrow \infty \quad J_3 \to P_3 \quad \left|\begin{array}{c} l \\ m \end{array}\right\rangle \to \left|\begin{array}{c} p \\ m \end{array}\right\rangle$$

$$(p/a)\beta = l\beta = x \text{ finite}$$

a. Compute the matrix elements of the operators $P_\pm$ in the algebra $iso(2)$ and show

$$\left\langle\begin{array}{c} l \\ m' \end{array}\right| aJ_\pm \left|\begin{array}{c} l \\ m \end{array}\right\rangle \xrightarrow{\lim} \left\langle\begin{array}{c} p \\ m' \end{array}\right| P_\pm \left|\begin{array}{c} p \\ m \end{array}\right\rangle$$

$$a\sqrt{(l \mp m)(l \pm m + 1)}\, \delta_{m', m\pm 1} \xrightarrow{\lim} p\, \delta_{m', m\pm 1}$$

b. Compute the contracted limit of the Jacobi polynomials and show that

$$\lim P^l_{mn}(\cos(x/l)) = (-)^{m-n} J_{m-n}(x)$$

where $J_k(x)$ is the kth Bessel function (Arecchi et al., 1972; Gilmore, 1974b).

c. Contract the spherical harmonics and show that

$$\lim \sqrt{\frac{2\pi}{l}} Y^l_m(\beta = x/l) \to J_m(x)$$

d. Contract the Legendre polynomials and show that

$$\lim P^l(\cos(\beta = x/l)) \to J_0(x)$$

e. In the generating function expression

$$e^{\alpha J_+} Y^l_m(\theta, \phi) = \sum_{k \geq 0} A^l_k Y^l_{m+k}(\theta, \phi) = Y^l_m(\theta', \phi')$$

compute the coefficients A^l_k and the arguments θ', ϕ' explicitly. Contract these results to construct the classical generating functions for Bessel functions.

f. Show that the operator $\mathbf{L} \cdot \mathbf{L}$ contracts to ∇^2 in the plane.

g. Show that the Casimir invariant operator for $SO(3)$ becomes the Laplace–Beltrami operator on $S^2 = SO(3)/SO(2)$ when restricted to the sphere surface, and this operator contracts to the Bessel equation.

2. Under the contraction $u(2) \to \mathfrak{h}_4$ the representations of the unitary group $U(2)$ contract to representations of the noncompact Heisenberg group H_4. Since H_4 is noncompact it has no faithful finite-dimensional unitary irreducible representations. We

therefore contract through a series of representations of $U(2)$ of ever increasing dimensions, as follows:

$$\lim \epsilon \to \infty \qquad \epsilon J_\pm \to h_\pm \qquad 2j\epsilon^2 \to 1$$

$$j \to +\infty, m \to -\infty \quad J_3 + \tfrac{1}{2\epsilon^2} \to h_3 \quad \left|\begin{array}{c} j \\ m \end{array}\right\rangle \to \left|\begin{array}{c} \infty \\ n \end{array}\right\rangle$$

$$j + m = n \text{ (finite)} \quad \theta \to \tfrac{\pi}{2} - \sqrt{2\epsilon}x$$

a. Compute the matrix elements

$$\left\langle \begin{array}{c} j \\ m' \end{array} \right| \epsilon J_\pm \left| \begin{array}{c} j \\ m \end{array} \right\rangle \xrightarrow{\lim} \left\langle \begin{array}{c} \infty \\ n' \end{array} \right| h_\pm \left| \begin{array}{c} \infty \\ n \end{array} \right\rangle$$

$$\epsilon\sqrt{(j \mp m)(j \pm m + 1)}\, \delta_{m',m\pm 1} \xrightarrow{\lim} \begin{array}{c} \sqrt{n+1}\, \delta_{n',n+1} \\ \sqrt{n}\, \delta_{n',n-1} \end{array}$$

b. Contract the spherical harmonics and show

$$l^{1/4} P^l_{n-l,0}\left(\tfrac{\pi}{2} - \sqrt{2\epsilon}x\right) \xrightarrow{\lim} \psi_n(x) = N_n H_n(x) e^{-x^2/2}$$

where $\psi_n(x)$ is the nth excited state wavefunction for the harmonic oscillator, $H_n(x)$ is the nth Hermite polynomial, and N_n is the usual normalization coefficient, $N_n = 1/\sqrt{2^n n! \sqrt{\pi}}$.

c. Carry out steps **c–f** of the previous problem. The results are obtained by making the following replacements:

Bessel function $\to$ harmonic oscillator eigenfunction
Bessel equation $\to$ Schrödinger equation for harmonic oscillator

3. Contract the Lie algebra $\mathfrak{su}(2)$ spanned by J_3, $J_\pm$ ($[J_3, J_\pm] = \pm J_\pm$, $[J_+, J_-] = 2J_3$) with respect to the subalgebra J_-. Use a simple Inönü–Wigner contraction to show

$$\lim_{\epsilon \to 0} \epsilon(2J_3) \to P \quad P' = \partial_x$$
$$\lim_{\epsilon \to 0} \epsilon(J_+) \to T \quad T' = \partial_t$$
$$\lim_{\epsilon \to 0} (J_-) \to V \quad V' = t\partial_x$$

Construct the commutation relations of the contracted operators and show that the operators on the right (P', T', V') satisfy an isomorphic set of commutation relations. The operators ∂_x, ∂_t, $t\partial_x$ generate the Galilean group in one dimension. Conclude that if the Lie algebra $\mathfrak{a}_1$ is contracted with respect to one of its shift operators the Galilean algebra $\mathfrak{gal}(1)$ results.

4. Contract $SO(n+1)$ with respect to the subgroup $SO(n)$ and show how the invariant metric and measure on the sphere $S^n = SO(n+1)/SO(n)$ reduce to the familiar metric and measure on $R^n = ISO(n)/SO(n)$.

13.6 Problems

5. Disentangling formulas can also be contracted.
   a. Use the defining 2×2 matrix representation for $\mathfrak{su}(2)$ to construct the disentangling theorem
   $$e^{\zeta J_+ - \zeta^* J_-} = e^{\tau J_+} e^{\log(1+\tau^*\tau) J_3} e^{-\tau^* J_-}$$
   and show $\tau = (\zeta/|\zeta|)\tan(|\zeta|)$.

   b. Use a faithful matrix representation of the Lie algebra $\mathfrak{h}_4$ to construct the disentangling theorem
   $$e^{\alpha a^\dagger - \alpha^* a} = e^{\alpha a^\dagger} e^{-\frac{1}{2}\alpha^* \alpha \, I} e^{-\alpha^* a}$$

   c. Use the contraction relation Eq. (13.30) for $\mathfrak{u}(2) \to \mathfrak{h}_4$ to show that the $\mathfrak{u}(2)$ disentangling theorem contracts to the $\mathfrak{h}_4$ disentangling theorem in the limit $\alpha = \lim_{c \to 0} \zeta/c$.

6. Thermal expectation values of the operator X are constructed by taking the trace: $\langle X \rangle = \text{tr } X e^{-\beta \mathcal{H}} / \text{tr } e^{-\beta \mathcal{H}}$, and a generating function for expectation values is $\langle e^{\alpha X} \rangle = \text{tr } e^{\alpha X} e^{-\beta \mathcal{H}} / \text{tr } e^{-\beta \mathcal{H}}$. When the operators X and $\mathcal{H}$ are elements in a finite dimensional Lie algebra these expectation values can often be computed rather simply.
   a. Assume $\mathcal{H} = \epsilon J_3$ and X is in the Lie algebra $\mathfrak{su}(2)$. Show that in the 2×2 defining matrix representation
   $$e^{\theta \cdot J} \to \begin{bmatrix} \cosh(\theta/2) + (\theta_z/\theta)\sinh(\theta/2) & (\theta_x - i\theta_y)/\theta \, \sinh(\theta/2) \\ (\theta_x + i\theta_y)/\theta \, \sinh(\theta/2) & \cosh(\theta/2) - (\theta_z/\theta)\sinh(\theta/2) \end{bmatrix}$$
   $$e^{-\beta \mathcal{H}} \to \begin{bmatrix} e^{-\beta\epsilon/2} & 0 \\ 0 & e^{+\beta\epsilon/2} \end{bmatrix}$$

   b. Show that the trace of this product is
   $$2\cosh(\theta/2)\cosh(\beta\epsilon/2) - 2(\theta_z/\theta)\sinh(\theta/2)\sinh(\beta\epsilon/2) \quad (= 2\cosh(\psi/2))$$

   c. Show that in the 2×2 matrix representation with $j = \frac{1}{2}$ and $2j+1 = 2$,
   $$\langle e^{\theta \cdot J} \rangle = (\sinh \psi / \sinh(\psi/2)) / (\sinh \beta\epsilon / \sinh(\beta\epsilon/2))$$

   d. Show that in the $(2j+1) \times (2j+1)$ dimensional representation,
   $$\langle e^{\theta \cdot J} \rangle = \frac{\sinh((2j+1)\psi/2)/\sinh(\psi/2)}{\sinh((2j+1)\beta\epsilon/2)/\sinh(\beta\epsilon/2)}$$

   e. As j becomes large, show that this ratio simplifies to
   $$\langle e^{\theta \cdot J} \rangle \xrightarrow{j \to \infty} \sinh(j\psi)/\sinh(j\beta\epsilon)$$

   f. Contract this generating function to the Heisenberg algebra.

7. One real form of D_3 is the conformal group $SO(4, 2)$.
   a. Write down the quadratic, cubic, and quartic Casimir operators for $SO(4, 2)$. These are analytic continuations of $C^2 = \sum_{ij} X_{ij}^2$, $C^3 = \epsilon^{abcdef} X_{ab} X_{cd} X_{ef}$, and $C^4 = \sum_{ij} Y_{ij}^2$, where $Y_{ij} = \epsilon^{ijcdef} X_{cd} X_{ef}$ of the group $SO(6)$.
   b. Contract $SO(4, 2)$ with respect to the subgroup $SO(4) \otimes SO(2)$.
   c. Construct the quadratic, cubic, and quartic Casimir operators of the contracted group. These are analytic continuations of the contractions of the three operators of part **a**. If we define $A_i = \lim_{\epsilon \to 0} \epsilon X_{i5}$ and $B_i = \lim_{\epsilon \to 0} \epsilon X_{i6}$, then show that the Casimir operators contract to

   $$C^2 \to \mathbf{A} \cdot \mathbf{A} + \mathbf{B} \cdot \mathbf{B}$$
   $$C^3 \to \epsilon^{ijkl} X_{ij} A_k B_l$$
   $$C^4 \to \sum_{ij}(\epsilon^{ijkl} A_k B_l)^2$$

   In these expressions the indices range from 1 to 4.
   d. Write down the Laplace–Beltrami operators in the eight-dimensional spaces $SO(4, 2)/[SO(4) \otimes SO(2)]$ and $I[SO(4) \otimes SO(2)]/[SO(4) \otimes SO(2)]$.

8. Riemannian symmetric spaces have been classified using the Cartan decomposition of simple Lie algebras:

   $$\mathfrak{g} = \mathfrak{h} + \mathfrak{p} \qquad \begin{array}{l} [\mathfrak{h}, \mathfrak{h}] \subseteq \mathfrak{h} \\ [\mathfrak{h}, \mathfrak{p}] = \mathfrak{p} \\ [\mathfrak{p}, \mathfrak{p}] \subseteq \mathfrak{h} \end{array}$$

   Operators X_i span $\mathfrak{h}$ and X_α span $\mathfrak{p}$.
   a. Show that the metric on $\mathfrak{p}$ is

   $$g_{\alpha,\beta} = C_{\alpha,\gamma}^{i} C_{\beta,i}^{\gamma} + C_{\alpha,i}^{\gamma} C_{\beta,\gamma}^{i}$$

   b. Show that in the contracted limit $Y_\alpha = \lim_{\epsilon \to 0} \epsilon X_\alpha$ a metric tensor on $\mathfrak{p}$ is well defined by

   $$g(\mathfrak{p}')_{\alpha,\beta} = \lim_{\epsilon \to 0}(Y_\alpha, Y_\beta)/\epsilon^2 = (X_\alpha, X_\beta)$$

   Use the structure constants to show this.
   c. Show that this metric is unchanged on the contracted space $P' = G'/H$, as opposed to the metric on $P = G/H$, which varies from place to place on the space.

14
Hydrogenic atoms

Many physical systems exhibit symmetry. When a symmetry exists it is possible to use group theory to simplify both the treatment and the understanding of the problem. Central two-body forces, such as the gravitational and Coulomb interactions, give rise to systems exhibiting spherical symmetry (two particles) or broken spherical symmetry (planetary systems). In this chapter we see how spherical symmetry has been used to probe the details of the hydrogen atom. We find a hierarchy of symmetries and symmetry groups. At the most obvious level is the geometric symmetry group, $SO(3)$, which describes invariance under rotations. At a less obvious level is the dynamical symmetry group, $SO(4)$, which accounts for the degeneracy of the levels in the hydrogen atom with the same principal quantum number. At an even higher level are the spectrum generating groups, $SO(4, 1)$ and $SO(4, 2)$, which do not maintain energy degeneracy at all, but rather map any bound (scattering) state of the hydrogen atom into linear combinations of all bound (scattering) states. We begin with a description of the fundamental principles underlying the application of group theory to the study of physical systems. These are the principle of relativity (Galileo) and the principle of equivalence (Einstein).

14.1 Introduction

Applications of group theory in physics start with two very important principles. These are Galileo's principle of relativity (of observers) and Einstein's principle of equivalence (of states). We show how these principles are used to establish the standard framework for the application of geometric symmetry groups to the treatment of quantum mechanical systems that possess some geometric symmetry. For the hydrogen atom the **geometric symmetry group** is $SO(3)$ and one prediction is that states occur in multiplets with typical angular momentum degeneracy: $2l + 1$. This is seen when we solve the Schrödinger and Klein–Gordon equations for the hydrogen atom – more specifically for the spinless electron in the Coulomb potential of a proton.

Invariance of a hamiltonian under a group action implies degeneracy of the energy eigenvalues. It is observed that in the nonrelativistic case the energy degeneracy is larger than required by invariance under the rotation group $SO(3)$. If we believe that the greater the symmetry, the greater the degeneracy, we would expect that the Hamiltonian is invariant under a larger group than the geometric symmetry group $SO(3)$. The larger group is called a **dynamical symmetry group**. This group is $SO(4)$ for the hydrogen bound states. Its infinitesimal generators include the components of two three-vectors: the angular momentum vector and the Laplace–Runge–Lenz vector.

When the dynamical symmetry is broken, as in the case of the Klein–Gordon equation, the classical orbit is a precessing ellipse and the bound states with a given principle quantum number N are slightly split according to their orbital angular momentum values l.

This suggests that we could look for even larger groups that do not pretend to preserve (geometric or dynamical) symmetry and do not maintain energy degeneracy. In fact, they map any bound (scattering) state into linear combinations of all other bound (scattering) states. Such groups exist. They are called **spectrum generating groups**. For the hydrogen atom the first spectrum generating group that was discovered was the deSitter group $SO(4, 1)$. A larger spectrum generating group is the conformal group $SO(4, 2)$. We illustrate how spectrum generating groups have been used to construct eigenfunctions and energy eigenvalues. We also describe how analytic continuations between two qualitatively different types of representations of a noncompact group lead to relations between the bound state spectrum, on the one hand, and the phase shifts of scattering states, on the other.

14.2 Two important principles of physics

There are two principles of fundamental importance that allow group theory to be used in profoundly important ways in physics. These are the principle of relativity and the principle of equivalence. We give a brief statement of both using a variant of Dirac notation.

Principle of relativity (of observers) Two observers, S and S', describe a physical state $|\psi\rangle$ in their respective coordinate systems. They describe the state by mathematical functions $\langle S|\psi\rangle$ and $\langle S'|\psi\rangle$. The two observers know the relation between their coordinate systems. The mathematical prescription for transforming functions from one coordinate system to the other is $\langle S'|S\rangle$. The set of transformations among observers forms a group. If observer S' wants to determine what observer S has seen, he applies the appropriate transformation, $\langle S|S'\rangle$, to his mathematical functions $\langle S'|\psi\rangle$ to determine how S has described the system:

$$\langle S|\psi\rangle = \langle S|S'\rangle \langle S'|\psi\rangle \tag{14.1}$$

The principle of relativity of observers is a statement that *the functions determined by S' in this fashion are exactly the functions used by S to describe the state $|\psi\rangle$*.

Principle of equivalence (of states): Two observes S and S' observe a system, as above. If

the rest of the universe looks the same

to both S and S', then S can use the mathematical functions $\langle S'|\psi\rangle$ written down by S' to describe a *new* physical state $|\psi'\rangle$

$$\langle S|\psi'\rangle = \langle S'|\psi\rangle \tag{14.2}$$

and that state must exist.

In this notation, the transformation of a hamiltonian under a group operation (for example, a rotation in $SO(3)$) is expressed by $\langle S'|H|S'\rangle = \langle S'|S\rangle\langle S|H|S\rangle\langle S|S'\rangle$, the invariance under the transformation $\langle S'|S\rangle$ is represented by $\langle S'|H|S'\rangle = \langle S|H|S\rangle$, and the existence of a $2p_z$ state in a system with spherical symmetry implies the existence (by the Principle of Equivalence) of $2p_x$ and $2p_y$ states, as well as arbitrary linear combinations of these three states.

14.3 The wave equations

Schrödinger's derivation of a wave equation for a particle of mass m began with the relativistic dispersion relation for the free particle: $p^\mu p_\mu = g_{\mu\nu} p^\mu p^\nu = (mc)^2$. In terms of the energy E and the three-momentum $\mathbf{p}$ this is

$$E^2 - (\mathbf{p}c)^2 = (mc^2)^2 \tag{14.3}$$

Interaction of a particle of charge q with the electromagnetic field is described by the principle of minimal electromagnetic coupling: $p_\mu \to \pi_\mu = p_\mu - (q/c)A_\mu$, where the four-vector potential A consists of the scalar potential Φ and the vector potential $\mathbf{A}$. These obey $\mathbf{B} = \nabla \times \mathbf{A}$ and $\mathbf{E} = -\nabla\Phi - (1/c)(\partial \mathbf{A}/\partial t)$. For an electron $q = -e$, where e is the charge on the proton, positive by convention. In the Coulomb field established by a proton, $\Phi = e/r$ and $\mathbf{A} = 0$, so that $E \to E + e^2/r$. Here r is the proton–electron distance. The Schrödinger prescription for converting a dispersion relation to a wave equation is to replace $\mathbf{p} \to (\hbar/i)\nabla$ and allow the resulting equation to act on a spacial function $\psi(\mathbf{x})$. This prescription results in the following wave equation, the Klein–Gordon equation:

$$\left\{ E^2 - (mc^2)^2 + 2E\left(\frac{e^2}{r}\right) + \left(\frac{e^2}{r}\right)^2 - (-i\hbar\nabla)^2 \right\} \psi(x) = 0 \qquad (14.4)$$

This equation exhibits spherical symmetry in the sense that it is unchanged (invariant) in form under rotations: $\langle S'|H|S'\rangle = \langle S|H|S\rangle$, where $\langle S'|S\rangle \in SO(3)$. Schrödinger solved this equation, compared its predictions with the spectral energy measurements on the hydrogen atom, was not convinced his theory was any good, and buried this approach in his desk drawer.

Sometime later he reviewed this calculation and took its nonrelativistic limit. Since the binding energy is about 13.6 eV and the electron rest energy mc^2 is about 510 000 eV, it makes sense to write $E = mc^2 + W$, where the principal part of the relativistic energy E is the electron rest energy and the nonrelativistic energy W is a small perturbation of either ($\simeq 0.0025\%$). Under this substitution, and neglecting terms of order $(W + e^2/r)^2/mc^2$, we obtain the nonrelativistic form of Eq. (14.4), the Schrödinger equation:

$$\left\{ \frac{\mathbf{p}\cdot\mathbf{p}}{2m} - \frac{e^2}{r} - W \right\} \psi(\mathbf{x}) = \left\{ -\frac{\hbar^2}{2m}\nabla^2 - \frac{e^2}{r} - W \right\} \psi(\mathbf{x}) = 0 \qquad (14.5)$$

Equation (14.4) is now known as the Klein–Gordon equation and its nonrelativistic limit Eq. (14.5) is known as the Schrödinger equation, although the former was derived by Schrödinger before he derived his namesake equation.

Remark Schrödinger began his quest for a theory of atomic physics with Maxwell's equations, in particular, the eikonal form of these equations. It is no surprise that his theory inherits key characteristics of electromagnetic theory: solutions that are amplitudes, the superposition principle for solutions, and interference effects that come about by squaring amplitudes to obtain intensities. Had he started from classical mechanics, there would be no amplitude-intensity relation and the only superposition principle would have been the superposition of forces or their potentials. The elegant but forced relation between Poisson brackets and commutator brackets ($[A, B]/i\hbar = \{A, B\}$) is an attempt to fit quantum mechanics into the straitjacket of classical mechanics.

14.4 Quantization conditions

The standard approach to solving partial differential equations is to separate variables. Since the two equations derived above have spherical symmetry, it is useful to introduce spherical coordinates (r, θ, ϕ). In this coordinate system the

14.4 Quantization conditions

Laplacian is

$$\nabla^2 = \left(\frac{1}{r}\frac{\partial}{\partial r}r\right)^2 + \frac{\mathcal{L}^2(S^2)}{r^2} \tag{14.6}$$

$$\mathcal{L}^2(S^2) = \frac{1}{\sin\theta}\frac{\partial}{\partial\theta}\sin\theta\frac{\partial}{\partial\theta} + \frac{1}{\sin^2\theta}\frac{\partial^2}{\partial\phi^2} \tag{14.7}$$

The second order differential operator $\mathcal{L}^2(S^2)$ is the Laplacian on the sphere S^2. Its eigenfunctions are the spherical harmonics $Y_m^l(\theta, \phi)$ and its spectrum of eigenvalues is $\mathcal{L}^2(S^2)Y_m^l(\theta, \phi) = -l(l+1)Y_m^l(\theta, \phi)$. The integers (l, m) satisfy $l = 0, 1, 2, \ldots$ and $-l \leq m \leq +l$. The negative sign and discrete spectrum characteristically indicate that S^2 is compact.

The partial differential equations (14.4) and (14.5) are reduced to ordinary differential equations by substituting the ansatz

$$\psi(r, \theta, \phi) \to \frac{1}{r}R(r)Y_m^l(\theta, \phi) \tag{14.8}$$

into these equations, replacing the angular part of the Laplacian by the eigenvalue $-l(l+1)$, and multiplying by r on the left. This gives the simple second order ordinary differential equation

$$\left(\frac{d^2}{dr^2} + \frac{A}{r^2} + \frac{B}{r} + C\right)R(r) = 0 \tag{14.9}$$

The values of the coefficients A, B, C that are obtained for the Klein–Gordon equation and the Schrödinger equation are as follows:

Equation	A	B	C	
Klein–Gordon	$-l(l+1)+(e^2/\hbar c)^2$	$2Ee^2/(\hbar c)^2$	$[E^2-(mc^2)^2]/(\hbar c)^2$	(14.10)
Schrödinger	$-l(l+1)$	$2me^2/\hbar^2$	$2mW/\hbar^2$	

There is a standard procedure for solving simple ordinary differential equations of the type presented in Eq. (14.9). This is the Frobenius method. The steps involved in this method, and the result of each step, are summarized in Table 14.1.

The energy eigenvalues for the bound states of both the relativistic and non-relativistic problems are expressed in terms of the radial quantum number $n = 0, 1, 2, \ldots$ and the angular momentum quantum number $l = 0, 1, 2, \ldots$, mass m of the electron, or more precisely the reduced mass of the proton–electron pair $m_{red}^{-1} = m_e^{-1} + M_p^{-1}$, and the fine structure constant (Gabrielse et al., 2006)

$$\alpha = \frac{e^2}{\hbar c} = \frac{1}{137.035\ 999\ 796(70)} = 0.007\ 297\ 352\ 531\ \underline{3}(3\ 8) \tag{14.11}$$

Table 14.1. *Left column lists the steps followed in the Frobenius method for finding the square-integrable solutions of simple ordinary differential equations, the right column shows the result of applying the step to Eq. (14.9)*

	Procedure	Result
1	Locate singularities	$0, \infty$
2	Determine analytic behavior at singular points	$r \to 0: R \simeq r^\gamma, \gamma(\gamma-1) + A = 0$ $r \to \infty: R \simeq e^{\lambda r}, \lambda^2 + C = 0$
3	Keep only $\mathcal{L}^2$ solutions	$\gamma = \frac{1}{2} + \sqrt{(\frac{1}{2})^2 - A}, \ \lambda = -\sqrt{-C}$
4	Look for solutions with proper asymptotic behavior	$R = r^\gamma e^{\lambda r} f(r)$
5	Construct differential equation for $f(r)$	$\left[(rD^2 + 2\gamma D) + (2\lambda\gamma + B + 2\lambda r D)\right] f(r) = 0$
6	Construct recursion relation	$f_{j+1} = -\dfrac{2\lambda(j+\gamma) + B}{j(j+1) + 2\gamma(j+1)} f_j$
7	Look at asymptotic behavior	$f \simeq e^{-2\lambda r}$ if series does not terminate $\simeq e^{+\lambda r}$ if series does terminate $(\lambda < 0)$
8	Construct quantization condition	$2\lambda(n+\gamma) + B = 0$ or $n + \dfrac{1}{2} + \sqrt{(\frac{1}{2})^2 - A} = \dfrac{B}{2\sqrt{-C}}$
9	Construct explicit solutions	$E = \dfrac{mc^2}{\sqrt{1+(\alpha/N')^2}}, \quad W = -\dfrac{1}{2}mc^2\alpha^2 \dfrac{1}{N^2}$ $N' = n + \frac{1}{2} + \sqrt{(l+\frac{1}{2})^2 - \alpha^2}, \ N = n + l + 1$

This is a dimensionless ratio of three physical constants that are fundamental in three "different" areas of physics: e (electromagnetism), $\hbar$ (quantum mechanics), and c (relativity). It is one of the most precisely measured of the physical constants. The bound state energy eigenvalues are

Klein–Gordon equation   Schrödinger equation

$$E(n, l) = \frac{mc^2}{\sqrt{1+(\alpha/N')^2}} \qquad W(n, l) = -\frac{1}{2}mc^2\alpha^2 \frac{1}{N^2} \qquad (14.12)$$

$$N' = n + \frac{1}{2} + \sqrt{\left(l+\frac{1}{2}\right)^2 - \alpha^2} \qquad N = n + l + 1$$

Both the nonrelativistic and relativistic energies have been plotted in Fig. 14.1. The nonrelativistic energies for the hydrogen atom appear as the darker lines. The nonrelativistic energy has been normalized by dividing by the hydrogen atom

14.5 Geometric symmetry $SO(3)$

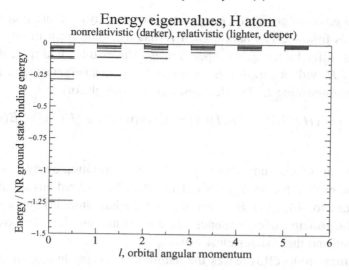

Figure 14.1. Spectrum of the hydrogen atom, normalized by the energy of the nonrelativistic ground state. The nonrelativistic spectrum is darker. The relativistic spectrum has been computed for $Z = 50$. These energies are computed by replacing $\alpha \rightarrow Z\alpha$ everywhere.

ground state energy $|W_1| = \frac{1}{2}mc^2\alpha^2$. These normalized energy levels decrease to zero like $1/N^2$, where $N = n + l + 1$ is the principal quantum number. The energies are displayed as a function of the orbital angular momentum l. The relativistic energies of the bound states for the proton–electron system converge to the rest energy mc^2 as N' increases. When this limit is removed these energies (also rescaled by dividing by $\frac{1}{2}mc^2\alpha^2$) can be plotted on the same graph. At the resolution shown, the two sets of rescaled energies are indistinguishable. To illustrate the difference, we have instead computed and plotted the bound state spectrum for a single electron in a potential with positive charge Z. The energies in this case are obtained by the substitution $\alpha \rightarrow Z\alpha$ everywhere. The energies of these bound states have been renormalized by subtracting the limit mc^2 and dividing by the nonrelativistic energy for the same ion: $\frac{1}{2}mc^2(Z\alpha)^2$. The energy difference between the 1s ground states is pronounced; this difference decreases rapidly as the principal quantum number increases.

14.5 Geometric symmetry $SO(3)$

Symmetry implies degeneracy.

To see this, assume $g_i \in G$ are group operations that leave a hamiltonian H invariant (unchanged in form)

$$g_i H g_i^{-1} = H \quad \text{or} \quad g_i H = H g_i \qquad (14.13)$$

When G is a group of geometric transformations the physical interpretation of this equation is as follows. The hamiltonian H has the same form in two coordinate systems that differ by the group operation g_i. Under this condition, if $|\psi\rangle$ is an eigenstate of H with eigenvalue E, then $g_i|\psi\rangle$ is also an eigenstate of H with the same energy eigenvalue E. The demonstration is straightforward:

$$H(g_i|\psi\rangle) = (Hg_i)|\psi\rangle = (g_iH)|\psi\rangle = g_i(H|\psi\rangle) = g_i(E|\psi\rangle) = E(g_i|\psi\rangle) \tag{14.14}$$

To illustrate this idea, assume that $|\psi\rangle = \psi_{2p_z}(\mathbf{x})$. A rotation by $\pi/2$ radians about the y-axis maps this state to $\psi_{2p_x}(\mathbf{x})$ and a rotation by $\pi/2$ radians about the x-axis maps this state to $-\psi_{2p_y}(\mathbf{x})$. By invariance (of the hamiltonian) under the rotation group and the principle of equivalence, these new functions describe possible states of the system, and these states must exist.

The rotation group $O(3)$ leaves the hamiltonian of the hydrogen atom invariant in both the nonrelativistic and relativistic cases. In the nonrelativistic case, $H = \mathbf{p}\cdot\mathbf{p}/2m - e^2/r$. The scalar $\mathbf{p}\cdot\mathbf{p} = -\hbar^2\nabla^2$ is invariant under rotations, as is also the potential energy term $-e^2/r$. Rotation operators can be expressed in terms of the infinitesimal generators of rotations about axis i: $\epsilon_{ijk}x_j\partial_k$. These geometric operators are proportional to the physical angular momentum operators $L_i = (\mathbf{r}\times\mathbf{p})_i = (\hbar/i)\epsilon_{ijk}x_j\partial_k$. Finite rotations can be expressed as exponentials as follows:

$$R(\theta) = e^{\epsilon_{ijk}\theta_i x_j\partial_k} = e^{i\theta\cdot\mathbf{L}/\hbar} \tag{14.15}$$

The angular momentum operators $\mathbf{L} = \mathbf{r}\times\mathbf{p}$ share the same commutation relations as the infinitesimal generators of rotations $\mathbf{r}\times\nabla$, up to the proportionality factor $\hbar/i$. The commutation relations are

$$[L_i, L_j] = i\hbar\epsilon_{ijk}L_k \tag{14.16}$$

It is useful to construct linear combinations of these operators that have canonical commutation relations of the type described in Chapter 10. To this end we define the raising (L_+) and lowering (L_-) operators by $L_\pm = L_x \pm iL_y$. The commutation relations are

$$[L_z, L_\pm] = \pm\hbar L_\pm \tag{14.17}$$

$$[L_+, L_-] = 2\hbar L_z \tag{14.18}$$

These angular momentum operators are related to the two boson operators as follows: $L_z = \hbar\frac{1}{2}(a_1^\dagger a_1 - a_2^\dagger a_2)$, $L_+ = \hbar a_1^\dagger a_2$, $L_- = \hbar a_2^\dagger a_1$. As a result, the angular momentum operators have matrix representations with basis vectors $|n_1\,n_2\rangle = |{}^j_m\rangle$, with $n_1 = 0, 1, 2, \ldots$, $n_2 = 0, 1, 2, \ldots$, $n_1 + n_2 = 2j$, $n_1 - n_2 = 2m$,

14.5 Geometric symmetry $SO(3)$

$-j \leq m \leq +j$. These basis vectors describe the finite-dimensional irreducible representations of the covering group $SU(2)$ of $SO(3)$. The subset of representations with $j = l$ (integer) describes representations of $SO(3)$.

To see this we construct a coordinate representation of the angular momentum operators. In spherical coordinates $((x, y, z) \to (r, \theta, \phi)$ with $x = r \sin\theta \cos\phi)$ these operators are

$$L_z = \frac{\hbar}{i} \frac{\partial}{\partial \phi}$$

$$L_\pm = \hbar \left(\pm \frac{\partial}{\partial \theta} + i \frac{\cos\theta}{\sin\theta} \frac{\partial}{\partial \phi} \right)$$
(14.19)

The functions on R^3 that transform under the angular momentum operators can be constructed from the mixed basis argument:

$$\langle \theta\phi | L_- \left| \begin{matrix} l \\ m \end{matrix} \right\rangle$$
$$\downarrow \qquad\qquad\qquad \downarrow$$
$$\langle \theta\phi | L_- | \theta'\phi' \rangle \langle \theta'\phi' \left| \begin{matrix} l \\ m \end{matrix} \right\rangle = \langle \theta\phi \left| \begin{matrix} l' \\ m' \end{matrix} \right\rangle \left\langle \begin{matrix} l' \\ m' \end{matrix} \right| L_- \left| \begin{matrix} l \\ m \end{matrix} \right\rangle$$
(14.20)

As usual, the intermediate arguments (with primes) are dummy arguments that are summed or integrated over. The symbols in Eq. (14.20) have the following meanings.

$\langle \theta\phi | L_- | \theta'\phi' \rangle$ Matrix element of the angular momentum shift down operator in the coordinate representation: $\hbar(-\partial/\partial\theta + i(\cos\theta/\sin\theta)(\partial/\partial\phi)) \delta(\cos\theta' - \cos\theta)\delta(\phi' - \phi)$.

$\left\langle \begin{matrix} l' \\ m' \end{matrix} \right| L_- \left| \begin{matrix} l \\ m \end{matrix} \right\rangle$ Matrix element of the angular momentum shift down operator in the algebraic representation: $\hbar\sqrt{(l'-m')(l+m)}$ $\delta_{l'l} \delta_{m', m-1}$.

$\langle \theta\phi \left| \begin{matrix} l \\ m \end{matrix} \right\rangle$ Matrix element of the similarity transformation between the coordinate representation and algebraic representation. Also called spherical harmonic: $Y^l_m(\theta, \phi)$.

This relation can be used to show that there are no geometric functions associated with values of the quantum number j that are half integral. It can also be used to construct the extremal function $Y^l_{-l}(\theta, \phi)$ by solving the equation $L_- Y^l_{-l}(\theta, \phi) = 0$ in the coordinate representation (Problem 14.12). Finally, the action of the shift up operators can be used to constuct the remaining functions $Y^l_m(\theta, \phi)$ through the recursion relation involving both the coordinate and the algebraic representations

Table 14.2. *Spherical harmonics $Y^l_m(\theta, \phi)$ for low values of l and m*

m	$l=0$	$l=1$	$l=2$	$l=3$
0	$\sqrt{\frac{1}{4\pi}}$	$\sqrt{\frac{3}{4\pi}}\cos\theta$	$\sqrt{\frac{5}{16\pi}}(3\cos^2\theta - 1)$	$\sqrt{\frac{7}{16\pi}}(5\cos^3\theta - 3\cos\theta)$
± 1		$\mp\sqrt{\frac{3}{8\pi}}\sin\theta\, e^{\pm i\phi}$	$\mp\sqrt{\frac{15}{8\pi}}\cos\theta\sin\theta\, e^{\pm i\phi}$	$\mp\sqrt{\frac{21}{64\pi}}\sin\theta(5\cos^2\theta - 1)\, e^{\pm i\phi}$
± 2			$\sqrt{\frac{15}{32\pi}}\sin^2\theta\, e^{\pm 2i\phi}$	$\sqrt{\frac{105}{32\pi}}\sin^2\theta\cos\theta\, e^{\pm 2i\phi}$
± 3				$\mp\sqrt{\frac{35}{64\pi}}\sin^3\theta\, e^{\pm 3i\phi}$

of the shift up operator L_+

$$L_+ Y^l_m(\theta, \phi) = Y^l_{m+1}(\theta, \phi)\sqrt{(l-m)(l+m+1)} \tag{14.21}$$

The lowest spherical harmonics ($l = 0, 1, 2, 3$) are collected in Table 14.2.

Remark The spectrum of the Casimir invariant for the rotation group $SO(3)$, or more specifically the Laplace–Beltrami operator constructed from its infinitesimal generators acting on the sphere parameterized by coordinates (θ, ϕ), is $-l(l+1)$, $l = 0, 1, 2, \ldots$. The fact that the spectrum is negative means that the space, S^2, on which these operators act, is compact. By the same token, the spectrum of the square of the angular momentum operator, $\mathbf{L} \cdot \mathbf{L}$, is $\hbar^2 l(l+1)$. This means physically that the inner product of the angular momentum operator with itself is never negative, and is quantized by integer angular momentum values, measured in units of Planck's constant $\hbar$.

14.6 Dynamical symmetry $SO(4)$

Symmetry implies degeneracy.

The greater the symmetry, the greater the degeneracy.

The states of the nonrelativistic hydrogen atom with fixed principal quantum number $N = n + l + 1$ are degenerate, with energy $E_N = -\frac{1}{2}mc^2\alpha^2 \frac{1}{N^2}$. There are $\sum_{l=0}^{l=N-1}(2l+1) = N^2$ states with this energy. This N^2-fold degeneracy is larger than the $2l+1$-fold degeneracy required by rotational invariance of the hamiltonian. If we believe the converse, that degeneracy implies symmetry, then we might be led to expect that the hydrogen atom exhibits more symmetry than meets the eye.

In fact this symmetry, called a dynamical symmetry (Schiff, 1968), exists and is related to a constant of motion that is peculiar to $1/r^2$ force laws. This constant of motion is known as the Laplace–Runge–Lenz vector. It is a constant of unperturbed planetary motion, for which the force law has the form $d\mathbf{p}/dt = -K\mathbf{r}/r^3$, where

14.6 Dynamical symmetry $SO(4)$

$K = GMm$, G is the universal gravitational constant, M and m are the two attracting masses, and $\mathbf{r} = x\hat{\mathbf{i}} + y\hat{\mathbf{j}} + z\hat{\mathbf{k}}$ is the vector from one mass to the other. The time derivative of the vector $\mathbf{p} \times \mathbf{L}$ is

$$\frac{d}{dt}(\mathbf{p} \times \mathbf{L}) = \underset{\downarrow}{\frac{d\mathbf{p}}{dt} \times \mathbf{L}} + \underset{\downarrow}{\mathbf{p} \times \frac{d\mathbf{L}}{dt}}$$

$$= -K\frac{\mathbf{r}}{r^3} \times (\mathbf{r} \times m\dot{\mathbf{r}}) + 0$$

$$= -mK\frac{\mathbf{r}(\mathbf{r} \cdot \dot{\mathbf{r}}) - \dot{\mathbf{r}}(\mathbf{r} \cdot \mathbf{r})}{r^3} = mK\frac{d}{dt}\left(\frac{\mathbf{r}}{r}\right) \qquad (14.22)$$

In going from the first line in Eq. (14.22) to the second, we use the fact that $\mathbf{L}$ is a constant of motion in any spherically symmetric potential. We also use the force law for a $1/r$ potential. In going from the second line to the third, we express the cross product $\mathbf{r} \times \mathbf{L}$ in terms of (generally) nonparallel vectors $\mathbf{r}$ and $\dot{\mathbf{r}}$. We also use the identity $(d/dt)(\mathbf{r}/r) = \dot{\mathbf{r}}/r - (\dot{\mathbf{r}} \cdot \mathbf{r})\,\mathbf{r}/r^3$. The result is that the Laplace–Runge–Lenz vector $\mathbf{M}$ is a constant of motion: $d\mathbf{M}/dt = 0$, where

$$\mathbf{M} = \frac{\mathbf{p} \times \mathbf{L}}{m} - K\frac{\mathbf{r}}{r} \qquad (14.23)$$

In the transition from classical to quantum mechanics the operator obtained from the classical operator in Eq. (14.23) is not hermitian. Pauli (1926) symmetrized it properly, defining the hermitian quantum mechanical operator

$$\hat{\mathbf{M}} = \frac{\hat{\mathbf{p}} \times \hat{\mathbf{L}} - \hat{\mathbf{L}} \times \hat{\mathbf{p}}}{2m} - K\frac{\hat{\mathbf{r}}}{r} \qquad (14.24)$$

where the $\hat{\ }$ over the classical symbol indicates a quantum mechanical operator. We will dispense with the $\hat{\ }$ over operators, in part to simplify notation, in part to prevent uncertainties in interpretation of the operator $\mathbf{r}$.

The hermitian operator $\mathbf{M}$ in Eq. (14.24) is a constant of motion, as it commutes with the nonrelativistic hamiltonian: $[H, \mathbf{M}] = 0$. The six operators L_i, M_j obey the following commutation relations

$$\begin{aligned} [L_i, L_j] &= i\hbar\epsilon_{ijk}\,L_k \\ [L_i, M_j] &= i\hbar\epsilon_{ijk}\,M_k \\ [M_i, M_j] &= \left(-\frac{2H}{m}\right)i\hbar\epsilon_{ijk}\,L_k \end{aligned} \qquad (14.25)$$

These are the commutation relations for the Lie algebra of the group $SO(4)$ for bound states ($E < 0$) or $SO(3, 1)$ for excited states ($E > 0$). The operators $\mathbf{L}$ and

M also obey

$$\mathbf{L} \cdot \mathbf{M} = \mathbf{M} \cdot \mathbf{L} = 0$$

$$\mathbf{M} \cdot \mathbf{M} = \frac{2H}{m} \left(\mathbf{L} \cdot \mathbf{L} + \hbar^2 \right) + K^2 \quad (14.26)$$

In order to simplify the discussion to follow, and make this discussion as independent of the principal quantum number N as possible, we renormalize the Laplace–Runge–Lenz vector by a scale factor as follows: $\mathbf{M}' = (-m/2H)^{1/2}\mathbf{M}$. (For $E > 0$ change $-\to +$ and $SO(4) \to SO(3,1)$.) The commutation relations of these operators are now

$$[L_i, L_j] = i\hbar \epsilon_{ijk} L_k$$
$$[L_i, M'_j] = i\hbar \epsilon_{ijk} M'_k \quad (14.27)$$
$$[M'_i, M'_j] = i\hbar \epsilon_{ijk} L_k$$

The Lie algebra $\mathfrak{so}(4)$ is the direct sum of two Lie algebras of type $\mathfrak{so}(3)$ (see Figs. 10.3, 10.8(b)). It is useful to introduce two vector operators **A** and **B** as follows

$$\mathbf{A} = \tfrac{1}{2}(\mathbf{L} + \mathbf{M}')$$
$$\mathbf{B} = \tfrac{1}{2}(\mathbf{L} - \mathbf{M}') \quad (14.28)$$

The operators **A** and **B** have angular momentum commutation relations. Further, they mutually commute. Finally, their squares have the same spectrum.

It is useful at this point to introduce the Schwinger representation for the angular momentum operators **A** in terms of two independent boson modes: $A_3 = \tfrac{1}{2}(a_1^\dagger a_1 - a_2^\dagger a_2)$, $A_+ = a_1^\dagger a_2$, $A_- = a_2^\dagger a_1$ (for simplicity, set $\hbar \to 1$). A similar representation of the angular momentum operators **B** in terms of two independent boson operators b_1, b_2 and their creation operators is also introduced.

Basis states for a representation of the algebra spanned by the operators **A** have the form $|p_1, p_2\rangle$, with $p_1 + p_2 = 2j_a$ constant and $p_1 - p_2 = m_a$. The $2j_a + 1$ basis states correspond to $p_1 = 2j_a, p_2 = 0$; $p_1 = 2j_a - 1, p_2 = 1$; etc. For **B** the basis states are $|q_1, q_2\rangle$, with $q_1 + q_2 = 2j_b$ constant and $q_1 - q_2 = m_b$. The invariant operators are $\mathbf{A} \cdot \mathbf{A} = j_a(j_a + 1)$ and $\mathbf{B} \cdot \mathbf{B} = j_b(j_b + 1)$. Since $\mathbf{A} \cdot \mathbf{A} = \mathbf{B} \cdot \mathbf{B}$ (cf. Problem 14.15), $j_a = j_b$ and the set of states related by the shift operators is $(2j + 1)^2$ fold degenerate, where $2j + 1 = N = n + l + 1$.

States with good l and m quantum numbers can be constructed from these states using Clebsch-Gordon coefficients:

$$\left| \begin{array}{c} l \\ m \end{array} \right\rangle = \left| \begin{array}{cc} j/2 & j/2 \\ m_a & m_b \end{array} \right\rangle \left\langle \begin{array}{cc} j/2 & j/2 \\ m_a & m_b \end{array} \right| \left. \begin{array}{c} l \\ m \end{array} \right\rangle \quad (14.29)$$

The action of the Laplace–Runge–Lenz shift operators on these states, and the spherical harmonics, is determined in a straightforward way. For example, $M'_+ = A_+ - B_+ = a_1^\dagger a_2 - b_1^\dagger b_2$, so that

$$M'_+ Y^l_m = \langle \theta\phi | \left(\begin{vmatrix} j/2 & j/2 \\ m_a+1 & m_b \end{vmatrix} \begin{vmatrix} j/2 & j/2 \\ m_a & m_b \end{vmatrix} \begin{vmatrix} l \\ m \end{vmatrix} \right) \times \sqrt{(j/2 - m_a)(j/2 + m_a + 1)}$$
$$- \begin{vmatrix} j/2 & j/2 \\ m_a & m_b+1 \end{vmatrix} \begin{vmatrix} j/2 & j/2 \\ m_a & m_b \end{vmatrix} \begin{vmatrix} l \\ m \end{vmatrix} \times \sqrt{(j/2 - m_b)(j/2 + m_b + 1)} \right)$$
(14.30)

In general, the Laplace–Runge–Lenz operators shift the values of l and m by ± 1 or 0, while the angular momentum shift operators change only m by ± 1. However, for certain stretched values of the Clebsch–Gordon coefficients, the Laplace–Runge–Lenz vectors act more simply, for example (Burkhardt and Leventhal, 2004)

$$M'_z \left| N \begin{matrix} l \\ \pm l \end{matrix} \right\rangle = D_1 \left| N \begin{matrix} l+1 \\ \pm l \end{matrix} \right\rangle \qquad D_1 = \frac{1}{N}\sqrt{\frac{N^2 - (l+1)^2}{2l+3}}$$

$$M'_\pm \left| N \begin{matrix} l \\ \pm l \end{matrix} \right\rangle = \pm D_2 \left| N \begin{matrix} l+1 \\ \pm(l+1) \end{matrix} \right\rangle \qquad D_2 = \frac{1}{N}\sqrt{\frac{2l+2}{2l+3}} \left[N^2 - (l+1)^2 \right]$$
(14.31)

14.7 Relation with dynamics in four dimensions

The operators $\mathbf{L}$ and $\mathbf{M}'$ are infinitesimal generators for the orthogonal group $SO(4)$. The relation between motion in the presence of a Coulomb or gravitational potential and motion in four (mathematical) dimensions was clarified by Fock (1935). Motion of a particle in a $1/r$ potential is equivalent to motion of a free particle in the sphere $S^3 \subset R^4$.

It is useful first to establish an orthogonal coordinate system in R^3. It is natural to do this in terms of the constant physical vectors that are available. These include the vectors $\mathbf{L}$ and $\mathbf{M}$. Their cross product $\mathbf{W} = \mathbf{L} \times \mathbf{M}$ is orthogonal to both and also a constant of motion. These classical vectors obey:

$$\mathbf{L} = \mathbf{r} \times \mathbf{p} \qquad\qquad \mathbf{L} \cdot \mathbf{L} = L^2$$

$$\mathbf{M} = \frac{\mathbf{p} \times \mathbf{L}}{m} - K\frac{\mathbf{r}}{r} \qquad \mathbf{M} \cdot \mathbf{M} = M^2 = \frac{2E}{m}L^2 + K^2 \qquad (14.32)$$

$$\mathbf{W} = \frac{\mathbf{p}}{m}L^2 - K\frac{\mathbf{L} \times \mathbf{r}}{r} \qquad \mathbf{W} \cdot \mathbf{W} = L^2 M^2$$

The particle moves in a plane perpendicular to the angular momentum vector $\mathbf{L}$, since $\mathbf{r} \cdot \mathbf{L} = 0$. The momentum vector moves in the same plane, since $\mathbf{p} \cdot \mathbf{L} = 0$.

While **r** moves in an ellipse, the momentum vector moves on a circle. For simplicity we choose the z-axis in the direction of **L** and the x- and y-axes in the directions of **M** and **W**. In this coordinate system $p_z = 0$, $p_x = \mathbf{p} \cdot \mathbf{M}/\sqrt{\mathbf{M} \cdot \mathbf{M}}$ and $p_y = \mathbf{p} \cdot \mathbf{W}/\sqrt{\mathbf{W} \cdot \mathbf{W}}$. The two nonzero components of the momentum vector are not independent, but obey the constraint

$$p_x^2 + \left(p_y - \frac{mM}{L}\right)^2 = \left(\frac{mK}{L}\right)^2 \tag{14.33}$$

This is the equation of a circle in the plane containing the motion. As the particle moves in the plane of motion on an elliptical orbit with one focus at the source, its momentum moves in the same plane on a circular orbit (radius mK/L) with the center displaced from the origin by mM/L.

The circle in R^3 is lifted to a circle in $S^3 \subset R^4$ by a projective transformation. We extend coordinates from R^3 to R^4 as follows:

$$\begin{aligned}(x, y, z) \in R^3 &\to (w, x, y, z) \in R^4 \\ (p_x, p_y, p_z) \in R^3 &\to (p_w, p_x, p_y, p_z) \in R^4\end{aligned} \tag{14.34}$$

With $p_0 = \sqrt{-2E/m}$, define the unit vector $\hat{\mathbf{u}} \in S^3 \subset R^4$ by the projective transformation T:

$$\hat{\mathbf{u}} \stackrel{T}{=} \frac{\mathbf{p} \cdot \mathbf{p} - p_0^2}{\mathbf{p} \cdot \mathbf{p} + p_0^2} \hat{\mathbf{w}} + \frac{2p_0}{\mathbf{p} \cdot \mathbf{p} + p_0^2} \mathbf{p} \tag{14.35}$$

Here $\hat{\mathbf{w}}$ is a unit vector in R^4 that is orthogonal to all vectors in the physical space R^3. The transformation in Eq. (14.35) is a stereographic projection. It is invertible and preserves angles (conformal). It is a simple matter to check that $\hat{\mathbf{u}}$ is a unit vector. The circular trajectory in R^3 (Eq. (14.33)) lifts to a circle in S^3. Reversibly, circles in S^3 project down to circles in the physical R^3 space under the reverse transformation.

Rotations in $SO(4)$ rigidly rotate the sphere S^3 into itself. They rotate circles into circles, which then project down to circular momentum trajectories in the physical space R^3:

$$\text{circle in } R^3 \xrightarrow{T} \text{circle in } S^3 \xrightarrow{SO(4)} \text{circle in } S^3 \xrightarrow{T^{-1}} \text{circle in } R^3 \tag{14.36}$$

The subgroup $SO(3)$ of rotations around the $\hat{\mathbf{w}}$ axis acts only on the physical space R^3. In this subgroup, the subgroup $SO(2)$ of rotations around the **L** axis leaves **L** fixed and simply rotates **M** in the plane of motion. The coset representatives $SO(3)/SO(2)$ act to reorient the plane of motion by rotating the angular momentum vector **L** while keeping the magnitude of **M** fixed. Rotations in the coset $SO(4)/SO(3)$ act to change the lengths of both **L** and **M**. All group operations in $SO(4)$ keep p_0 fixed. In this way the group $SO(4)$ maps states with principal quantum number N into (linear combinations of) states with the same principal quantum number N. In short, $SO(4)$ acts on the bound hydrogen atom states through unitary irreducible representations of dimension $N^2 = (n+l+1)^2$.

14.8 DeSitter symmetry $SO(4, 1)$

The dynamical symmetry group $SO(4)$ that rotates bound states to bound states does not change their energy; the dynamical symmetry group $SO(3, 1)$ that rotates scattering states to scattering states does not change their energy either. It would be nice to find a set of transformations that rescales the energy. If such a group could be found, it would be possible, for example, to map the $1s$ ground state into any other bound state. Such a group exists: it is the deSitter group $SO(4, 1)$ (Malkin and Man'ko, 1965; Ogievetskii and Polubarinov, 1960).

That such a group might exist is strongly suggested by the appearance of the hydrogen atom spectrum, as replotted in Fig. 14.2. In this figure we have multiplied each energy eigenvalue by $-N^3$, where N is the principal quantum number. The rescaled energies have been plotted as a function of N (vertically) and orbital angular momentum quantum number l (horizontally). In this format, the eigenvalue spectrum bears a strong resemblance to the spectrum of states that supports finite-dimensional representations of $\mathfrak{su}(2)$ (Fig. 6.1) and the infinite-dimensional representations of $\mathfrak{su}(1, 1)$ (Fig. 11.2).

We begin with a group that preserves inner products in some N-dimensional linear vector space: $\mathbf{x}' = M\mathbf{x}$, with M a transformation in the group and the inner product defined by $(\mathbf{x}, \mathbf{x})_N = \mathbf{x}^t g \mathbf{x} = x_i g_{ij} x_j$. As always, the metric-preserving condition leads to $M^t G M = G$.

It is useful to define a new N-vector $\mathbf{y}$ as a scaled version of the original vector: $\mathbf{y} = \lambda \mathbf{x}$. We introduce two additional coordinates by defining $z_1 = \lambda$ and $z_2 = \lambda(\mathbf{x}, \mathbf{x})_N$. With these definitions we find the conformal condition

$$(\mathbf{y}, \mathbf{y})_N - z_1 z_2 = (\lambda \mathbf{x}, \lambda \mathbf{x})_N - \lambda [\lambda(\mathbf{x}, \mathbf{x})_N] = 0 \tag{14.37}$$

The conformal condition defines an inner product in the $N+2$ dimensional linear vector space that is nondiagonal in the coordinates $\mathbf{y}, z_1, z_2$ but diagonal in the coordinates $\mathbf{y}, y_{N+1}, y_{N+2}$, with $y_{N+1} = \frac{1}{2}(z_1 + z_2)$ and $y_{N+2} = \frac{1}{2}(z_1 - z_2)$:

$$\begin{bmatrix} G & & \mathbf{y} \\ \hline & -\frac{1}{2} & z_1 \\ -\frac{1}{2} & & z_2 \end{bmatrix} \quad \begin{bmatrix} G & & \mathbf{y} \\ \hline & -1 & y_{N+1} \\ & +1 & y_{N+2} \end{bmatrix} \tag{14.38}$$

The conformal condition Eq. (14.37) defines a cone in the enlarged $N+2$ dimensional space. If the group that preserves the metric G in R^N is $SO(p, q)$, the group

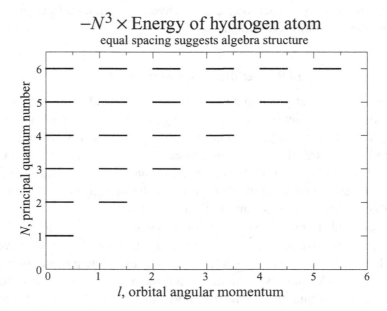

Figure 14.2. Nonrelativistic spectrum of the hydrogen atom, replotted to emphasize the possibility of a dynamical algebra.

that preserves the metric in R^{N+2} is $SO(p+1, q+1)$. We seek to construct a spherical or hyperbolic slice of this cone.

The connection with the Kepler problem is made as follows. The momenta **p** are lifted to the coordinates on a sphere $S^3 \subset R^4$ ($E < 0$) or a two-sheeted hyperboloid $H^3 \subset R^4$ ($E > 0$) by the following projective transformations:

$$\hat{\mathbf{u}} = \frac{\frac{1}{2}(p_0^2 - \mathbf{p} \cdot \mathbf{p})}{\frac{1}{2}(p_0^2 + \mathbf{p} \cdot \mathbf{p})} \mathbf{w} + \frac{p_0 \mathbf{p}}{\frac{1}{2}(p_0^2 + \mathbf{p} \cdot \mathbf{p})} \quad E < 0$$

$$\hat{\mathbf{u}} = \frac{\frac{1}{2}(p_0^2 + \mathbf{p} \cdot \mathbf{p})}{\frac{1}{2}(p_0^2 - \mathbf{p} \cdot \mathbf{p})} \mathbf{w} + \frac{p_0 \mathbf{p}}{\frac{1}{2}(p_0^2 - \mathbf{p} \cdot \mathbf{p})} \quad E > 0$$

(14.39)

For the four-vectors **u** the metric G that appears in Eq. (14.38) is determined from the denominators in Eq. (14.39):

$$\mathbf{u}^t G \mathbf{u} = u_0^2 \pm \sum_{i=1}^{3} u_i^2 \qquad \begin{array}{l} + \text{ for } E < 0 \\ - \text{ for } E > 0 \end{array} \tag{14.40}$$

The algebraic surfaces on which the projective vector **u** lies is defined by the condition $\mathbf{u}^t G \mathbf{u} = 1$.

14.8 DeSitter symmetry $SO(4,1)$

The connection with the conformal transformations introduced above is as follows. The group that leaves invariant the conformal metric $\text{diag}(1, \pm I_3, -1, +1)$ is $SO(5, 1)$ for $E < 0$ and $SO(2, 4)$ for $E > 0$. On the surfaces (sphere, hyperboloid) the condition $\mathbf{u}^t G \mathbf{u} = 1$ is satisfied, so that $z_1 = z_2$, $y_4 = \lambda$ and $y_5 = 0$ (the six coordinates are labeled $(y_0, \mathbf{y} = \lambda \mathbf{u}, y_4 = \frac{1}{2}(z_1 + z_2), y_5 = \frac{1}{2}(z_1 - z_2))$. Transformations that map the algebraic surface to itself must map $y_5 = 0$ to $y_5 = 0$. It is a simple matter to verify that this is the matrix subgroup of the 6×6 matrix group $SO(5, 1)$ or $SO(2, 4)$ of the form $\begin{bmatrix} M & 0 \\ 0 & 1 \end{bmatrix}$, with M a 5×5 matrix that preserves the metric $\text{diag}(1, \pm I_3, -1)$ in R^5. This is $SO(4, 1)$ for $E < 0$ and $SO(1, 4)$ for $E > 0$.

It remains to show that this group maps these algebraic surfaces into themselves. To this end we write the linear transformation in R^5 as follows

$$\begin{bmatrix} \lambda \mathbf{u} \\ \lambda \end{bmatrix}' = \begin{bmatrix} A & B \\ C & D \end{bmatrix} \begin{bmatrix} \lambda \mathbf{u} \\ \lambda \end{bmatrix} \tag{14.41}$$

where A is a 4×4 matrix, etc. From this we determine

$$\mathbf{u}' = \frac{A(\lambda \mathbf{u}) + B\lambda}{C(\lambda \mathbf{u}) + D\lambda} \tag{14.42}$$

The inner product of $\mathbf{u}'$ with itself satisfies

$$(\mathbf{u}')^t G \mathbf{u}' - 1 = \frac{(A\lambda \mathbf{u} + B\lambda)^t G (A\lambda \mathbf{u} + B\lambda) - (C\lambda \mathbf{u} + D\lambda)^t (C\lambda \mathbf{u} + D\lambda)}{(C\lambda \mathbf{u} + D\lambda)^t (C\lambda \mathbf{u} + D\lambda)} \tag{14.43}$$

By using the relations among the submatrices required by the metric preserving condition (e.g., $A^t G A - C^t C = G$, etc.) it is a simple matter to show that this reduces to

$$(\mathbf{u}', \mathbf{u}')_N - 1 = \frac{(\mathbf{u}, \mathbf{u})_N - 1}{(C\mathbf{u} + D)^t (C\mathbf{u} + D)} \tag{14.44}$$

In short, the algebraic surface is invariant under this transformation group.

Remark The subgroup $SO(4)$ rigidly rotates the sphere $S^3 \subset R^4$ into itself while the subgroup $SO(3, 1)$ "rigidly rotates" the hyperboloid into itself. In the latter case this is less intuitive. This means that the coordinates of the hyperboloid are mapped into themselves by a *linear* transformation in R^4. The group $SO(4, 1)$ maps coordinates in these spaces to themselves through a *nonlinear* transformation in R^4: in this case a simple projective transformation. It is a linear transformation in R^5.

The infinitesimal generators of this nonlinear transformation are constructed as follows (Bander and Itzykson, 1966a, 1966b). For $E < 0$ introduce a four-vector u as usual ($u_0 \to u_4$)

$$\begin{aligned} \mathbf{u} &= 2 p_4 (\mathbf{p} \cdot \mathbf{p} + p_4^2)^{-1} \mathbf{p} \\ u_4 &= (\mathbf{p} \cdot \mathbf{p} - p_4^2)(\mathbf{p} \cdot \mathbf{p} + p_4^2)^{-1} \end{aligned} \tag{14.45}$$

Define the four-vector B in terms of the four-vector u and the angular momentum vector **L** and the scaled (by $1/\sqrt{2m|E|}$) Runge–Lenz vector **M**' as follows:

$$\mathbf{B} = \mathbf{M}'u_4 + \mathbf{L} \times \mathbf{u} - \tfrac{3}{2}i\mathbf{u} = \tfrac{i}{2}\left[\mathbf{u}, \mathbf{L}^2 + \mathbf{M}'^2\right]$$

$$B_4 = \mathbf{M}' \cdot \mathbf{u} + \tfrac{3}{2}iu \qquad = \tfrac{i}{2}\left[u_4, \mathbf{L}^2 + \mathbf{M}'^2\right]$$

The operators L_i, M'_i, and B_μ are the infinitesimal generators of $SO(4, 1)$ as follows, for $E < 0$.

$$\begin{bmatrix} 0 & L_3 & -L_2 & M_1 & B_1 \\ -L_3 & 0 & L_1 & M_2 & B_2 \\ L_2 & -L_1 & 0 & M_3 & B_3 \\ -M_1 & -M_2 & -M_3 & 0 & B_4 \\ B_1 & B_2 & B_3 & B_4 & 0 \end{bmatrix} \begin{matrix} + \\ + \\ + \\ + \\ - \end{matrix}$$

14.9 Conformal symmetry $SO(4, 2)$

The largest set of transformations that leave the states of the hydrogen atom invariant, in some sense, is the conformal group $SO(4, 2)$. Several different ways have been developed to prove this point. We review three here.

14.9.1 Schwinger representation

The algebra of the dynamical symmetry group has infinitesimal generators **L** and **M'**. Their linear combinations given two sets of vector operators **A** and **B** that mutually commute and have angular momentum commutation relations on bound states. It is possible to represent these operators using the boson representation. That is, for the operators **A** we introduce annihilation and creation operators a_i, $a_j^\dagger$ for two independent modes, and similarly we introduce operators b_i, $b_j^\dagger$ to describe **B**. Basis states on which these operators act have the form $|m_1, m_2; n_1, n_2\rangle$ where, for example

$$a_1^\dagger a_2 |m_1, m_2; n_1, n_2\rangle = |m_1 + 1, m_2 - 1; n_1 n_2\rangle \sqrt{m_1 + 1}\sqrt{m_2}$$

$$b_1^\dagger b_1 |m_1, m_2; n_1, n_2\rangle = |m_1, m_2; n_1 n_2\rangle (\sqrt{n_1})^2$$

The orthogonality of **L** and **M** leads to the orthogonality of **A** and **B**, and this leads directly to the condition $j_a = j_b$, where $j_a = \tfrac{1}{2}(m_1 + m_2)$ and $j_b = \tfrac{1}{2}(n_1 + n_2)$.

From the previous section we know there is a group that maps bound states into (linear combinations of) bound states. We determine an algebra of operators that performs the same function on bound states as follows. Operators that change the

14.9 Conformal symmetry $SO(4,2)$

principal quantum number $N = 2j_a + 1 = 2j_b + 1 = (j_a + j_b) + 1$ must change $j_a = j_b$. Operators that change j_a have the form $a_i^\dagger$ or $a_i^\dagger a_j^\dagger$, but they do not simultaneously change j_b. Only operators that simultaneously add or subtract one excitation to the subsystems A and B simultaneously maintain the constraint $j_a = j_b$. The largest set of operators bilinear in the boson operators that map hydrogen atom bound states to bound states consists of the operators

$$
\begin{array}{lcccc}
\text{operators} & a_i^\dagger a_j & b_i^\dagger b_j & a_i^\dagger b_j^\dagger & a_i b_j \\
\text{subalgebra} & u(2) & u(2) & & \\
\text{number} & 4 & 4 & 4 & 4
\end{array}
\tag{14.46}
$$

What is this algebra? Among these 16 operators, the maximal number of mutually commuting operators that can be found is four. These are conveniently chosen as the number operators for the four boson modes: $(H_1, H_2, H_3, H_4) = (a_1^\dagger a_1, a_2^\dagger a_2, b_1^\dagger b_1, b_2^\dagger b_2)$. The remaining twelve operators have eigenoperator commutation relations with this set:

$$
\begin{array}{llll}
a_1^\dagger a_2 \ (+1,-1,0,0) & a_1^\dagger b_1^\dagger \ (+1,0,+1,0) & a_1 b_1 \ (-1,0,-1,0) & \\
a_2^\dagger a_1 \ (-1,+1,0,0) & a_1^\dagger b_2^\dagger \ (+1,0,0,+1) & a_1 b_2 \ (-1,0,0,-1) & \\
b_1^\dagger b_2 \ (0,0+1,-1) & a_2^\dagger b_1^\dagger \ (0,+1,+1,0) & a_2 b_1 \ (0,-1,-1,0) & (14.47) \\
b_2^\dagger b_1 \ (0,0,-1,+1) & a_2^\dagger b_2^\dagger \ (0,+1,0,+1) & a_2 b_2 \ (0,-1,0,-1) &
\end{array}
$$

All these roots have equal length, and inner products among these roots are all $\pm\frac{1}{2}$ or 0. The operator

$$(a_1^\dagger a_1 + a_2^\dagger a_2) - (b_1^\dagger b_1 + b_2^\dagger b_2)$$

commutes with all operators in this set. It is a constant of motion, and in fact vanishes on all hydrogen atom bound states. As a result the algebra is the direct sum of an abelian invariant subalgebra spanned by this operator, and a rank-three simple Lie algebra, all of whose roots have equal lengths and are either orthogonal or make angles of $\pi/4$ or $3\pi/4$ radians with each other. The algebra is uniquely a real form of $A_3 = D_3$.

Which real form? It is possible to form a number of subalgebras of type A_1 from these operators:

$$
\begin{array}{lll}
a_1^\dagger a_2 \quad a_2^\dagger a_1 & \frac{1}{2}(a_1^\dagger a_1 - a_2^\dagger a_2) & su(2) \\
b_1^\dagger b_2 \quad b_2^\dagger b_1 & \frac{1}{2}(b_1^\dagger b_1 - b_2^\dagger b_2) & su(2) \\
a_i^\dagger b_j^\dagger \quad a_i b_j & \frac{1}{2}(a_i^\dagger a_i + b_j^\dagger b_j + 1) & su(1,1)
\end{array}
$$

The first two are compact, the last four are not compact. The maximal compact subalgebra is spanned by the two compact subalgebras together with the diagonal operator $a_1^\dagger a_1 + a_2^\dagger a_2 + b_1^\dagger b_1 + b_2^\dagger b_2$. This is the algebra $\mathfrak{so}(4) + \mathfrak{so}(2)$. The fifteen-dimensional Lie algebra that maps bound states to bound states is therefore $\mathfrak{so}(4, 2) = \mathfrak{su}(2, 2)$. This is the conformal algebra.

14.9.2 Dynamical mappings

Although the classical Kepler problem is analytically solvable, analyticity disappears under perturbation. In this case classical orbits must be computed numerically. At points of very close approach the velocity of the particles increases greatly, so it is prudent to slow down the integration time step to preserve accuracy. This procedure has been implemented formally through a canonical transformation (Kustaanheimo and Stiefel, 1965; Stiefel and Scheifele, 1971), and is now widely known as the Kustaanheimo–Stiefel transformation. Under this transformation time is stretched out when the distance R between the interacting particles becomes small. In addition the (relative) coordinates are projected from R^3 to a fictitious space R^4. Under this transformation, and a constraint, the Kepler hamiltonian is transformed into a four-dimensional harmonic oscillator hamiltonian.

Coordinates (q_1, q_2, q_3, q_4) in the fictitious space R^4 are related to coordinates (Q_1, Q_2, Q_3) in the real space by the 4×4 transformation

$$\begin{bmatrix} Q_1 \\ Q_2 \\ Q_3 \\ Q_4 \end{bmatrix} = M_{KS} \begin{bmatrix} q_1 \\ q_2 \\ q_3 \\ q_4 \end{bmatrix} = \begin{bmatrix} q_1 & -q_2 & -q_3 & q_4 \\ q_2 & q_1 & -q_4 & -q_3 \\ q_3 & q_4 & q_1 & q_2 \\ q_4 & -q_3 & q_2 & -q_1 \end{bmatrix} \begin{bmatrix} q_1 \\ q_2 \\ q_3 \\ q_4 \end{bmatrix} \qquad (14.48)$$

The transformation is constructed so that the "fourth" real coordinate Q_4 is identically zero. This transformation is invertible provided $q_1^2 + q_2^2 + q_3^2 + q_4^2 \neq 0$. The distance $R = \sqrt{Q_1^2 + Q_2^2 + Q_3^2}$ in R^3 and the distance $q = \sqrt{q_1^2 + q_2^2 + q_3^2 + q_4^2}$ in R^4 are related by $R = q^2$.

The other half of the canonical transformation, involving the momenta in the real and fictitious spaces, is

$$(P_1, P_2, P_3, P_4)^t = \frac{1}{2R} M_{KS} (p_1, p_2, p_3, p_4)^t$$

A constraint condition must be applied to force $P_4 = 0$. This condition is

$$\zeta = -2R P_4 = (q_1 p_4 - q_4 p_1) + (q_3 p_2 - q_2 p_3) = 0 \qquad (14.49)$$

With this constraint we find $P^2 = P_1^2 + P_2^2 + P_3^2 = (1/4R p^2) - (\zeta^2 / 4R^2) \to (1/4R)(p_1^2 + p_2^2 + p_3^2 + p_4^2)$. With these transformations the hamiltonian in the

14.9 Conformal symmetry $SO(4, 2)$

real space can be transformed to a hamiltonian in the fictitious space by

$$\frac{P^2}{2m} - \frac{e^2}{R} = E \xrightarrow{\times R} \frac{RP^2}{2m} - e^2 = ER \xrightarrow{KS} \frac{p^2}{8m} - e^2 = Eq^2 \quad (14.50)$$

This is the hamiltonian for a four-dimensional harmonic oscillator when $E < 0$, as easily seen by rearranging the terms

$$\frac{p^2}{2m} - 4Eq^2 = 4e^2 \quad (14.51)$$

The angular momentum operators in the real and fictitious spaces are bilinear products of the position and momentum coordinates, as follows:

$$(Q_1, Q_2, Q_3, Q_4) \begin{bmatrix} 0 & \theta_3 & -\theta_2 & * \\ -\theta_3 & 0 & \theta_1 & * \\ \theta_2 & -\theta_1 & 0 & * \\ -* & -* & -* & 0 \end{bmatrix} \begin{bmatrix} P_1 \\ P_2 \\ P_3 \\ P_4 \end{bmatrix}$$

$$\frac{1}{2}(q_1, q_2, q_3, q_4) \begin{bmatrix} 0 & \theta_3 & -\theta_2 & \theta_1 \\ -\theta_3 & 0 & \theta_1 & \theta_2 \\ \theta_2 & -\theta_1 & 0 & \theta_3 \\ -\theta_1 & -\theta_2 & -\theta_3 & 0 \end{bmatrix} \begin{bmatrix} p_1 \\ p_2 \\ p_3 \\ p_4 \end{bmatrix} \quad (14.52)$$

Similar expressions can be given for the Runge–Lenz vector. However, these are quadratic in the position and momentum operators. As a result they must be expressed in matrix form using 8×8 matrices acting on the vector $(q_1, q_2, q_3, q_4; p_1, p_2, p_3, p_4)$ on the left and its transpose on the right (Sadovskií and Žhilinskií, 1998).

We now ask: what is the largest group of transformations on the coordinates and momenta that

(i) is linear,
(ii) is canonical, and
(iii) preserves $\zeta = 0$.

We address this question in the usual way. Linear transformations allow us to use matrices. These are 8×8 matrices acting on the four coordinates and four momenta. Preserving the Poisson brackets requires that the matrices satisfy a symplectic metric-preserving condition: $M^t G_1 M = G_1$. Preserving the condition $\zeta = 0$ requires these transformations to satisfy another metric-preserving condition: $M^t G_2 M = G_2$.

The matrices G_i have the form

$$G_i = \begin{bmatrix} 0 & M_i \\ -M_i & 0 \end{bmatrix}$$

where

$$M_1 = \begin{bmatrix} 1 & 0 & 0 & 0 \\ 0 & 1 & 0 & 0 \\ 0 & 0 & 1 & 0 \\ 0 & 0 & 0 & 1 \end{bmatrix} \quad M_2 = \begin{bmatrix} 0 & 0 & 0 & 1 \\ 0 & 0 & -1 & 0 \\ 0 & 1 & 0 & 0 \\ -1 & 0 & 0 & 0 \end{bmatrix} \quad (14.53)$$

$$M_1^t = +M_1 \quad G_1^t = -G_1 \quad M_2^t = -M_2 \quad G_2^t = +G_2$$

The metric G_1 is antisymmetric and the metric G_2 is symmetric, with signature $(+4, -4)$. The group that preserves the antisymmetric metric is $Sp(8; \mathbb{R})$ and the group that preserves the symmetric metric is $SO(4, 4)$. The group that satisfies both metric-preserving conditions is their intersection:

$$Sp(8; \mathbb{R}) \cap SO(4, 4) = SU(2, 2) \simeq SO(4, 2) \quad (14.54)$$

The simplest way to see this result is to perform a canonical transformation from coordinates (q, p) to coordinates (s, r):

$$\begin{bmatrix} s_1 \\ r_4 \end{bmatrix} = \tfrac{1}{\sqrt{2}} \begin{bmatrix} 1 & 1 \\ -1 & 1 \end{bmatrix} \begin{bmatrix} q_1 \\ p_4 \end{bmatrix} \quad \begin{bmatrix} s_2 \\ r_3 \end{bmatrix} = \tfrac{1}{\sqrt{2}} \begin{bmatrix} -1 & 1 \\ -1 & -1 \end{bmatrix} \begin{bmatrix} q_2 \\ p_3 \end{bmatrix}$$

$$\begin{bmatrix} s_3 \\ r_2 \end{bmatrix} = \tfrac{1}{\sqrt{2}} \begin{bmatrix} 1 & 1 \\ -1 & 1 \end{bmatrix} \begin{bmatrix} q_3 \\ p_2 \end{bmatrix} \quad \begin{bmatrix} s_4 \\ r_1 \end{bmatrix} = \tfrac{1}{\sqrt{2}} \begin{bmatrix} -1 & 1 \\ -1 & -1 \end{bmatrix} \begin{bmatrix} q_4 \\ p_1 \end{bmatrix} \quad (14.55)$$

Since the new coordinates are already canonical, only the condition $\zeta = 0$ remains to be satisfied. It is a simple matter to verify that

$$z_1 = \tfrac{1}{\sqrt{2}}(s_1 + is_2) \quad z_2 = \tfrac{1}{\sqrt{2}}(r_1 + ir_2)$$

$$z_3 = \tfrac{1}{\sqrt{2}}(s_3 + is_4) \quad z_4 = \tfrac{1}{\sqrt{2}}(r_3 + ir_4)$$

$$z_1^* z_1 - z_2^* z_2 + z_3^* z_3 - z_4^* z_4 = \zeta$$

$$(14.56)$$

The noncompact group $U(2, 2)$ preserves the constraint Eq. (14.49).

14.9.3 Lie algebra of physical operators

A number of workers have shown that the hamiltonian describing the interaction of a charged particle interacting with an external Coulomb field ($V(r) = -e^2/r$) can be expressed in terms of operators that close under commutation. The Lie algebra that these operators span is isomorphic with the Lie algebra of a noncompact orthogonal group.

14.10 Spin angular momentum

Three vector operators and a scalar operator

$$\mathbf{J} = \mathbf{r} \times \mathbf{p} \qquad \text{angular momentum}$$

$$\mathbf{M} = \frac{1}{2m}(\mathbf{p} \times \mathbf{L} - \mathbf{L} \times \mathbf{p}) - K\frac{\mathbf{r}}{r} \qquad \text{Laplace–Runge–Lenz vector}$$

$$\mathbf{A} = \frac{1}{2m}(\mathbf{p} \times \mathbf{L} - \mathbf{L} \times \mathbf{p}) + K\frac{\mathbf{r}}{r} \qquad \text{dual vector} \qquad (14.57)$$

$$A_4 = \mathbf{r} \cdot \mathbf{p} + \frac{3}{2}\frac{\hbar}{i} \qquad \text{dual scalar}$$

close under commutation to span a Lie algebra that is isomorphic with $\mathfrak{so}(4, 1)$.

Five additional operators can be introduced that extend the algebra to $\mathfrak{so}(4, 2)$. These include one vector operator and two additional operators:

$$\Gamma_i = r p_i$$
$$\Gamma_4 = \tfrac{1}{2}(r\mathbf{p} \cdot \mathbf{p} - r) \qquad (14.58)$$
$$\Gamma_5 = \tfrac{1}{2}(r\mathbf{p} \cdot \mathbf{p} + r)$$

The commutation relations that these 15 operators satisfy are summarized by the 6×6 matrix

$$\left[\begin{array}{ccc|ccc}
0 & J_3 & -J_2 & M_1 & A_1 & \Gamma_1 \\
-J_3 & 0 & J_1 & M_2 & A_2 & \Gamma_2 \\
J_2 & -J_1 & 0 & M_3 & A_3 & \Gamma_3 \\
\hline
-M_1 & -M_2 & -M_3 & 0 & A_4 & \Gamma_4 \\
A_1 & A_2 & A_3 & A_4 & 0 & \Gamma_5 \\
\Gamma_1 & \Gamma_2 & \Gamma_3 & \Gamma_4 & -\Gamma_5 & 0
\end{array}\right] \begin{array}{c} + \\ + \\ + \\ + \\ - \\ - \end{array}$$

The four triplets J_i, M_i, A_i, Γ_i ($i = 1, 2, 3$) have transformation properties of three-vectors under rotations. The three additional operators A_4, Γ_4, Γ_5 close under commutation and span a Lie algebra that is isomorphic with $\mathfrak{so}(2, 1)$.

The Schrödinger and Klein–Gordon hamiltonians for an electron of charge $-e$ in the Coulomb field $\Phi(r) = e/r$ of a proton can be expressed in terms of operators of type A_4, Γ_4, and Γ_5. These operators are displayed in Table 14.3, along with the hamiltonians and the algebraic representation of the wave equations.

14.10 Spin angular momentum

The interaction of the electron with the electromagnetic field is properly described by the Dirac equation. The electromagnetic field $(\mathbf{E}, \mathbf{B})$ is described by the four-vector potential $A_\mu = (\phi, \mathbf{A})$. The electron has charge $q = -e$ (where e is the charge on the proton) and spin $\tfrac{1}{2}$. The Dirac equation $H_D \psi = E\psi$ is a matrix

Table 14.3. *Nonrelativistic and relativistic hamiltonians for a spinless particle, operator representation of the operators A_4, Γ_4, and Γ_5, expression of the hamiltonians and wave equations in terms of these operators, and explicit values of the coefficients in these equations*

H	$\dfrac{p^2}{2m} - \dfrac{\alpha}{r}$	$\sqrt{p^2 + m^2} - \dfrac{\alpha}{r}$
A_4	$\mathbf{r}\cdot\mathbf{p} - i$	$\mathbf{r}\cdot\mathbf{p} - i$
Γ_4	$\tfrac{1}{2}(r\mathbf{p}\cdot\mathbf{p} - r)$	$\dfrac{1}{2}\left(r\mathbf{p}\cdot\mathbf{p} - r - \dfrac{\alpha^2}{r}\right)$
Γ_5	$\tfrac{1}{2}(r\mathbf{p}\cdot\mathbf{p} + r)$	$\dfrac{1}{2}\left(r\mathbf{p}\cdot\mathbf{p} + r - \dfrac{\alpha^2}{r}\right)$
Θ	$r(H_S - W)$	$r\left\{\left(H_{KG} + \dfrac{\alpha}{r}\right)^2 - \left(E + \dfrac{\alpha}{r}\right)^2\right\}$
	$A(\Gamma_5 + \Gamma_4) + B(\Gamma_5 - \Gamma_4) + C$	$A(\Gamma_5 + \Gamma_4) + B(\Gamma_5 - \Gamma_4) + C$
A	$1/2m$	1
B	$-W$	$m^2 - E^2$
C	$-\alpha$	$-2\alpha E$

In the event a magnetic field **B** is present, the momentum operators **p** should be replaced by $\pi = \mathbf{p} - \tfrac{q}{c}\mathbf{A}$. Under this condition the operators still close under commutation.

differential equation of first order:

$$H_D = -e\phi(r) + \beta mc^2 + \gamma \cdot (c\mathbf{p} + e\mathbf{A}) \tag{14.59}$$

The 4×4 matrices β and γ_i can be chosen as

$$\beta = \begin{bmatrix} I_2 & 0 \\ 0 & -I_2 \end{bmatrix} \qquad \gamma_i = \begin{bmatrix} 0 & \sigma_i \\ \sigma_i & 0 \end{bmatrix} \tag{14.60}$$

Here σ_i are the standard Pauli 2×2 spin matrices (cf., Eq. (3.39), Problem 3.1).

The fifteen-dimensional Lie algebra for the Dirac equation is spanned by the operators **J**, **M**, **A**, **Γ** as given in Eq. (14.57), and the three operators A_4, Γ_4, Γ_5. The latter two are modified to allow a treatment of the Dirac operator along the same lines as the treatment of the Schrödinger and Klein–Gordon operators given in Section 14.9.3. We define operators

$$\begin{aligned} M_4 &= \mathbf{r}\cdot\mathbf{p} - i \\ \Gamma_4 &= \frac{1}{2}\left\{\left(r\mathbf{p}\cdot\mathbf{p} - r - \frac{\alpha^2}{r} - \frac{i\alpha\gamma\cdot\mathbf{r}}{r^2}\right)\right\} \\ \Gamma_5 &= \frac{1}{2}\left\{\left(r\mathbf{p}\cdot\mathbf{p} + r - \frac{\alpha^2}{r} - \frac{i\alpha\gamma\cdot\mathbf{r}}{r^2}\right)\right\} \end{aligned} \tag{14.61}$$

As before, the substitution $\mathbf{p} \to \pi = \mathbf{p} - \frac{q}{c}\mathbf{A}$ is in order in the event there is a nonzero magnetic field $\mathbf{B}$. These operators close under commutation to form an $\mathfrak{so}(2,1)$ Lie algebra. These operators also close under commutation with the four three-vectors J_i, M_i, A_i, Γ_i defined in Table 14.3. The Dirac hamiltonian is expressed in terms of these generators as follows:

$$\Theta = r\left\{\left(H_D + \frac{\alpha}{r}\right)^2 - \left(E + \frac{\alpha}{r}\right)^2\right\}$$
$$= A(\Gamma_5 + \Gamma_4) + B(\Gamma_5 - \Gamma_4) + C \qquad (14.62)$$

where the coefficients A, B, C have exactly the same values as for the Klein–Gordon operator (see Table 14.3). In short, the operators Γ_4, Γ_5 are modified but the relation among these operators in the algebraic representation of the relativistic wave equations is not.

14.11 Spectrum generating group

The physics of the hydrogenic problem is determined primarily by the radial equation Eq. (14.9). It is possible to determine solutions of this equation using operators that close under commutation. These are the generators of a Lie algebra. The corresponding group is called a spectrum generating group.

To construct a set of operators that close under commutation, we first simplify the radial equation by multiplying on the left by r

$$\left(rD^2 + \frac{A}{r} + B + Cr\right)R(r) = 0 \qquad (14.63)$$

with $D = d/dr$. The operators r and D behave under commutation like the boson creation and annihilation operators $a^\dagger$ and a. In fact, the nonzero commutation relations are

$$[rD, r] = +r \qquad [a^\dagger a, a^\dagger] = +a^\dagger$$
$$[rD, rD^2] = -rD^2 \quad [a^\dagger a, a^\dagger aa] = -a^\dagger aa \qquad (14.64)$$
$$[r, rD^2] = -2rD \quad [a^\dagger, a^\dagger aa] = -2a^\dagger a$$

The linear combinations $rD^2 + r$ and $rD^2 - r$ are compact and noncompact, respectively. In order to model the differential operator Eq. (14.63) with a set of operators that close under commutation to form a finite-dimensional Lie algebra, we must be careful, as

$$\left[rD, \frac{1}{r}\right] = -\frac{1}{r}$$

$$\left[rD^2, \frac{1}{r}\right] = \frac{2}{r^2} - \frac{1}{r} D$$

We choose as operators in the Lie algebra $\mathfrak{so}(2, 1)$ the three differential operators

$$\Gamma_5 = \frac{1}{2}\left(rD^2 + \frac{a}{r} - r\right)$$

$$\Gamma_4 = \frac{1}{2}\left(rD^2 + \frac{a}{r} + r\right) \tag{14.65}$$

$$M_4 = rD$$

The Casimir operator for this algebra is $C^2 = \Gamma_5^2 - \Gamma_4^2 - M_4^2 = -a$. The representations of this algebra have been described in Problem 11.6.

The radial equation Eq. (14.63) is expressed in terms of the three operators as follows ($a \to A$)

$$((\Gamma_5 + \Gamma_4) + B + C(\Gamma_4 - \Gamma_5)) R(r) = 0 \tag{14.66}$$

Next, we rotate the generators of the algebra according to

$$e^{\theta M_4}\begin{pmatrix}\Gamma_5 \\ \Gamma_4\end{pmatrix}e^{-\theta M_4} = \begin{bmatrix}\cosh\theta & -\sinh\theta \\ -\sinh\theta & \cosh\theta\end{bmatrix}\begin{pmatrix}\Gamma_5 \\ \Gamma_4\end{pmatrix} \tag{14.67}$$

When this similarity transformation is applied to Eq. (14.66) we obtain the following result:

$$\left[(e^{-\theta} - C e^{\theta})\Gamma_5 + (e^{-\theta} + C e^{\theta})\Gamma_4 + B\right] e^{\theta M_4} R(r) = 0 \tag{14.68}$$

The rotation angle θ can be chosen to eliminate either the noncompact generator Γ_4 or the compact generator Γ_5, depending on the sign of the parameter C.

14.11.1 Bound states

If $C < 0$ we can choose $e^{-\theta} + C e^{\theta} = 0$, so that the resulting equation becomes

$$\left(2\sqrt{-C}\, \Gamma_5 + B\right) u(r) = 0 \tag{14.69}$$

where $u(r) = e^{\theta M_4} R(r)$. If A is the Casimir invariant of this representation of $\mathfrak{su}(1, 1)$, the discrete spectrum of the compact operator Γ_5 is $N = -\frac{1}{2} + \sqrt{(\frac{1}{2})^2 - A + 1} + n$, $n = 0, 1, 2, \ldots$. This result leads directly to the eigenvalue

14.11 Spectrum generating group

spectrum for the nonrelativistic and the relativistic hydrogen atom (no spin) obtained in Eq. (14.12).

Remark The spectrum generating algebra Eq. (14.65) acts in Hilbert spaces that carry unitary irreducible representations of the noncompact group $SO(2, 1)$. These representations are indexed by an integer l that has an interpretation as angular momentum. The energy spectrum that we have computed has the behavior (in the nonrelativistic case) $W = -\frac{1}{2}mc^2\alpha^2(1/N^2)$, where $N = l + 1 + k$, $k = 0, 1, 2, \ldots$. Here N is the principal quantum number. The result is that this algebra acts to change the principal quantum number while keeping l constant. Since the three operators in the spectrum generating algebra commute with the angular momentum operators, the quantum number m_l (eigenvalue of L_z) is also invariant under the action of these operators. The states connected by the operators of this $\mathfrak{so}(2, 1)$ algebra are $|N, lm\rangle \leftrightarrow |N \pm 1, lm\rangle$. The states on which these operators act are organized in "angular momentum towers." These states are organized vertically in Fig. 14.2.

Remark The angular momentum operators L_z, $L_\pm$ act on multiplets shown as a single horizontal line in Figs. 14.1 and 14.2. The operators M_z, $M_\pm$ associated with the Laplace–Runge–Lenz vector act horizontally on the levels shown in these two figures. The operators Γ_z, $\Gamma_\pm = \Gamma_4 \pm iM_4$ act vertically on the levels shown in these figures. Since $[\mathbf{L}, \mathbf{\Gamma}] = 0$, the operators Γ do not change the m values of hydrogenic states.

Remark The shift down operator Γ_- annihilates the ground state in a given angular momentum tower: $\Gamma_- \langle r|N^{l=N-1}_m\rangle = 0$. Since the differential operators are known, this relation can be used, as was the relation $L_- Y^l_{m=-l}(\theta, \phi) = 0$, to determine the radial wavefunction $\langle r|N, l = N-1\rangle$.

14.11.2 Scattering states

If $C > 0$ we can choose θ so that $e^{-\theta} - C e^\theta = 0$. Equation (14.66) reduces to

$$\left(2\sqrt{C}\,\Gamma_4 + B\right) u(r) = 0 \tag{14.70}$$

where as before $u(r) = e^{\theta M_4} R(r)$. Since the generator Γ_4 is noncompact, it has a continuous spectrum. The energy can be written in terms of the scaling factor $k \simeq e^{-\theta}$ with $E = \hbar^2 k^2 / 2m$. The asymptotic form of the wave function is (Gilmore et al., 1993; Kais and Kim, 1986)

$$R_{k,l}(r) \sim \sqrt{\frac{2}{\pi}} \sin\left(kr - \frac{\pi}{2} j + \frac{\alpha}{k}(\log(2kr) + \delta(j))\right) \tag{14.71}$$

where $\delta(j) = \arg[\Gamma(j+1-i(\alpha/k)]$ is part of the scattering phase shift, and the expression for j is given by $j = -\tfrac{1}{2} + \sqrt{(\tfrac{1}{2})^2 - A}$.

14.11.3 Quantum defect

Multielectron atoms are complicated objects. If one of the electrons is promoted to a high lying level, it is on average far from the nucleus and the core electrons. Some simplifications can then be made in the description of its excited state spectrum. As the "Rydberg" electron approaches the core, the positive nuclear charge is less completely screened by the core electrons, and the electron is more strongly attracted than a simple $-1/r$ potential suggests. It is possible to represent this extra attraction by adding a term of the form $-1/r^2$ to the potential to represent penetration of the core electrons. To this end the potential used in the Schrödinger and Klein–Gordon equations is $V(r) = -e^2/r \to -e^2/r - \mu_l(\hbar^2/2m)/r^2$. This perturbation produces a modification in the radial equation. The modification is encapsulated entirely in the change

$$A \to A' = A + \mu_l \tag{14.72}$$

This change produces a change in the value of $j \to j' = j + \Delta j$, where $\Delta j = -\mu_l/(2l+1)$ in the nonrelativistic case. This change produces a change in the bound state energy spectrum:

$$E_{N=n+l+1} = -\frac{mc^2\alpha^2}{2N^2} \to -\frac{mc^2\alpha^2}{2(N+\Delta j)^2} \tag{14.73}$$

The quantum defect Δj causes the Rydberg states to be bound more strongly than in a pure hydrogenic atom (without screening). The same change occurs in scattering states. There is an additional phase shift due to the stronger attraction in the core. The excess phase shift is

$$\Delta\phi = -\frac{\pi}{2}\Delta j + \frac{\alpha}{\pi}\arg\left(\Gamma[j+1+\Delta j - i(\alpha/k)] - \Gamma[j+1-i(\alpha/k)]\right) \tag{14.74}$$

Remark More accurate calculations of bound state spectra and scattering phase shifts employ more accurate representations of core screening (than $-1/r^2$). Nevertheless, the results are the same: a quantum defect in the bound state energies translates, through analytic continuation, to a corresponding excess phase shift in the scattering states (Seaton, 1966a, 1966b).

14.12 Conclusion

Group theory entered physics in two distinct ways. On one level the set of transformations from one coordinate system (or observer) to another forms a group. Observers are related by the Galilean principle of relativity. On another level, some physical systems exhibit symmetry. This symmetry allows us to predict new states on the basis of states that are already observed, together with the application of some symmetry transformation. This is done through Einstein's principle of equivalence.

We have exploited these principles to describe the quantum mechanical properties, particularly the energy level structure, of hydrogenic atoms. Initially, we exploited a *geometric* symmetry, the symmetry of the hamiltonian under rotations. The symmetry group is $SO(3)$ or the disconnected group $O(3)$. This symmetry requires that states occur in multiplets with angular momentum degeneracy $2l + 1$. It is surprising that hydrogenic states have a larger degeneracy than required by the rotation group $SO(3)$.

We believe that symmetry implies degeneracy, and the greater the symmetry, the greater the degeneracy. If we also believe that the N^2-fold degeneracy of the hydrogen states with principal quantum number N is due to invariance under some group, we are prodded to search for a larger group $G \supset SO(3)$ that explains the N^2-fold degeneracy. This *dynamical* symmetry group is $SO(4)$: its six infinitesimal generators include both the angular momentum operators and the components of the Laplace–Runge–Lenz vector.

Why stop here? Why not search for a "symmetry" that breaks the degeneracy but maps any state of the hydrogen atom to linear combinations of all other states? Such *spectrum generating* groups include $SO(4)$. The largest such group is the conformal group $SO(4, 2)$. Before this group was discovered, the deSitter group $SO(4, 1)$ was employed as a spectrum-generating group. A simple noncompact subgroup of these groups, isomorphic with $SO(2, 1)$, was used to illustrate explicitly how the generators of a Lie algebra are used to determine eigenstates and energy eigenvalues. In addition, representations that describe bound states can be analytically continued to representations that describe scattering states. This analytic continuation relates bound state energies to phase shifts of scattering states. In the case that the Coulomb potential is perturbed by core shielding effects, the energy eigenvalue spectrum is often simply represented by a quantum defect that depends on the angular momentum. The phase shift of scattering states with angular momentum l is related to the quantum defect with the same angular momentum.

In applications to the hydrogen atom, the role and scope of group theory in physics is seen to extend far beyond applications depending on simple geometric symmetry.

14.13 Problems

1. **a. Principle of relativity** Assume two observers S and S' are locked in the hold of a boat without windowports, so they cannot perceive the exterior world. Galilean relativity is founded on two assumptions: (1) it is impossible to determine whether a noninertial frame is at rest or in uniform relative motion with respect to its surroundings; (2) a body in an inertial frame will move with uniform velocity unless acted on by a force. Special relativity is also founded on two assumptions: (1) the laws of physics are the same in all inertial frames; (2) the speed of light is the same in all inertial frames. The first of the Galilean assumptions is implicit in the special theory of relativity. Show that the existence of the 3^{deg} microwave background radiation is incompatible with the first of Galileo's assumptions. Does this create a problem for the Special Theory of Relativity?

    b. Equivalence principle Assume two observers S and S' are locked inside elevators without windows, so they cannot perceive the exterior world. Assume one elevator is sitting on the surface of the Earth, so that the observer S experiences a gravitational force $\mathbf{F} = m\mathbf{g}$ in the "down" direction. Assume that the other elevator is in "interstellar space" so that external gravitational forces "vanish," but that his elevator experiences an acceleration $\mathbf{g}$ in the "up" direction. If the "rest of the universe" "looks the same" to both observers, argue that you can represent a gravitational field by a local acceleration. This use of the equivalence principle is one of the foundations of the general theory of relativity.

2. In the presence of a uniform magnetic field $\mathbf{B}$ show that the vector potential $\mathbf{A}$ can be taken as $\mathbf{A} = \frac{1}{2}\mathbf{B} \times \mathbf{r}$, so that $\mathbf{B} = \nabla \times \mathbf{A}$. Derive the Klein–Gordon equation for an electron in a Coulomb potential and a uniform magnetic field. Take the nonrelativistic limit of this and derive the Schrödinger equation for an electron in the presence of these two fields.

3. Make the ansatz $E = mc^2 + W$ in the Klein–Gordon equation and exhibit the terms in this equation that must be neglected in order to recover the nonrelativistic approximation, the Schrödinger equation.

4. Introduce spherical coordinates as follows: $(r, \theta, \phi) = (\theta_3, \theta_2, \theta_1)$ and

$$z = x_3 = \theta_3 \cos\theta_2$$
$$y = x_2 = \theta_3 \sin\theta_2 \cos\theta_1$$
$$x = x_1 = \theta_3 \sin\theta_2 \sin\theta_1$$

Show that $\mathcal{L}^2(S^1) = \partial^2/\partial\theta_1^2$. Show that

$$\sin^2\theta_2 \, \mathcal{L}^2(S^2) = \left(\sin\theta_2 \frac{\partial}{\partial\theta_2}\right)^2 + \mathcal{L}^2(S^1)$$

Generalize this result to $\mathcal{L}^2(S^3)$ recursively using $\mathcal{L}^2(S^2)$ and $(\partial/\partial \cos\theta_3)^2$. Do this more generally for $\mathcal{L}^2(S^n)$.

5. This problem carries through the steps indicated in Table 14.1.
   a. Show that the singular points of Eq. (14.9) occur at $r = 0$ and $r \to \infty$.
   b. Show that in the neighborhood of the singular points

   $$r \to 0 \quad \left(\frac{d^2}{dr^2} + \frac{A}{r^2} + \frac{B}{r} + C\right) R(r) \to \left(\frac{d^2}{dr^2} + \frac{A}{r^2}\right) R(r) = 0$$
   $$R(r) \simeq r^\gamma \qquad \gamma(\gamma - 1) + A = 0$$

   $$r \to \infty \quad \left(\frac{d^2}{dr^2} + \frac{A}{r^2} + \frac{B}{r} + C\right) R(r) \to \left(\frac{d^2}{dr^2} + C\right) R(r) = 0$$
   $$R(r) \simeq e^{\lambda r} \qquad \lambda^2 + C = 0$$

   Show that $\gamma = \frac{1}{2} \pm \sqrt{(\frac{1}{2})^2 - A}$ and $\lambda = \pm\sqrt{-C}$.

   c. Show that if $\sqrt{(\frac{1}{2})^2 - A}$ is real, the solution with the positive sign is always square integrable in the neighborhood of $r = 0$. Under what conditions is the solution with the negative sign square integrable? Show that if $C < 0$ the solution $\pm\sqrt{-C}$ with the negative sign is square integrable. What happens if $C > 0$?
   d. Show that a solution of the form $R(r) = r^\gamma e^{\lambda r} f(r)$ can be found where the function $f(r)$ is a simple polynomial function.
   e. Find the equation that the function $f(r)$ satisfies. Show that it is equivalent to the equation given in Table 14.1.
   f. Represent the function $f(r)$ as an ascending power series: $f(r) = \sum_{j=0}^{\infty} f_j r^j$. Find the two-term recursion relation satisfied by the coefficients f_j. Show that the recursion relation is

   $$[(j+1)j + 2\gamma(j+1)] f_{j+1} + (2\lambda\gamma + 2\lambda j + B) f_j = 0$$

   Use this relation to show

   $$f(r) = \sum_{j=0}^{\infty} \frac{\Gamma(j + \gamma + (B/2\lambda))}{\Gamma(\gamma + (B/2\lambda))} \frac{\Gamma(2\gamma)}{\Gamma(j + 2\gamma)} \frac{(-2\lambda r)^j}{j!}$$

   g. If this series does not terminate, show that its asymptotic behavior as $r \to \infty$, determined from the behavior of f_j as $j \to \infty$, is $f(r) \to e^{-2\lambda r}$. Since $\lambda < 0$ this solution is not square integrable.
   h. Conclude that the function $f(r)$ must be a polynomial of finite degree. If the highest nonzero degree term present is r^n, so that $f_n \neq 0$ but $f_{n+1} = 0 \, (\Rightarrow f_{n+2} = f_{n+3} = \cdots = 0)$, show that the quantization conditon $2\lambda(n + \gamma) + B = 0$ must be satisfied. Show that this leads to the quantization condition in terms of the three parameters A, B, C that appear in Eq. (14.9):

   $$n + \frac{1}{2} + \sqrt{\left(\frac{1}{2}\right)^2 - A} = \frac{B}{2\sqrt{-C}}$$

252   Hydrogenic atoms

i. Use the values of the parameters A, B, C given in Eq. (14.10) to solve for the energy eigenvalues of the Klein–Gordon and Schrödinger equations:

$$E(n,l) = \frac{mc^2}{\sqrt{1 + \dfrac{\alpha^2}{(n + \frac{1}{2} + \sqrt{(l + \frac{1}{2})^2 - \alpha^2})^2}}}$$

$$W(n,l) = -\frac{1}{2} mc^2 \alpha^2 \frac{1}{(n + l + 1)^2}$$

Show that the polynomial solution is

$$f(r) = \sum_{j=0}^{n} \frac{\Gamma(2\gamma)}{\Gamma(j + 2\gamma)} \frac{n!}{(n-j)! j!} (2\lambda r)^j$$

The radial part of the wavefunction $\frac{1}{r} r^\gamma f(r) e^{\lambda r}$ has exactly n nodes in the open interval $(0, \infty)$.

6. For a highly ionized atom with Z protons in its nucleus and a single remaining electron, show that the potential is Ze/r and the solutions of the relativistic and nonrelativistic equations are obtained by the replacement $\alpha \to Z\alpha$. How large can Z become before the relativistic solution is clearly incorrect? (Hint: set $l = 0$.)

7. Expand the relativistic energy in ascending powers of the fine structure constant to determine the relativistic corrections to the nonrelativistic energy. Show that, with $N' = n + \frac{1}{2} + \sqrt{(l + \frac{1}{2})^2 - \alpha^2}$ and $N = n + l + 1$

$$E(n,l) = \frac{mc^2}{\sqrt{1 + (\frac{\alpha}{N'})^2}} \to mc^2 - mc^2 \frac{1}{2N^2} \alpha^2 + mc^2 \left(\frac{3}{8N^4} - \frac{1}{N^3(2l+1)}\right) \alpha^4$$

$$+ mc^2 \left(-\frac{5}{16N^6} + \frac{3}{2N^5(2l+1)} - \frac{2N + 3(2l+1)}{2N^4(2l+1)^3}\right) \alpha^6 + mc^2$$

$$\times \left(\frac{35}{128N^8} - \frac{15}{8N^7(2l+1)} + \frac{6N + 9(2l+1)}{4N^6(2l+1)^3}\right.$$

$$\left. - \frac{2N^2 + 3N(2l+1) + 2(2l+1)^2}{N^5(2l+1)^5}\right) \alpha^8 + \mathcal{O}(\alpha^{10})$$

8. The radial part of the wavefunction dies off like $e^{\lambda r}$ for large r, where $\lambda < 0$ for bound states. The parameter λ^{-1} has the dimensions of length, and $a \simeq 1/|\lambda|$ characterizes the size of a bound state orbit. Show that bound states with quantum numbers (n, l) ($N = n + l + 1$ is the principal quantum number) have size scales

relativistic $a(n, l) = \sqrt{(N')^2 + \alpha^2}\, a_B$ $N' = n + \frac{1}{2} + \sqrt{(l + \frac{1}{2})^2 - \alpha^2}$
nonrelativistic $a(n, l) = N a_B$ $N = n + l + 1$

14.13 Problems

Table 14.4. *Some particles that can be used to form hydrogen-like atoms*

Particle	Rest energy (MeV)
electron $e^{\pm}$	0.511
mu meson $\mu^{\pm}$	105.7
tau meson $\tau^{\pm}$	1784.0
proton, antiproton $p^{\pm}$	938.26
deuteron d^{+}	1875.6
tritium t^{+}	2809.4
pi meson $\pi^{\pm}$	139.6
sigma meson $\Sigma^{\pm}$	1385.0
cascade meson Ξ^{-}	1533.0
omega Ω^{-}	1672.0

In these expressions $a_B = \hbar^2/me^2 = 0.529 \times 10^{-8}$ cm is the Bohr radius: the characteristic size of the hydrogen atom in its ground state. By what percentage do the sizes of the atoms in the (n, l) states differ between the relativistic and nonrelativistic treatments?

9. Many charged particles can form hydrogen-like atoms through their electrostatic interaction. Compute the energy spectrum for bound states of neutral atoms formed from a positively charged particle and a negatively charged particle drawn from this list of particles in Table 14.4. For each particle the mass is given in terms of the particle rest energy. Recall that the mass, m, that appears in the expression for the binding energy $W = -\frac{1}{2}mc^2\alpha^2/N^2$ is the reduced mass: $1/m = 1/m_1 + 1/m_2$ of the two particles.

10. The motion of a classical nonrelativistic particle in a $1/r^2$ radial force field is a conic section: an elliptical orbit for bound states ($E < 0$); hyperbolic for scattering states ($E > 0$); and parabolic at the separatrix ($E = 0$). If the radial force field includes a radial $1/r^3$ perturbation

$$f = -\frac{K}{r^2} + \frac{C}{r^3}$$

the trajectory has the form (Goldstein, 1950)

$$r = \frac{a(1-\epsilon^2)}{1+\epsilon\cos(\alpha\theta)}$$

where $\alpha = \sqrt{1-\eta}$, $\eta = C/Ka$. This can be treated as an ellipse that is slowly rotating, $\alpha \simeq 1$. In this case the parameters a and ϵ have their usual meanings for elliptical orbits: a is the semimajor axis and ϵ is the eccentricity. The ratio η is a measure of the strength of the perturbation to the strength of the Coulomb potential.

254   Hydrogenic atoms

   a. Expand the relativistic energy $E = \sqrt{(mc^2)^2 + (pc)^2} - K/r$ to fourth order in $\mathbf{p}$
   and show $E = (mc^2) + (p^2/2m) - (p^2/2m)^2/(2mc^2) - K/r = mc^2 + W$.
   b. Replace the quartic term $-(p^2/2m)^2/(2mc^2)$ by $-(W + K/r)^2/(2mc^2)$ and expand. Show that the classical hamiltonian for the motion of the (special) relativistic particle is

$$H = mc^2 + \frac{p^2}{2m} - \frac{K'}{r} + \frac{C'}{r^2}$$

   Evaluate K' and C' and show $K' = K(1 + W/mc^2)$ and $C' = -K^2/(2mc^2)$.
   c. Argue that the classical motion involves a renormalized coupling $K \to K'$ as well as a $1/r^3$ component to the force, with $C = 2C'$.
   d. Show that the advance in the perihelion of the orbit is $\delta\theta \simeq \eta/2$ per period.
   e. Evaluate η for the planet Mercury, for which $\epsilon = 0.206$ and the period is $T = 0.24$ year. Show that this amounts to about 7" per century. The general relativistic correction is larger by a factor of 6, and accounts for the observed advance in Mercury's perihelion of 42" per century.
   f. The existence of precessing elliptical orbits is due to the "relativistic mass velocity" correction. This can be viewed from two perspectives. (1) Newton's equations are correct and the mass of the particle varies with its state of motion according to $m = m_0/\sqrt{1-(v/c)^2}$. (2) The mass of a particle is a constant of nature and Newton's (nonrelativistic) equations of motion are not correct for relativistic particles, and must be modified. The author feels the second interperetation is far superior to the first.

11. When the attracting potential is central and nearly $1/r$, the motion of a bound particle is nearly elliptical. It is useful to describe this motion as if it were elliptical, with the semimajor axis of the ellipse precessing in the plane of motion. Assume that the force has the form $\mathbf{F}(r) = (-K/r^2 + p(r))\hat{\mathbf{r}}$, where $p(r)$ is a small perturbation. The rate at which the Runge–Lenz vector precesses is

$$\omega = \frac{\partial}{\partial L}\left(\frac{1}{T}\int_0^T p(r)\,dt\right) = \frac{\partial}{\partial L}\left(\frac{m}{LT}\oint r^2 p(r)\,d\theta\right)$$

with $1/r = (mK/L^2)(1 + (M/mK)\cos\theta)$. Here L is the particle's orbital angular momentum and T is its period. If the perturbing term is of the form C/r^3 the integral is $C \times 2\pi \frac{mK}{L^2}$. The perturbations due to special and General Relativity are

$$\text{special relativity}\quad C = \frac{KL^2}{2m^2c^2} \qquad \omega = \frac{\pi K^2}{TL^2c^2}$$

$$\text{general relativity}\quad C = 6 \times \frac{KL^2}{2m^2c^2} \qquad \omega = \frac{6\pi K^2}{TL^2c^2}$$

For planetary motion $K = GMm$. When $M \gg m$, ω is (almost) independent of m. Why? Determine how the relativistic precession ω scales (cf. Problem 16.3) with

14.13 Problems

planetary distance from the Sun. What is the precession for the Earth? Use $\omega = 42''$ per century for Mercury and the following distance ratios:

Mercury	Venus	Earth	Mars	Jupiter	Saturn	Uranus	Neptune
0.39	0.72	1.00	1.52	5.20	9.54	19.18	30.06

12. The action of the angular momentum shift down operator L_- on the lowest m-value spherical harmonic for a given value of l is zero: $L_- Y^l_{m=-l}(\theta, \phi) = 0$. Use the coordinate representation for L_- to compute this function.

   a. Write $Y^l_m(\theta, \phi) = P^l_{-l}(\theta) e^{-il\phi}$ and show

   $$\left(-\frac{\partial}{\partial \theta} + i\frac{\cos\theta}{\sin\theta}\frac{\partial}{\partial \phi}\right) P^l_{-l}(\theta) e^{-il\phi} = e^{-il\phi} \left(-\frac{\partial}{\partial \theta} + l\frac{\cos\theta}{\sin\theta}\frac{\partial}{\partial \phi}\right) P^l_{-l}(\theta)$$

   b. Show $P^l_{-l} = (\sin\theta)^l$ satisfies this equation.

   c. This function is not normalized to unity over the sphere. Normalize it by introducing a normalization coefficient N_l and enforcing the condition

   $$\int_0^\pi d\theta \sin\theta \int_0^{2\pi} d\phi |N_l \sin^l \theta e^{-il\phi}|^2 = 1$$

   d. Show that

   $$N_l = \sqrt{\frac{1}{4\pi}} \sqrt{\frac{(2l+1)!!}{(2l)!!}}$$

   e. This leads the the simple recursion relation for normalization coefficients for the $Y^l_{\pm l}(\theta, \phi)$:

   $$N_l = \sqrt{\frac{2l+1}{2l}} N_{l-1}$$

   Compare these results with Table 14.2 using initial condition $N_0 = \sqrt{1/4\pi}$. Compute N_3.

   f. Use the numerical value of the matrix elements $\langle ^l_{m'} |L_+| ^l_m \rangle = \sqrt{(l+m')(l-m)}\, \delta_{m',m+1}$ and the coordinate representation of the shift up operator L_+ to construct the correctly normalized spherical harmonics $Y^l_m(\theta, \phi)$.

13. Use methods similar to those described in Problem 12 to construct the radial wavefunctions for hydrogenic atoms with extreme orbital angular momentum quantum numbers: $l = N - 1$, where in general the principal quantum number $N = n + l + 1$. These functions have no nodes in the interval $(0, \infty)$ (since $n = 0$).

14. Show

   $$\frac{d}{dt}\left(\frac{\mathbf{r}}{r}\right) = \frac{\dot{\mathbf{r}}(\mathbf{r}\cdot\mathbf{r}) - \mathbf{r}(\mathbf{r}\cdot\dot{\mathbf{r}})}{r^3} = -\frac{\mathbf{r} \times (\mathbf{r} \times \dot{\mathbf{r}})}{r^3}$$

15. $\mathbf{r}$ is the position vector from the sun to a planet, or from the proton to the electron in the hydrogen atom, $\mathbf{L} = \mathbf{r} \times \mathbf{p}$ is the orbital angular momentum, and $\mathbf{M}$ is the Laplace–Runge–Lenz vector.

a. $\mathbf{M} \cdot \mathbf{L} = 0$.
b. $\mathbf{M} \cdot \mathbf{M} = (2\mathbf{L} \cdot \mathbf{L}/m)(\mathbf{p} \cdot \mathbf{p}/2m - K/r) + K^2$.
c. $\mathbf{M} \cdot \mathbf{r} = \mathbf{L} \cdot \mathbf{L}/m - Kr$.
d. $\mathbf{M} \cdot \mathbf{r} = Mr\cos\theta$.
e. $r = (\mathbf{L} \cdot \mathbf{L}/mK/1 + (M/K)\cos\theta)$.
f. Compare this result to the standard solution of the trajectory equations for motion in a $1/r$ potential to conclude that L^2/mK is the semimajor axis of the elliptical orbit and $\epsilon = M/K$ is the eccentricity of the orbit.
g. Conclude that the Laplace–Runge–Lenz vector is a constant of motion that points to the perihelion of the elliptical orbit.

16. Show that $\mathbf{A} \cdot \mathbf{A} = (-1/4\hbar^2)(\mathbf{L} \cdot \mathbf{L} + \mathbf{M}' \cdot \mathbf{M}' + \mathbf{L} \cdot \mathbf{M}' + \mathbf{M}' \cdot \mathbf{L})$. Show that $\mathbf{B} \cdot \mathbf{B}$ has a similar expression. Show that the two expressions are equal since $\mathbf{L} \cdot \mathbf{M} = \mathbf{M} \cdot \mathbf{L} = 0$.

17. Show that the inverse of the stereographic projection given in Eq. (14.35) is

$$\frac{\mathbf{p}}{p_0} = \frac{\mathbf{u}}{1 - \mathbf{u} \cdot \mathbf{w}}$$

18. Compute $p_x = \mathbf{p} \cdot \mathbf{M}/M$ and $p_y = \mathbf{p} \cdot \mathbf{W}/W$. Show $p_x^2 + (p_y - a)^2 = r^2$. Explicitly compute the displacement vector a (i.e., $(0, a)$) and the radius r of circular motion. Show that circles in R^3 lift to circles in $S^3 \subset R^4$ under the stereographic projection of Eq. (14.35). Show that circles in S^3 project back down to circles in R^3 under the inverse transformation.

19. Show that the number of independent monomials of the form $x^a y^b z^c$, with a, b, c nonnegative integers and $a + b + c = l$ is $N(l, 3) = (l + 3 - 1)/l!(3 - 1)!$. In N-dimensional space show that the number of homogeneous polynomials of degree l in $x_1, x_2, \ldots, x_N$ is obtained by replacing $3 \to N$ in this expression. This is the Bose–Einstein counting statistic.
a. Show that the functions $r^l Y_m^l(\theta, \phi)$ are homogeneous polynomials in x, y, z of degree l.
b. Show that the number of independent spherical harmonics of degree l is the difference between the number of homogeneous polynomials of degree l and $l - 2$ on three variables: $\dim \{Y_m^l\} = N(l, 3) - N(l - 2, 3) = 2l + 1$.
c. After stereographic transformation into four dimensions, the hydrogen wavefunctions in the momentum representation are spherical harmonics in four variables (Bander and Itzykson, 1966a). Show that the number of spherical harmonics of degree n is $\dim \{\mathcal{Y}_{lm}^n\} = N(n, 4) - N(n - 2, 4) = (n + 1)^2$.
d. Construct homogeneous polynomials of degree 0, 1, 2 and the spherical harmonics associated with these homogeneous polynomials. Take the inverse Fourier transform of these spherical harmonics to obtain the hydrogen atom wavefunctions $\psi(\mathbf{x})_{nlm}$ for $n = 0, 1, 2; l = 0, \ldots, n - 1$; and $-l \le m \le +l$.

e. Show that the recursive relation used to build up a Pascal triangle can be written in the symmetric form

$$\frac{(a+b+1)!}{(a+\frac{1}{2})!(b+\frac{1}{2})!} = \frac{(a+b)!}{(a-\frac{1}{2})!(b+\frac{1}{2})!} + \frac{(a+b)!}{(a+\frac{1}{2})!(b-\frac{1}{2})!}$$

where a and b are half odd integers: $\frac{1}{2}, \frac{3}{2}, \frac{5}{2}, \ldots$.

f. Show homogeneous polynomials satisfy the recursion relation: $N(l, d) = N(l, d-1) + N(l-1, d)$.

g. Use this result to derive the following recursion relation for the dimensions of the spaces of spherical harmonics on spheres S^n and S^{n-1}:

$$\dim \mathcal{Y}^l(S^n) = \dim \mathcal{Y}^{l-1}(S^n) + \dim \mathcal{Y}^l(S^{n-1})$$

For the case $n = 3$ this gives $(l+1)^2 = l^2 + (2l+1)$. The initialization for all n is $\mathcal{Y}^0(S^n) = 1 = \dim \mathcal{Y}^0(S^n)$.

h. $\dim \mathcal{Y}^l(S^n) = \frac{(l+n-2)!}{l!(n-1)!}(2l+n-1)$.

20. **D-dimensional Coulomb problem** In D-dimensional space the Schrödinger equation for the Kepler problem is Eq. (14.4) in the relativistic case and Eq. (14.5) in the nonrelativistic case. The only difference is that the Laplacian ∇^2 is on D coordinates rather than three. In this case the Laplacian operator is

$$\nabla^2 = \left(\frac{1}{r^{D-1/2}} \frac{\partial}{\partial r} r^{D-1/2}\right)^2 + \frac{\mathcal{L}^2}{r^2}$$

The angular part of the Laplacian operator, $\mathcal{L}^2$, acts on spherical harmonics on S^{D-1}, $\mathcal{Y}^l(S^{D-1})$. These spherical harmonics are eigenfunctions of this (Laplace–Beltrami) operator with eigenvalue $-\left[(l+\alpha)^2 - \alpha^2\right]$, and α is a quantity that depends on the Lie algebra of $SO(D)$: it is half the sum over all positive roots of the algebra. For the Lie algebras of the orthogonal roots the coefficient of the sum that is important is $\alpha = D - 2$.

a. Show that $\psi(\mathbf{x}) = (1/r^{(D-1)/2})\mathcal{Y}^l(\text{angles})$ is a clever ansatz that reduces the Schrödinger equation in D dimensions to the form of Eq. (14.4) in the relativistic case and Eq. (14.5) in the nonrelativistic case.

b. Show that the only change in Eq. (14.10) is the replacement

$$\left(l+\frac{1}{2}\right)^2 - \left(\frac{1}{2}\right)^2 \to \left(l+\frac{D-2}{2}\right)^2 - \left(\frac{D-2}{2}\right)^2$$

in column A.

c. Show that the relativistic and nonrelativistic energies shown in Eq. (14.12) change as follows:

relativistic $\quad N' \to n + \frac{1}{2} + \sqrt{(l+\frac{1}{2})^2 + l(D-3) - \alpha^2}$

nonrelativistic $\quad N \to n + \frac{1}{2} + \sqrt{(l+\frac{1}{2})^2 + l(D-3)}$

21. Compute the quantum defect in heavy atoms by using the Klein–Gordon equation and a $-1/r^2$ perturbation. Show that the bound state energy and scattering phase shifts are given by the substitution $l(l+1) \to l(l+1) - \mu_l$. Argue that electrons in the s state penetrate the core much more deeply (on average) and p-state electrons (than d-state electrons, ...) so that $\mu_0 \gg \mu_1 > \cdots$.

22. The isotropic harmonic oscillator in n dimensions has hamiltonian

$$H = \sum_{i=1}^{n} \hbar\omega \left(a_i^\dagger a_i + \frac{1}{2} \right)$$

   a. Show that the Lie algebra of its geometric symmetry group is spanned by the angular momentum operators $L_{ij} = a_i^\dagger a_j - a_j^\dagger a_i = -L_{ji}$.

   b. Show that the Lie algebra of its dynamical symmetry group is spanned by the angular momentum operators together with the quadrupole tensor operators $Q_{ij} = a_i^\dagger a_j + a_j^\dagger a_i = +Q_{ji}$.

   c. Show that one spectrum generating algebra includes the operators $\mathbf{L}$ and $\mathbf{Q}$ as well as the single boson operators $a_i^\dagger$ and a_j, as well as their commutator $[a_i, a_j^\dagger] = 1$. Show that this algebra is nonsemisimple and describe its structure.

   d. Show that another spectrum generating algebra consists of the operators $\mathbf{L}$ and $\mathbf{Q}$ as well as the two boson creation operators $a_i^\dagger a_j^\dagger$ and two boson annihilation operators $a_i a_j$. Show that this algebra is simple and describe its structure. Show that this spectrum generating algebra does not couple all the states that exist: "parity" is an invariant, where "parity" is even or odd according to whether the number of excitations in the spectrum is even or odd.

15

Maxwell's equations

The electromagnetic field $\mathbf{E}(\mathbf{x}, t)$, $\mathbf{B}(\mathbf{x}, t)$ is determined by Maxwell's equations. These equations are linear in the space and time derivatives. In the momentum representation, obtained by taking a Fourier transform of the electric and magnetic fields, Maxwell's equations impose a set of four linear constraints on the six amplitudes $\mathbf{E}(k)$, $\mathbf{B}(k)$. Why? At a more fundamental level, the electromagnetic field is described by photons. For each photon momentum state there are only two degrees of freedom, the helicity (polarization) states, corresponding to an angular momentum 1 aligned either in or opposite to the direction of propagation. Thus, the classical description of the electromagnetic field is profligate, introducing six amplitudes for each k when in fact only two are independent. The remaining four degrees must be absent in any description of a physically allowed field. The equations that annihilate these four nonphysical linear combinations are the equations of Maxwell. We derive these equations, in the absence of sources, by comparing the transformation properties of the helicity and classical field states for each four-momentum.

15.1 Introduction

The electromagnetic field has been described in two different ways. Following the nineteenth century approach (pre quantum mechanics), a field is introduced having appropriate transformation properties. The price one pays is that not every field represents a physically allowed state: such fields must be annihilated by appropriate equations. Following the twentieth century approach, a Hilbert space is introduced. An arbitrary superposition of states in this space represents a physically allowed field. The price one pays is that the field so constructed does not have obvious transformation properties.

In the older approach a field is defined at every point in space time. It is required to be "manifestly covariant." That is, it transforms as a tensor under homogeneous

Table 15.1. *Comparison of descriptions of the electromagnetic field*

Time period	Approach	Strengths	Weaknesses
Nineteenth century	Manifestly covariant	Fields have elegant transformation properties	Many fields represent nonphysical states
Twentieth century	Hilbert space	All linear superpositions represent physical states	Transformation properties are complicated

Lorentz transformations. This requires there to be a certain number of field components at every space-time point, or more conveniently, for every allowed momentum vector. In the Hilbert space formulation the number of independent components is just the allowed number of spin or helicity states. The number of components is never greater than the number of components required to define the "manifestly covariant" field; however, it may be less than this number. In this case there are linear combinations of the components of the manifestly covariant field that cannot represent physically allowed states. These linear combinations must be suppressed. It is the function of the field equations to suppress those linear combinations of components that do not correspond to physical states. These two approaches are compared in Table 15.1.

Maxwell's equations fulfill this function. The classical description involves six field components for each allowed mementum state. These are the classical electric and magnetic fields, $\mathbf{E}(\mathbf{x}, t)$ and $\mathbf{B}(\mathbf{x}, t)$, or their components after Fourier transformation, $\mathbf{E}(k)$ and $\mathbf{B}(k)$, where k is a four-vector that obeys $k \cdot k = \mathbf{k} \cdot \mathbf{k} - k_4 k_4 = 0$. Here $\mathbf{k}$ is essentially a three-momentum vector and k_4 is essentially an energy. The quantum description involves arbitrary superpositions of two helicity components for each momentum vector. The helicity states involve an angular momentum aligned along the direction of motion (helicity $+1$ and right-handed polarization) and opposite to the direction of propagation (helicity -1 and left-handed polarization). There are four $(6 - 2)$ linear combinations of classical field components that must be suppressed for each k-vector, and that are annihilated by Maxwell's equations. We derive these equations by comparing the transformation properties of the basis vectors for the "manifestly covariant" but nonunitary representations of the inhomogeneous Lorentz group with the basis vectors for its unitary irreducible representations, which are not manifestly covariant. The set of constraints so derived reduce, for $j = 1$, to Maxwell's equations. This derivation is carried out for free fields (no sources) only. When sources are present the photon four-vector k no longer obeys $k \cdot k = 0$. In this case the manifestly covariant equations provide a beautiful prescription for describing the coupling to source terms.

15.2 Review of the inhomogeneous Lorentz group

15.2.1 Homogeneous Lorentz group

The wavefront for a light signal expanding from a source at the origin of coordinates for observers S and S' obeys the equation

$$x^2 + y^2 + z^2 - (ct)^2 = x'^2 + y'^2 + z'^2 - (ct')^2 = 0 \qquad (15.1)$$

This requires that the coordinates (x, y, z, ict) and $(x, y, z, ict)'$ for observers S and S' be related by a homogeneous Lorentz transformation

$$\begin{bmatrix} x \\ y \\ z \\ ict \end{bmatrix} = \begin{bmatrix} \Lambda \end{bmatrix} \begin{bmatrix} x \\ y \\ z \\ ict \end{bmatrix}' \qquad (15.2)$$

The 4×4 matrix transformations Λ belong to the Lie group $O(3, 1)$. The infinitesimal generators of a group operation in $SO(3, 1)$ are

$$\Lambda \to I_4 + \epsilon \begin{bmatrix} 0 & +\theta_3 & -\theta_2 & ib_1 \\ -\theta_3 & 0 & +\theta_1 & ib_2 \\ +\theta_2 & -\theta_1 & 0 & ib_3 \\ -ib_1 & -ib_2 & -ib_3 & 0 \end{bmatrix} = I_4 + \epsilon \left(\boldsymbol{\theta} \cdot \mathbf{J} + \mathbf{b} \cdot \mathbf{K} \right) \qquad (15.3)$$

Homogeneous Lorentz transformations leave invariant inner products: $k \cdot a = \Lambda k \cdot \Lambda a$, where k and a are four vectors and $\Lambda \in O(3, 1)$. The infinitesimal generators $\mathbf{J}, \mathbf{K}$ satisfy the following commutation relations:

$$\begin{aligned}{} [J_i, J_j] &= -\epsilon_{ijk} J_k \\ [J_i, K_j] &= -\epsilon_{ijk} K_k \\ [K_i, K_j] &= +\epsilon_{ijk} J_k \end{aligned} \qquad (15.4)$$

15.2.2 Inhomogeneous Lorentz group

Intervals are preserved by the inhomogeneous Lorentz group:

$$(x_2 - x_1)^2 + (y_2 - y_1)^2 + (z_2 - z_1)^2 - (ct_2 - ct_1)^2 = \text{invariant} \qquad (15.5)$$

The inhomogeneous Lorentz group consists of homogeneous Lorentz transformations, Λ, together with displacements of the origin. The general group transformation can be written as a 5×5 matrix, in terms of the 4-vector $a = (x, y, z, ct)$:

$$\{\Lambda, a\} = \begin{bmatrix} & & & & x \\ & \Lambda & & & y \\ & & & & z \\ & & & & ct \\ 0 & 0 & 0 & 0 & 1 \end{bmatrix} \quad (15.6)$$

as shown. The group composition law is matrix multiplication. The following results are immediate:

$$\{\Lambda_2, a_2\}\{\Lambda_1, a_1\} = \{\Lambda_2\Lambda_1, a_2 + \Lambda_2 a_1\}$$

$$\{I, a\}\{\Lambda, 0\} = \{\Lambda, a\} = \{\Lambda, 0\}\{I, \Lambda^{-1}a\} \quad (15.7)$$

The inhomogeneous Lorentz group is the semidirect product of the homogeneous Lorentz group and the commutative invariant subgroup of translations of the origin of coordinates in space and time. The infinitesimal generators for this invariant subgroup are $(\partial/\partial x, \partial/\partial y, \partial/\partial z, i\partial/\partial(ct))$.

15.3 Subgroups and their representations

The group of inhomogeneous Lorentz transformations has two important subgroups. These are the subgroup of homogeneous Lorentz transformations $\{\Lambda, 0\}$ and the invariant subgroup of translations $\{I, a\}$. Both their representations play a role in the derivation of the relativistically covariant field equations.

15.3.1 Translations $\{I, a\}$

The translation subgroup $\{I, a\}$ is abelian (commutative). All of its unitary irreducible representations are one dimensional, and in fact

$$\Gamma^k(\{I, a\}) = e^{ik \cdot a} \quad (15.8)$$

where k is a four-vector that parameterizes the one-dimensional representations. We may define a basis state for the one-dimensional representation Γ^k of $\{I, a\}$ as $|k\rangle$:

$$\{I, a\}|k\rangle = |k'\rangle\langle k'|\{I, a\}|k\rangle = |k'\rangle\delta(k' - k)\Gamma^k(\{I, a\}) = |k\rangle e^{ik \cdot a} \quad (15.9)$$

Physically, k has a natural interpretation as the four-momentum of the photon.

15.3 Subgroups and their representations

15.3.2 Homogeneous Lorentz transformations

The Lie algebra $D_2 = A_1 + A_1$ is semisimple: it is the direct sum of two simple Lie algebras of type A_1 (see Fig. 10.3). We can construct linear combinations of the infinitesimal generators $\mathbf{J}$, $\mathbf{K}$ of $SO(3, 1)$ that are mutually commuting and that satisfy angular momentum commutation relations. These are

$$\mathbf{J}^{(1)} = \frac{1}{2}(\mathbf{J} - i\mathbf{K})$$
$$\mathbf{J}^{(2)} = \frac{1}{2}(\mathbf{J} + i\mathbf{K})$$
(15.10)

These operators satisfy angular momentum commutation relations

$$\left[J_i^{(1)}, J_j^{(1)} \right] = -\epsilon_{ijk} J_k^{(1)}$$
$$\left[J_i^{(2)}, J_j^{(2)} \right] = -\epsilon_{ijk} J_k^{(2)}$$
$$\left[J_i^{(1)}, J_j^{(2)} \right] = 0$$
(15.11)

The algebra $\mathbf{J}^{(1)}$ has $2j + 1$ dimensional irreducible representations D^j while $\mathbf{J}^{(2)}$ has $2j' + 1$ dimensional irreducible representations $D^{j'}$. Any element in $SO(3, 1)$ can be expressed in a $(2j + 1)(2j' + 1)$ dimensional representation $D^{jj'}$ as follows

$$\text{EXP}(\boldsymbol{\theta} \cdot \mathbf{J} + \mathbf{b} \cdot \mathbf{K}) = \text{EXP}\left[(\boldsymbol{\theta} + i\mathbf{b}) \cdot \mathbf{J}^{(1)} + (\boldsymbol{\theta} - i\mathbf{b}) \cdot \mathbf{J}^{(2)} \right]$$
$$= D^j \left[(\boldsymbol{\theta} + i\mathbf{b}) \cdot \mathbf{J}^{(1)} \right] D^{j'} \left[(\boldsymbol{\theta} - i\mathbf{b}) \cdot \mathbf{J}^{(2)} \right] \quad (15.12)$$

15.3.3 Representations of $SO(3, 1)$

The Lie algebra $\mathfrak{so}(3, 1)$ is isomorphic to the Lie algebra for the group of 2×2 matrices $SL(2; \mathbb{C})$. We have the following two isomorphisms

$$\begin{array}{ll} \mathbf{J} = \frac{i}{2}\sigma & \mathbf{J} = \frac{i}{2}\sigma \\ \mathbf{K} = -\frac{1}{2}\sigma & \mathbf{K} = +\frac{1}{2}\sigma \end{array}$$
(15.13)

These two isomorphisms give rise to the following two inequivalent sets of representations

$$\begin{array}{cc} D^{j0} & D^{0j} \\ \mathbf{K}^{(j)} = i\mathbf{J}^{(j)} & \mathbf{K}^{(j)} = -i\mathbf{J}^{(j)} \end{array}$$
(15.14)

where $\mathbf{J}^{(j)}$ are the three $(2j+1) \times (2j+1)$ angular momentum matrices. The following matrices are associated with these representations

$$D^{j0}[\boldsymbol{\theta}\cdot\mathbf{J} + \mathbf{b}\cdot\mathbf{K}] = \mathrm{EXP}\left[\boldsymbol{\theta}\cdot\mathbf{J}^{(j)} + \mathbf{b}\cdot(+i\mathbf{J})^{(j)}\right] = \mathrm{EXP}\left[(\boldsymbol{\theta} + i\mathbf{b})\cdot\mathbf{J}^{(j)}\right]$$
$$D^{0j}[\boldsymbol{\theta}\cdot\mathbf{J} + \mathbf{b}\cdot\mathbf{K}] = \mathrm{EXP}\left[\boldsymbol{\theta}\cdot\mathbf{J}^{(j)} + \mathbf{b}\cdot(-i\mathbf{J})^{(j)}\right] = \mathrm{EXP}\left[(\boldsymbol{\theta} - i\mathbf{b})\cdot\mathbf{J}^{(j)}\right]$$
(15.15)

These representations are complex conjugates of each other. The most general representation of $SO(3, 1)$ is

$$D^{jj'}(\boldsymbol{\theta}\cdot\mathbf{J} + \mathbf{b}\cdot\mathbf{K}) = \mathrm{EXP}\left[(\boldsymbol{\theta} + i\mathbf{b})\cdot\mathbf{J}^{(j)}\right]\mathrm{EXP}\left[(\boldsymbol{\theta} - i\mathbf{b})\cdot\mathbf{J}^{(j')}\right] = D^{jj'}(\Lambda)$$
(15.16)

Basis states for the action of Λ through the representation $D^{jj'}(\Lambda)$ can be computed

$$\Lambda \left| \begin{matrix} j & j' \\ \mu & \mu' \end{matrix} \right\rangle = \left| \begin{matrix} j & j' \\ \nu & \nu' \end{matrix} \right\rangle D^{jj'}_{\nu\nu';\mu\mu'}(\Lambda) \tag{15.17}$$

Under restriction to the subgroup $SO(3) \subset SO(3, 1)$ this representation is reducible in a Clebsch–Gordan series

$$D^{jj'}(\Lambda) \overset{\Lambda \downarrow SO(3)}{\longrightarrow} D^{j}[SO(3)] \times D^{j'}[SO(3)] = \sum_{j''} D^{j''}[SO(3)]$$
$$|j - j'| \leq j'' \leq j + j' \tag{15.18}$$

This representation remains irreducible only if $j' = 0$ or $j = 0$.

15.4 Representations of the Poincaré group

We construct here two kinds of representations for the inhomogeneous Lorentz group. These are the manifestly covariant representations and the unitary irreducible representations.

15.4.1 Manifestly covariant representations

A field $T_{\mu\nu}(x)$ is said to be **manifestly covariant** (obviously covariant) under transformations of the homogeneous Lorentz group $\Lambda \in SO(3, 1)$ if

$$\Lambda T_{\mu\nu}(x) = T_{\mu'\nu'}(x\Lambda^{-1})\Lambda_{\mu'\mu}\Lambda_{\nu'\nu} \tag{15.19}$$

That is, the field components obviously form a basis on which the Lorentz transformation acts. The point at which the transformation acts is fixed, but since the coordinate system changes, the coordinates of the fixed point are changed by $x' = x\Lambda^{-1}$.

15.4 Representations of the Poincaré group

We construct manifestly covariant representations of the inhomogeneous Lorentz group by constructing direct products of basis vectors

$$|k\rangle \times \left|\begin{matrix} j & j' \\ \mu & \mu' \end{matrix}\right\rangle \tag{15.20}$$

for the subgroups $\{I, a\}$ and $\{\Lambda, 0\}$ of the inhomogeneous Lorentz group. We define the action of the inhomogeneous Lorentz group on these direct product states by defining the action of the two subgroups, of homogeneous Lorentz transformations and of translations, on the momentum states $|k\rangle$ and the field component states $\left|\begin{smallmatrix} j & j' \\ \mu & \mu' \end{smallmatrix}\right\rangle$ separately.

We define the action of $\{I, a\}$ on these states by

$$\{I, a\} |k\rangle = |k\rangle e^{ik \cdot a}$$
$$\{I, a\} \left|\begin{matrix} j & j' \\ \mu & \mu' \end{matrix}\right\rangle = \left|\begin{matrix} j & j' \\ \mu & \mu' \end{matrix}\right\rangle \tag{15.21}$$

The action of $\{\Lambda, 0\}$ on the momentum states follows from

$$\{I, a\} [\{\Lambda, 0\} |k\rangle] = \{\Lambda, 0\} \{I, \Lambda^{-1}a\} |k\rangle$$
$$= [\{\Lambda, 0\} |k\rangle] e^{ik \cdot \Lambda^{-1}a}$$
$$= [\{\Lambda, 0\} |k\rangle] e^{i\Lambda k \cdot a} = |\Lambda k\rangle e^{i\Lambda k \cdot a} \tag{15.22}$$

The action of $\{\Lambda, 0\}$ on the field component states is

$$\{\Lambda, 0\} \left|\begin{matrix} j & j' \\ \mu & \mu' \end{matrix}\right\rangle = \left|\begin{matrix} j & j' \\ \nu & \nu' \end{matrix}\right\rangle D^{jj'}_{\nu\nu';\mu\mu'}(\Lambda) \tag{15.23}$$

If the vector space that carries a manifestly covariant representation of the inhomogeneous Lorentz group has the states

$$|k\rangle \left|\begin{matrix} j & j' \\ \mu & \mu' \end{matrix}\right\rangle \tag{15.24}$$

then all states of the form

$$|\Lambda k\rangle \left|\begin{matrix} j & j' \\ \nu & \nu' \end{matrix}\right\rangle \tag{15.25}$$

are also present in the underlying vector space.

The action of the two subgroups on the two types of states is summarized by

$$\begin{array}{c|c}
 & |k\rangle \quad \left|\begin{array}{cc} j & j' \\ \mu & \mu' \end{array}\right\rangle \\
\hline
\{I, a\} & |k\rangle e^{ik\cdot a} \quad \left|\begin{array}{cc} j & j' \\ \nu & \nu' \end{array}\right\rangle \delta_{\nu\nu';\mu\mu'} \\
\{\Lambda, 0\} & |\Lambda k\rangle \quad \left|\begin{array}{cc} j & j' \\ \nu & \nu' \end{array}\right\rangle D^{jj'}_{\nu\nu';\mu\mu'}(\Lambda)
\end{array} \qquad (15.26)$$

15.4.2 Unitary irreducible representations

Suppose we have a representation of $\{\Lambda, a\}$ that is unitary and irreducible. Under restriction to the subgroup $\{I, a\}$ this reduces to a direct sum of irreducibles $\Gamma^k(\{I, a\})$ of $\{I, a\}$. The basis states are $|k; \xi\rangle$, where k is defined by the action of the translation $\{I, a\}$

$$\{I, a\} |k; \xi\rangle = |k; \xi\rangle e^{ik\cdot a} \qquad (15.27)$$

and ξ is a helicity index that distinguishes different states with the same four-momentum. A homogeneous Lorentz transformation maps the state $|k; \xi\rangle$ into a subspace of states parameterized by $k' = \Lambda k$

$$\begin{aligned}
\{I, a\}\{\Lambda, 0\} |k; \xi\rangle &= \{\Lambda, 0\}\{I, \Lambda^{-1}a\} |k; \xi\rangle \\
&= \{\Lambda, 0\} |k; \xi\rangle e^{ik\cdot \Lambda^{-1}a} \\
&= [\{\Lambda, 0\} |k; \xi\rangle] e^{i\Lambda k\cdot a}
\end{aligned} \qquad (15.28)$$

As a result

$$\{\Lambda, 0\} |k; \xi\rangle = |\Lambda k; \xi'\rangle M_{\xi'\xi}(\Lambda) \qquad (15.29)$$

where $M_{\xi'\xi}(\Lambda)$ is a matrix that remains to be determined.

This simple calculation shows that if the four-vector k parameterizes a state in an irreducible representation of the inhomogeneous Lorentz group, then the states k' with

$$k' = \Lambda k \qquad (15.30)$$

15.4 Representations of the Poincaré group

are present also. To construct the matrix $M(\Lambda)$, we choose one particular four-vector k^0 for each of the possible cases

$$
\begin{array}{llll}
\text{(i)} & k \cdot k > 0 & k^0 = (0, 0, 1, 0) & \\
\text{(ii)} & k \cdot k = 0 \quad k \neq 0 & k^0 = (0, 0, 1, +i) & \text{(a)} \\
& & k^0 = (0, 0, 1, -i) & \text{(b)} \\
\text{(iii)} & k \cdot k < 0 & k^0 = (0, 0, 0, +i) & \text{(a)} \\
& & k^0 = (0, 0, 0, -i) & \text{(b)} \\
\text{(iv)} & k \cdot k = 0 \quad k = 0 & k^0 = (0, 0, 0, 0) &
\end{array}
\tag{15.31}
$$

The states (a), (b) are related to each other by the discrete time reversal operator T. The vector k^0 is called the **little vector**.

The effect of a homogeneous Lorentz transformation on the state $|k^0; \xi\rangle$ is determined by writing each Λ as a product of two group operations

$$\Lambda = C_k H_{k^0} \tag{15.32}$$

where

$$
\begin{aligned}
H_{k^0} k^0 &= k^0 \\
C_k k^0 &= k
\end{aligned}
\tag{15.33}
$$

That is, H_{k^0} is the stability subgroup of the little vector k^0 and C_k is a coset representative that maps k^0 into k:

$$C_k k^0 = k = \Lambda k^0 \tag{15.34}$$

The little groups (stability groups) of the little vectors k^0 are

(i) $SO(2, 1)$
(ii) $ISO(2)$
(iii) $SO(3)$
(iv) $SO(3, 1)$

These are determined as follows.

Case (i) An arbitrary element in the Lie subgroup acting on k^0 must leave k^0 invariant. Linearizing, an element in the Lie algebra must annihilate k^0:

$$
\begin{bmatrix}
0 & +\theta_3 & -\theta_2 & ib_1 \\
-\theta_3 & 0 & +\theta_1 & ib_2 \\
+\theta_2 & -\theta_1 & 0 & ib_3 \\
-ib_1 & -ib_2 & -ib_3 & 0
\end{bmatrix}
\begin{bmatrix} 0 \\ 0 \\ 1 \\ 0 \end{bmatrix}
=
\begin{bmatrix} -\theta_2 \\ +\theta_1 \\ 0 \\ -ib_3 \end{bmatrix}
=
\begin{bmatrix} 0 \\ 0 \\ 0 \\ 0 \end{bmatrix}
\tag{15.35}
$$

The subalgebra leaving k^0 fixed is defined by $\theta_1 = \theta_2 = b_3 = 0$, θ_3, b_1, b_2 arbitrary. This is the three-dimensional subgroup $SO(2, 1)$ consisting of generators for rotations about the z-axis and boosts in the x- and y-directions.

Case (ii) Applying the same arguments, we find

$$\begin{bmatrix} 0 & +\theta_3 & -\theta_2 & ib_1 \\ -\theta_3 & 0 & +\theta_1 & ib_2 \\ +\theta_2 & -\theta_1 & 0 & ib_3 \\ -ib_1 & -ib_2 & -ib_3 & 0 \end{bmatrix} \begin{bmatrix} 0 \\ 0 \\ 1 \\ i \end{bmatrix} = \begin{bmatrix} -\theta_2 - b_1 \\ +\theta_1 - b_2 \\ -b_3 \\ -ib_3 \end{bmatrix} = \begin{bmatrix} 0 \\ 0 \\ 0 \\ 0 \end{bmatrix} \tag{15.36}$$

The stability subalgebra is defined by

$$\begin{aligned} b_3 &= 0 \\ b_2 &= +\theta_1 \\ b_1 &= -\theta_2 \end{aligned} \tag{15.37}$$

A general element in this subalgebra is

$$\begin{bmatrix} 0 & +\theta_3 & -\theta_2 & -i\theta_2 \\ -\theta_3 & 0 & +\theta_1 & i\theta_1 \\ +\theta_2 & -\theta_1 & 0 & 0 \\ i\theta_2 & -i\theta_1 & 0 & 0 \end{bmatrix} = \sum_i \theta_i Y_i \qquad \begin{aligned} Y_1 &= J_1 + K_2 \\ Y_2 &= J_2 - K_1 \\ Y_3 &= J_3 \end{aligned} \tag{15.38}$$

The operators Y_i obey the commutation relations

$$\begin{aligned} [Y_3, Y_1] &= -Y_2 \\ [Y_3, Y_2] &= +Y_1 \qquad\qquad ISO(2) \\ [Y_1, Y_2] &= 0 \end{aligned} \tag{15.39}$$

These are the commutation relations for the group $ISO(2)$, the group of inhomogeneous motions of the Euclidean plane R^2. Acting on the time-reversed little vector $(0, 0, 1, -i) = T(0, 0, 1, +i)$ the infinitesimal generators are $Y_1 = J_1 - K_2$, $Y_2 = J_2 + K_1$, $Y_3 = J_3$.

Case (iii) Proceeding as above

$$\begin{bmatrix} 0 & +\theta_3 & -\theta_2 & ib_1 \\ -\theta_3 & 0 & +\theta_1 & ib_2 \\ +\theta_2 & -\theta_1 & 0 & ib_3 \\ -ib_1 & -ib_2 & -ib_3 & 0 \end{bmatrix} \begin{bmatrix} 0 \\ 0 \\ 0 \\ i \end{bmatrix} = \begin{bmatrix} -b_1 \\ -b_2 \\ -b_3 \\ 0 \end{bmatrix} = \begin{bmatrix} 0 \\ 0 \\ 0 \\ 0 \end{bmatrix} \tag{15.40}$$

The subalgebra defined by $\mathbf{b} = 0$ is spanned by the angular momentum operators $\mathbf{J}$. It is $\mathfrak{su}(2)$.

15.4 Representations of the Poincaré group

Case (iv) This is the simplest case:

$$\begin{bmatrix} 0 & +\theta_3 & -\theta_2 & ib_1 \\ -\theta_3 & 0 & +\theta_1 & ib_2 \\ +\theta_2 & -\theta_1 & 0 & ib_3 \\ -ib_1 & -ib_2 & -ib_3 & 0 \end{bmatrix} \begin{bmatrix} 0 \\ 0 \\ 0 \\ 0 \end{bmatrix} = \begin{bmatrix} 0 \\ 0 \\ 0 \\ 0 \end{bmatrix} \quad (15.41)$$

The little group of this vector is the entire homogeneous Lorentz group $SO(3, 1)$. The action of the little group on the subspace of states $|k^0; \xi\rangle$ is

$$H_{k^0}|k^0; \xi\rangle = |H_{k^0}k^0; \xi'\rangle D_{\xi'\xi}(H_{k^0})$$
$$= |k^0; \xi'\rangle D_{\xi'\xi}(H_{k^0}) \quad (15.42)$$

The original representation of the inhomogeneous Lorentz group is unitary and irreducible if and only if the representation $D_{\xi'\xi}(H_{k^0})$ of the little group is unitary and irreducible.

The cases (i)–(iv) are discussed here.

Case (i) The unitary irreducible representations of the noncompact group $SO(2, 1)$ were described in Problem 5 of Chapter 11. Since $k \cdot k > 0$ describes negative mass particles, we will not need to discuss these representations here.

Case (ii) See below.

Case (iii) The unitary irreducible representations for the group $SU(2)$, which is the little group for a massive particle at rest, were described in Problem 2 of Chapter 6. They are described by an integer or half-integer: $j = 0, \frac{1}{2}, 1, \frac{3}{2}, \ldots$. The angular momentum j is a property of each massive particle.

Case (iv) The unitary irreducible representations of $SO(3, 1)$ are known but not interesting for the present discussion.

We consider the case of zero mass particles in more detail here. The unitary irreducible representations of $ISO(2)$ are constructed following the prescription we are using to study the unitary irreducible representations of the inhomogeneous Lorentz group – the method of the little group. Since $ISO(2)$ has a two-dimensional translation invariant subgroup, basis states in a unitary irreducible representation can be labeled by a vector $\kappa = (\kappa_1, \kappa_2)$ in a two-dimensional Euclidean space, $\kappa \in R^2$, $\kappa \cdot \kappa \geq 0$. If a state $|\kappa\rangle$ is in one such representation, so are all states $|\kappa'\rangle$ for which $\kappa' \cdot \kappa' = \kappa \cdot \kappa$. That is, $\kappa' = (\kappa'_1, \kappa'_2)$ is related to $\kappa = (\kappa_1, \kappa_2)$ by a rotation: $\kappa' = R(\theta)\kappa$. The invariant length $\kappa \cdot \kappa$ parameterizes the representation. As before, two cases occur (cf., Cases (i) or (iii) and Case (iv) above):

$$\begin{array}{lll} \text{(i)} & \kappa \cdot \kappa > 0 & \text{little group} = \text{Identity} \\ \text{(ii)} & \kappa \cdot \kappa = 0 & \text{little group} = ISO(2) \end{array} \quad (15.43)$$

The first case presents us with two problems. First, κ^2 is a continuous quantum number, and there are no known particles with a continuous spin index. Second, if $\kappa^2 > 0$ there must be an infinite number of states with this same continuous index, for each four-momentum value. Therefore we require $\kappa = 0$. This leaves us with the following physically allowable representations of the little group ($Y_1 \to 0$, $Y_2 \to 0$)

$$\text{EXP}(\theta_3 Y_3 + \theta_1 Y_1 + \theta_2 Y_2) = e^{i\xi\theta_3} \tag{15.44}$$

where ξ is an integer or half-integer.

The coset representatives C_k permute the four-vector subspaces:

$$C_k |k^0; \xi\rangle = |k; \xi\rangle \tag{15.45}$$

The action of an arbitrary element of the inhomogeneous Lorentz group on any state in this Hilbert space is

$$\begin{aligned}
\{\Lambda, a\} |k; \xi\rangle &= \{\Lambda, 0\} \{I, \Lambda^{-1} a\} |k; \xi\rangle \\
&= \{\Lambda, 0\} |k; \xi\rangle e^{ik \cdot \Lambda^{-1} a} \\
&= \{\Lambda, 0\} C_k |k^0; \xi\rangle e^{i\Lambda k \cdot a} \\
&= \{\Lambda C_k, 0\} |k^0; \xi\rangle e^{i\Lambda k \cdot a} \\
&= \{C_{k'} H_{k^0}, 0\} |k^0; \xi\rangle e^{i\Lambda k \cdot a} \\
&= |k'; \xi\rangle e^{i\xi\Theta} e^{i\Lambda k \cdot a}
\end{aligned} \tag{15.46}$$

where

$$C_{k'}^{-1} \Lambda C_k = H_{k^0} = \text{EXP}(\Theta J_3 + \theta_1 Y_1 + \theta_2 Y_2) \longrightarrow e^{i\xi\Theta} \tag{15.47}$$

15.5 Transformation properties

The Hilbert space that carries a unitary irreducible representation of a massless particle with helicity ξ contains all states of the form

$$\begin{aligned}|k; \xi\rangle \quad k &= \Lambda k^0 \\ k^0 &= (0, 0, 1, \pm i)\end{aligned} \tag{15.48}$$

The vector space that carries a manifestly covariant representation of a massless particle with transformation indices (j, j') contains all states of the form

$$|k\rangle \begin{vmatrix} j & j' \\ \mu & \mu' \end{vmatrix} \quad \begin{aligned} k &= \Lambda k^0 \\ k^0 &= (0, 0, 1, \pm i) \end{aligned} \tag{15.49}$$

To compare these two ways of describing a massless particle we compare transformation properties of their states.

15.5 Transformation properties

A. $\{H_{k^0}, 0\}$ on $|k^0; \xi\rangle$

$$\{H_{k^0}, 0\}|k^0; \xi\rangle = |k^0; \xi\rangle e^{i\xi\Theta} \qquad (15.50)$$

where $H_{k^0} = \text{EXP}(\Theta J_3 + \theta_1 Y_1 + \theta_2 Y_2)$.

B. $\{H_{k^0}, 0\}$ on $|k^0\rangle|{}^{j\ j'}_{\mu\ \mu'}\rangle$ The little group maps k^0 to k^0 but acts in a nontrivial way on the spin states

$$\{H_{k^0}, 0\}|k^0\rangle\left|{}^{j\ j'}_{\mu\ \mu'}\right\rangle = |k^0\rangle\left|{}^{j\ j'}_{v\ v'}\right\rangle D^{jj'}_{vv';\mu\mu'}(H_{k^0}) \qquad (15.51)$$

The direct product representation $D^{jj'}$ has the following form

$$D^{j0}(H_{k^0}) = \text{EXP}\left(\theta_3 J_3^{(j)} + \theta_1(J_1^{(j)} + iJ_2^{(j)}) + \theta_2(J_2^{(j)} - iJ_1^{(j)})\right)$$

$$= \text{EXP}\left(\theta_3 J_3^{(j)} + (\theta_1 - i\theta_2)(J_1^{(j)} + iJ_2^{(j)})\right)$$

$$= \begin{bmatrix} e^{ij\theta_3} & * & * & * & * \\ & e^{i(j-1)\theta_3} & * & * & * \\ & & \ddots & * & * \\ & & & & * \\ & & & & e^{-ij\theta_3} \end{bmatrix} \qquad (15.52)$$

$$D^{0j'}(H_{k^0}) = \text{EXP}\left(\theta_3 J_3^{(j')} + \theta_1(J_1^{(j')} - iJ_2^{(j')}) + \theta_2(J_2^{(j')} + iJ_1^{(j')})\right)$$

$$= \text{EXP}\left(\theta_3 J_3^{(j')} + (\theta_1 + i\theta_2)(J_1^{(j')} - iJ_2^{(j')})\right)$$

$$= \begin{bmatrix} e^{ij'\theta_3} & & & & \\ * & e^{i(j'-1)\theta_3} & & & \\ * & * & \ddots & & \\ * & * & * & \ddots & \\ * & * & * & * & e^{-ij'\theta_3} \end{bmatrix} \qquad (15.53)$$

By comparing Eq. (15.50) with Eq. (15.52) and Eq. (15.53) we reach the following conclusions.

The state $|k^0\rangle|{}^{j\ 0}_{j\ 0}\rangle$ transforms identically to $|k^0; \xi\rangle$ if $\xi > 0$ and $j = +\xi$.
The state $|k^0\rangle|{}^{0\ j'}_{0\ -j'}\rangle$ transforms identically to $|k^0; \xi\rangle$ if $\xi < 0$ and $j' = -\xi$.

272   Maxwell's equations

If $|\psi\rangle$ is any physical state, it can be expanded in terms of either the helicity basis states $|k;\xi\rangle$ or the direct product states $|k\rangle|{}^{j\ j'}_{\mu\ \mu'}\rangle$:

$$|\psi\rangle = \sum_{k,\xi} |k;\xi\rangle\langle k;\xi|\psi\rangle$$

$$|\psi\rangle = \sum_{k,\mu\mu'} |k\rangle\left|{}^{j\ j'}_{\mu\ \mu'}\right\rangle\left\langle k; {}^{j\ j'}_{\mu\ \mu'}\bigg|\psi\right\rangle$$

The amplitudes of the projection of $|\psi\rangle$ onto the basis states are $\langle k;\xi|\psi\rangle$ in the first case and $\langle k; {}^{j\ j'}_{\mu\ \mu'}|\psi\rangle$ in the second. In both cases the sum extends over all k vectors for which $\Lambda k \cdot \Lambda k = 0, k \neq 0$. In the first case the sum extends over the appropriate helicity states ξ ($\xi = \pm 1$ for photons). In the second case the sum extends over the appropriate values of μ, μ': $-j \leq \mu \leq +j$, $-j' \leq \mu' \leq +j'$.

We discuss the positive helicity state $\xi = j > 0$ first. The amplitude $\langle k^0; j|\psi\rangle$ of the state $|k^0; j\rangle$ in any physical state $|\psi\rangle$ may be arbitrary. This is simply the amplitude of the massless particle of helicity j in the state $|\psi\rangle$. The amplitude $\langle k^0; {}^{j\ 0}_{j\ 0}|\psi\rangle$ in the same physical state $|\psi\rangle$ is the same. The amplitudes of the states $\langle k^0; {}^{j\ 0}_{m\ 0}|\psi\rangle$, $m \neq j$, must all vanish. These states are all superfluous – allowed in the manifestly covariant representation but not present in the Hilbert space that carries the unitary irreducible representation. A simple linear way to enforce this condition on the superfluous amplitudes is to require

$$\left\{J_3^{(j)}k_3^0 - jk_4^0 I_{2j+1}\right\}\left\langle k^0; {}^{j\ 0}_{m\ 0}\bigg|\psi\right\rangle = 0 \tag{15.54}$$

The matrix within the bracket $\{\cdot\}$ is diagonal, with the coefficient $(j - j)k_3^0 = 0$ multiplying the allowed amplitude $\langle k^0; {}^{j\ 0}_{j\ 0}|\psi\rangle$ and nonzero coefficients $(m - j)k_3^0$ multiplying the amplitudes $\langle k^0; {}^{j\ 0}_{m\ 0}|\psi\rangle$. Since $(m - j)k_3^0 \neq 0$, the amplitudes that are absent in the description of a physical state $(m \neq j)$ must vanish.

For the negative helicity states $\xi = -j$ we have by a completely similar argument

$$\left\{J_3^{(j)}k_3^0 + jk_4^0 I_{2j+1}\right\}\left\langle k^0; {}^{0\ j'}_{0\ m'}\bigg|\psi\right\rangle = 0 \tag{15.55}$$

C. Other k-vector subspaces The coset operator C_k maps the state $|k^0;\xi\rangle$ into the state

$$C_k|k^0;\xi\rangle = |k;\xi\rangle \tag{15.56}$$

and the subspace $|k^0\rangle|{}^{j\ j'}_{\mu\ \mu'}\rangle$ into the subspace $|k\rangle|{}^{j\ j'}_{\nu\ \nu'}\rangle$ through the following nontrivial similarity transformation

$$C_k|k^0\rangle|{}^{j\ j'}_{\mu\ \mu'}\rangle = |k\rangle|{}^{j\ j'}_{\nu\ \nu'}\rangle D^{jj'}_{\nu\nu';\mu\mu'}(C_k) \tag{15.57}$$

The condition on the amplitude $\langle k; {}^{j\ j'}_{\mu\ \mu'}|\psi\rangle$ in the subspace $|k\rangle$ is related to the conditions (15.54) and (15.55) in the subspace $|k^0\rangle$ by a similarity transformation

$$M^{jj'}(k^0)\langle k^0; {}^{j\ j'}_{\mu\ \mu'}|\psi\rangle = 0$$

$$C_k M^{jj'}(k^0) C_k^{-1}\langle k; {}^{j\ j'}_{\mu\ \mu'}|\psi\rangle = 0 \tag{15.58}$$

For the positive helicity state $\xi = j$ the matrix $M^{jj'}(k^0) = M^{j0}(k^0)$ is given in (15.54). The coset representative may be taken as the product of a boost in the z-direction,

$$B_z(k)(0, 0, 1, i) = (0, 0, k, ik) \tag{15.59}$$

followed by a rotation

$$R(k)(0, 0, k, ik) = (k_1, k_2, k_3, ik_4) \qquad k_1^2 + k_2^2 + k_3^2 = k_4^2 = k^2 \tag{15.60}$$

For $j = 1$ the similarity transformation becomes

$$R(\mathbf{k}) B_z(k_4) \left\{ J_3^{(j)} - j I_{2j+1} \right\} B_z^{-1}(k_4) R^{-1}(\mathbf{k})$$

$$= \{\mathbf{J} \cdot \mathbf{k} - 1 k_4 I_3\} \langle k; {}^{1\ 0}_{\mu\ 0}|\psi\rangle = 0 \tag{15.61}$$

as the linear constraint that must be satisfied in the subspace $|k\rangle |{}^{1\ 0}_{\mu\ 0}\rangle$. The negative helicity states satisfy the constraint

$$\{\mathbf{J} \cdot \mathbf{k} + 1 k_4 I_3\} \langle k; {}^{0\ 1}_{0\ \mu'}|\psi\rangle = 0 \tag{15.62}$$

15.6 Maxwell's equations

The constraint equation is conveniently expressed in the coordinate rather than the momentum representation by inverting the original Fourier transform that brought us from the coordinate to the momentum representation

$$\langle k|x\rangle \left\{ \mathbf{J} \cdot \frac{1}{i}\nabla + 1\frac{1}{i}\frac{\partial}{\partial(ict)} I_3 \right\} \langle x|k\rangle \langle k; {}^{1\ 0}_{m\ 0}|\psi\rangle = 0 \tag{15.63}$$

If we define complex fields $\langle x|k\rangle\langle k; {}^{j\ 0}_{m\ 0}|\psi\rangle$ by $\psi_{jm}(x)$, ($j = 1$, $m = +1, 0, -1$ or x, y, z or $1, 2, 3$) then this equation simplifies to a differential equation. In the standard representation for the angular momentum operators for $j = 1$ we find

$$\begin{bmatrix} -\dfrac{i}{c}\dfrac{\partial}{\partial t} & +\partial_3 & -\partial_2 \\ -\partial_3 & -\dfrac{i}{c}\dfrac{\partial}{\partial t} & +\partial_1 \\ +\partial_2 & -\partial_1 & -\dfrac{i}{c}\dfrac{\partial}{\partial t} \end{bmatrix} \begin{bmatrix} B_1 + iE_1 \\ B_2 + iE_2 \\ B_3 + iE_3 \end{bmatrix} = 0 \qquad (15.64)$$

$$-\dfrac{i}{c}\dfrac{\partial}{\partial t}(B + iE)_1 + \partial_3(B + iE)_2 - \partial_2(B + iE)_3 = 0$$

$$-\partial_3(B + iE)_1 - \dfrac{i}{c}\dfrac{\partial}{\partial t}(B + iE)_2 + \partial_1(B + iE)_3 = 0 \qquad (15.65)$$

$$+\partial_2(B + iE)_1 - \partial_1(B + iE)_2 - \dfrac{i}{c}\dfrac{\partial}{\partial t}(B + iE)_3 = 0$$

These three equations are summarized as a vector equation by

$$-\dfrac{i}{c}\dfrac{\partial}{\partial t}(\mathbf{B} + i\mathbf{E}) - \nabla \times (\mathbf{B} + i\mathbf{E}) = 0 \qquad (15.66)$$

By taking the real and imaginary part of this complex equation we find

$$\begin{array}{ll} \text{Re} & +\dfrac{1}{c}\dfrac{\partial \mathbf{E}}{\partial t} - \nabla \times \mathbf{B} = 0 \\ \\ \text{Im} & -\dfrac{1}{c}\dfrac{\partial \mathbf{B}}{\partial t} - \nabla \times \mathbf{E} = 0 \end{array} \qquad (15.67)$$

These are Maxwell's equations for positive helicity $+1$ massless particles (photons):

$$\begin{aligned} \nabla \times \mathbf{B} - \dfrac{1}{c}\dfrac{\partial \mathbf{E}}{\partial t} &= 0 \\ \nabla \times \mathbf{E} + \dfrac{1}{c}\dfrac{\partial \mathbf{B}}{\partial t} &= 0 \end{aligned} \qquad (15.68)$$

The equations for negative helicity states are derived from the complex conjugate representation D^{01} and are

$$\begin{bmatrix} +\dfrac{i}{c}\dfrac{\partial}{\partial t} & +\partial_3 & -\partial_2 \\ -\partial_3 & +\dfrac{i}{c}\dfrac{\partial}{\partial t} & +\partial_1 \\ +\partial_2 & -\partial_1 & +\dfrac{i}{c}\dfrac{\partial}{\partial t} \end{bmatrix} \begin{bmatrix} B_1 - iE_1 \\ B_2 - iE_2 \\ B_3 - iE_3 \end{bmatrix} = 0 \qquad (15.69)$$

It is easily verified that the resulting equations are identical to Eq. (15.68).

15.7 Conclusion

In some sense, Maxwell's equations were a historical accident. Had the discovery of quantum mechanics preceeded the unification of electricity and magnetism, Maxwell's equations might not have loomed so large in the history of physics.

In the quantum description of the electromagnetic field, photons are the fundamental building blocks. Photons are described by a four-vector k that obeys $k \cdot k = 0$ in free space, and a helicity index indicating a projection of an angular momentum ± 1 along the direction of propagation of the photon. Every physical state is described by a superposition of the photon basis states, and every superposition describes a possible physical state. In this description of the electromagnetic field in free space no constraint equations are necessary.

The nineteenth century description of the electromagnetic field proceeds along somewhat different lines. A multicomponent field $(\mathbf{E}, \mathbf{B})$ is introduced at each point in space-time. The components of the field transform in a very elegant way under homogeneous Lorentz transformations (as a tensor). If the field is Fourier transformed from the coordinate to the momentum representation, then each four-momentum has six components associated with it. These are the components of a second order antisymmetric tensor. Since the quantum description has only two independent components associated with each four-momentum, there are four dimensions worth of linear combinations of the classical field components that do not describe physically allowed states, for each four-momentum. Some mechanism must be derived for annihilating these superpositions. This mechanism is the set of equations discovered by Maxwell. In this sense, Maxwell's equations are an expression of our ignorance.

It is ironic that the first truly powerful applications of group theory were to the solutions of equations. We now understand that group theory, by pointing to the appropriate Hilbert space for the electromagnetic field, allows us to relate physical states to arbitrary superpositions of basis states. Since no superpositions are forbidden, no equations are necessary.

15.8 Problems

1. So, where are the divergence equations? In the special frame with little vector $k^0 = (0, 0, 1, i)$ the only nonvanishing component of the field, $\langle k; {}^{j=1}_{m} {}^{0}_{0} | \psi \rangle$, is the component with $m = +1$ (cf., Eq. (15.54)). The coordinates are $-(v_x + iv_y)$. The vector $\mathbf{v} = (v_x, v_y, 0)$ represented by this coordinate is orthogonal to the spacial part of the little vector $\mathbf{k}^0 = (0, 0, 1)$: $\mathbf{k}^0 \cdot \mathbf{v} = 0$. Under boosts B_z and rotations, the nonvanishing component of the boosted field is orthogonal to the spacial part of the $\mathbf{k}$ vector: $\mathbf{k} \cdot \mathbf{v}(\mathbf{k}) = 0$. Backtransforming from the Fourier to the spacial representation,

show that

$$\mathbf{k} \cdot \mathbf{v}(\mathbf{k}) = 0 \xrightarrow{FT^{-1}} \nabla \cdot (\mathbf{B} + i\mathbf{E}) = 0$$

Taking the real and imaginary parts of this equation give the source-free divergence equations $\nabla \cdot \mathbf{E} = 0$ and $\nabla \cdot \mathbf{B} = 0$. Show this.

2. When sources are present the Maxwell equations are modified in a way that is most clearly expressed in the "manifestly covariant representation." If particle j at $\mathbf{x}(j)$ has electric charge e_j and magnetic charge m_j, the electric and magnetic charge densities and current densities are defined as follows.

	Electric	Magnetic
Charge density	$\rho_e(\mathbf{x}, t) = \sum_j e_j \mathbf{x}_j(t)$	$\rho_m(\mathbf{x}, t) = \sum_j m_j \mathbf{x}_j(t)$
Current density	$\mathbf{J}_e(\mathbf{x}, t) = \sum_j e_j \dfrac{d\mathbf{x}_j(t)}{dt}$	$\mathbf{J}_m(\mathbf{x}, t) = \sum_j m_j \dfrac{d\mathbf{x}_j(t)}{dt}$
Conservation law	$\nabla \cdot \mathbf{J}_e(\mathbf{x}, t) + \dfrac{\partial \rho_e(\mathbf{x}, t)}{\partial t} = 0$	$\nabla \cdot \mathbf{J}_m(\mathbf{x}, t) + \dfrac{\partial \rho_m(\mathbf{x}, t)}{\partial t} = 0$

The conservation equations enforce the conditions of charge conservation (both electric and magnetic, separately).

In order to extend Maxwell's equations to include sources, the source free (homogeneous) equations (15.66) must be coupled to the source terms in such a way that the symmetry properties on the left (the fields) match the symmetry properties of the sources. Thus, the right-hand side must include only vector terms, and these terms must have appropriate transformation properties under the discrete operations T, P, TP. The result is unique up to scale factor:

$$\left(\nabla \times + \frac{i}{c}\frac{\partial}{\partial t}\right)(\mathbf{B} + i\mathbf{E}) = \frac{1}{i}\frac{4\pi}{c}(\mathbf{J}_m + i\mathbf{J}_e) \qquad (15.70)$$

The factor 4π is the surface area of the unit sphere in R^3, and the factor $1/c$ on the right is determined by the system of units used (Gaussian).

a. Show that Maxwell's equations with sources are

$$\nabla \times \mathbf{B} - \frac{1}{c}\frac{\partial \mathbf{E}}{\partial t} = +\frac{4\pi}{c}\mathbf{J}_e$$

$$\nabla \times \mathbf{E} + \frac{1}{c}\frac{\partial \mathbf{B}}{\partial t} = -\frac{4\pi}{c}\mathbf{J}_m$$

b. Show that the Maxwell equations with sources are invariant under the simultaneous transformation

$$\mathbf{B} + i\mathbf{E} \to \mathbf{B}' + i\mathbf{E}' = e^{i\phi}(\mathbf{B} + i\mathbf{E})$$
$$\mathbf{J}_m + i\mathbf{J}_e \to \mathbf{J}'_m + i\mathbf{J}'_e = e^{i\phi}(\mathbf{J}_m + i\mathbf{J}_e)$$

In particular, show that for $\phi = \pi/2$ this is the dual transformation $(\mathbf{B}, \mathbf{E}) \to (\mathbf{E}, -\mathbf{B})$.

c. Take the divergence of both sides of Eq. (15.70). Use the vector identity div curl $(*) = 0$, for $* =$ anyvector. Show

$$\frac{i}{c}\frac{\partial}{\partial t}\{\nabla \cdot (\mathbf{B} + i\mathbf{E}) - 4\pi(\rho_m + i\rho_e)\} = 0$$

d. By taking real and imaginary parts and integrating over time, find the following:

$$\nabla \cdot \mathbf{B}(\mathbf{x}, t) = 4\pi \rho_m(\mathbf{x}, t) + C_m(\mathbf{x})$$
$$\nabla \cdot \mathbf{E}(\mathbf{x}, t) = 4\pi \rho_e(\mathbf{x}, t) + C_e(\mathbf{x})$$

e. Two "constants of integration" appear in these equations. They are functions of space but not of time. If these "constant functions of position" are zero the Maxwell divergence equations result. Provide arguments to show that these constants should be zero. These should take the form of investigating what the field looks like when all particles head towards "infinity."

Remark So far magnetic charges (monopoles) have not been observed, despite being predicted by supersymmetric theories and searched for actively by experimentalists. This means that the first divergence equation is $\nabla \cdot \mathbf{B} = 0$.

3. In order to describe gravitational waves in free space it is possible to use the representation $D^{jj'+j'j}(\Lambda)$, with $j - j' = \pm 2$. In the case with $(j, j') = (2, 0)$ a curl equation is introduced to suppress four nonphysical complex amplitudes. Show that the gravitational wave equations in free space are

$$-\frac{2i}{c}\frac{\partial}{\partial t}(\mathbf{G_m} + i\mathbf{G_e}) - \nabla \times (\mathbf{G_m} + i\mathbf{G_e}) = 0 \tag{15.71}$$

The real and imaginary parts of this complex equation are

$$\begin{aligned} \text{Re} &\quad +\frac{2}{c}\frac{\partial \mathbf{G_e}}{\partial t} - \nabla \times \mathbf{G_m} = 0 \\ \text{Im} &\quad -\frac{2}{c}\frac{\partial \mathbf{G_m}}{\partial t} - \nabla \times \mathbf{G_e} = 0 \end{aligned} \tag{15.72}$$

The fields $\mathbf{G_e}$ and $\mathbf{G_m}$ are called the gravitoelectric and gravitomagnetic fields. These fields can be treated in Cartesian coordinates as real symmetric 3×3 traceless matrices and in spherical coordinates as five-component rank-two spherical tensors. In the latter case the curl operator is $\mathbf{J} \cdot \nabla$, where $\mathbf{J}$ is the 5×5 angular momentum operator:

$$\mathbf{J} \cdot \nabla = \begin{bmatrix} +2\partial_0 & \sqrt{4}\partial_+ & 0 & 0 & 0 \\ \sqrt{4}\partial_- & +1\partial_0 & \sqrt{6}\partial_+ & 0 & 0 \\ 0 & \sqrt{6}\partial_- & 0\partial_0 & \sqrt{6}\partial_+ & 0 \\ 0 & 0 & \sqrt{6}\partial_- & -1\partial_0 & \sqrt{4}\partial_+ \\ 0 & 0 & 0 & \sqrt{6}\partial_- & -2\partial_0 \end{bmatrix}$$

In Cartesian coordinates the curl operator is slightly more complicated. The Maxwell-like equations for the gravitoelectric and gravitomagnetic field are

$$\begin{bmatrix} 0 & \partial_y & -\partial_x & 2\partial_z & 0 \\ -\partial_y & 0 & \partial_z & -\partial_x & -\sqrt{3}\partial_x \\ \partial_x & -\partial_z & 0 & -\partial_y & \sqrt{3}\partial_y \\ -2\partial_z & \partial_x & \partial_y & 0 & 0 \\ 0 & \sqrt{3}\partial_x & -\sqrt{3}\partial_y & 0 & 0 \end{bmatrix} \begin{pmatrix} F_1 \\ F_2 \\ F_3 \\ F_4 \\ F_5 \end{pmatrix} + \frac{2}{c}\frac{\partial}{\partial t} \begin{pmatrix} G_1 \\ G_2 \\ G_3 \\ G_4 \\ G_5 \end{pmatrix} = 0$$

$$\begin{bmatrix} 0 & \partial_y & -\partial_x & 2\partial_z & 0 \\ -\partial_y & 0 & \partial_z & -\partial_x & -\sqrt{3}\partial_x \\ \partial_x & -\partial_z & 0 & -\partial_y & \sqrt{3}\partial_y \\ -2\partial_z & \partial_x & \partial_y & 0 & 0 \\ 0 & \sqrt{3}\partial_x & -\sqrt{3}\partial_y & 0 & 0 \end{bmatrix} \begin{pmatrix} G_1 \\ G_2 \\ G_3 \\ G_4 \\ G_5 \end{pmatrix} - \frac{2}{c}\frac{\partial}{\partial t} \begin{pmatrix} F_1 \\ F_2 \\ F_3 \\ F_4 \\ F_5 \end{pmatrix} = 0$$

The relation between the five components of the rank-two spherical tensor and the nine matrix elements of a second order Cartesian tensor are (Ramos and Gilmore, 2006)

$$F_{ij} = \begin{pmatrix} F_{11} & F_{12} & F_{13} \\ F_{21} & F_{22} & F_{23} \\ F_{31} & F_{32} & F_{33} \end{pmatrix} = \begin{pmatrix} F_4 - \frac{1}{\sqrt{3}}F_5 & F_1 & F_3 \\ F_1 & -F_4 - \frac{1}{\sqrt{3}}F_5 & F_2 \\ F_3 & F_2 & +\frac{2}{\sqrt{3}}F_5 \end{pmatrix}$$

The matrix components obey $F_{ij} = F_{ji}$, $\sum_i F_{ii} = 0$, and $\partial^i F_{ij} = 0$. The gravitoelectric and gravitomagnetic tensors have the same discrete symmetries as the electric and magnetic fields.

4. Follow the outline of Problem 2 to show the following.
   a. The gravitoelectric and gravitomagnetic fields satisfy divergence conditions in free space. Write them down.
   b. In the presence of source terms (stationary and moving masses) the homogeneous equations are "dressed" with source terms on the right-hand side. In Cartesian coordinates the source term for the gravitoelectric field is $U_{ij} = \sum_k m_k (\mathbf{x}_k(t)\mathbf{x}_k(t))_{ij}$, and the form of the rank-two tensor is determined from the expression at the conclusion of Problem 3. What is the gravitational analog of the magnetic monopole?
   c. The coupled equations are invariant under a gauge transformation of the first kind of both the gravitoelectric and gravitomagnetic fields and the current terms: $\mathbf{G_m} + i\mathbf{G_e} \to e^{i\phi}(\mathbf{G_m} + i\mathbf{G_e})$ and $\mathbf{J_m} + i\mathbf{J_e} \to e^{i\phi}(\mathbf{J_m} + i\mathbf{J_e})$. Show this.
   d. What are the divergence equations in the presence of moving matter?

5. Construct the source-free field equations for gravitons for the $D^{jj}(\Lambda)$ representation, with $j = 1$. Show that there are seven constraints that correspond to (J, M) with $(J, M) = (0, 0), (1, 0), (1, \pm 1), (2, 0), (2 \pm 1)$. What are these equations in the stan-

dard differential representation? How are source terms (moving masses) coupled to these equations?

6. Observed redshifts are extremely important in interpreting the history of our universe. There appear to be four sources for redshifts (so far):
   (i) Döppler shift;
   (ii) gravitational redshift;
   (iii) universal expansion redshift;
   (iv) Mach redshift.

   The Döppler shift has been recognized since 1842. Radiation from a source is redshifted if the source and observer are moving away from each other, blueshifted if they are moving towards each other. The gravitational redshift is a consequence of the conservation of energy. As a photon climbs out of a gravitational potential it loses energy and its frequency is redshifted. The universal expansion redshift is a consequence of the expansion of the universe. Two points (e.g., a source and an observer) that are at rest with respect to the the COBE background radiation (the "aether") move apart due to the expansion of the universe. If a wave with N wavelengths connects the two (distance $N\lambda$), as time goes on and the distance increases the wavelength must also increase to $N\lambda'$. This redshift source is sometimes confused with the Döppler shift because the two points appear to be moving apart due to the expansion of the universe. The fourth redshift source is controversial. Mach proposed that the inertia (mass) of a particle depends on the distribution of mass in the universe. Field theory requires that this information is transmitted by the fields set up by charges (electric, magnetic (if they exist), and masses). In fact, the exchange of virtual gravitons provides information about the distribution of mass in the universe within our horizon and should contribute to the mass (inertia) of a particle in the same way that exchange of virtual photons contributes to the energy (mass) changes in the Lamb effect.

   a. Assume that the energy density in the universe has the form $\rho(\mathbf{x}, t) = \rho(t)$ (time dependent only). Assume that since recombination ($\sim$300 kY after the Big Bang) the horizon of the accessible universe has been uniformly expanding. Assume that the mass of the electron comes from two sources: interactions with electromagnetic radiation and interaction with graviational radiation. Compute how the mass changes with time.

   b. Estimate the mass dependence of the electron–proton mass ratio $m_e(t)/M_p(t)$.

   c. If the electron mass is increasing in time because of the expansion of the horizon with time, then the electron was less massive in the past. Radiation emitted from the hydrogen atom has frequency $v = \frac{1}{2}(mc^2/\hbar) \times |(1/n_1^2 - 1/n_2^2)|$ where n_1 and n_2 are the principal quantum numbers of the two states involved in the transition and m is the reduced mass of the electron–proton system. Show that H_α photons emitted from hydrogen at rest with the COBE background are redshifted because of the universal expansion *and* because the electron was less massive in the past. Disentangle these two effects and argue that the Mach shift aliases the universal expansion redshift.

Maxwell's equations

7. The locally flat metric of space-time and the metric representing a certain type of gravitational field are given by the matrices

$$g_{\text{flat}} = \begin{bmatrix} c^2 & & & \\ & -1 & & \\ & & -1 & \\ & & & -1 \end{bmatrix} \qquad g_{\text{grav}} = \begin{bmatrix} c^2\left(1+\dfrac{2\Phi(x)}{c^2}\right) & & & \\ & -1 & & \\ & & -1 & \\ & & & -1 \end{bmatrix}$$

Here $\Phi(x)$ is the local Newtonian gravitational field. Find a locally linear coordinate transformation S that brings the curved metric to flat form: $S^t g_{\text{grav}} S = g_{\text{flat}}$. Interpret S in terms of a locally free-falling coordinate transformation.

8. **Gauss' law on the sphere S^2** Gauss' law in R^3 states

$$\oint \mathbf{E} \cdot d\mathbf{S} = \int 4\pi \rho \, dV$$

The integral on the left is over the surface bounding the volume V over which the integral on the right extends, $\mathbf{E}$ is the electric field and ρ is the charge density. For a charge q at the origin of a sphere of radius a, $\rho(x) = q\delta(x)$, The $\mathbf{E}$ field is spherically symmetric, and Gauss' Law reduces to

$$4\pi a^2 |\mathbf{E}(a)| = 4\pi q$$

From this, and symmetry, we deduce the Coulomb/gravitational force law:

$$\mathbf{E}(a) = \frac{q}{a^2} \frac{\mathbf{a}}{|\mathbf{a}|}$$

By completely similar arguments Gauss' Law in the plane R^2 gives $|\mathbf{E}(a)| = q/|\mathbf{a}|$.

Assume a Gauss law ($\oint \mathbf{E} \cdot d\mathbf{S} = \int 2\pi\rho \, dA$) holds on the sphere S^2. Place a charge q on the north pole of a sphere of radius R (see Fig. 15.1).

a. An observation point subtends an angle θ when measured from the center of the sphere S^2 (c.f., Fig. 15.1). Show that its distance a from the north pole is $a = R\theta$

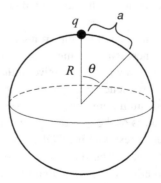

Figure 15.1. A charge q is placed on the north pole of a sphere of radius R.

15.8 Problems

and the circumference of a circle of latitude through this point is $2\pi R \sin\theta$. Use this information to deduce

$$|E| = \frac{q}{R\sin\theta} = \frac{q}{R\sin(a/R)}$$

Conclude that the field is stronger than the q/a form it would have in a plane.

b. Show that this effective strengthening is due to the relative compression of the E field lines (compared to the planar case) due to the positive curvature of the sphere.

c. Rewrite this result as

$$|E| = \frac{q}{R\sin(a/R)} = \frac{q(a)}{a} \qquad q(a) = q\left(\frac{a/R}{\sin(a/R)}\right)$$

where a ($a = R\theta$) is the distance from the charge to the observation point.

d. If the observer thinks (s)he is in a flat space, conclude (s)he will think the effective charge depends on the distance from the observation point. In particular, if $a = ct$, the further back in time the observer looks, the stronger (s)he will think the charge is.

9. **Gauss' law on rank-one homogeneous spaces** The invariant metric and measure on the three Riemannian symmetric spaces $H^n = SO(n, 1)/SO(n)$, $R^n = ISO(n)/SO(n)$, and $S^n = SO(n+1)/SO(n)$ are

$$ds^2 = \frac{dr^2}{1 - kr^2} + r^2 \sum_{j=2}^{n}(\sin\theta_2 \sin\theta_3 \cdots \sin\theta_{j-1} d\theta_j)^2$$

where $k = (-1, 0, +1)$ for H^n, R^n, S^n and radial coordinates are used:

$$\begin{aligned}x_1 &= r\cos\theta_2 \\ x_2 &= r\sin\theta_2 \cos\theta_3 \\ &\vdots \\ x_{n-1} &= r\sin\theta_2 \sin\theta_3 \cdots \sin\theta_{n-1} \cos\theta_n \\ x_n &= r\sin\theta_2 \sin\theta_3 \cdots \sin\theta_{n-1} \sin\theta_n\end{aligned}$$

a. Derive the metric for H^n, S^n from Eq. (12.9) and the coordinate transformation above.

b. Assume a Gauss Law of the form

$$\oint \mathbf{E} \cdot d\mathbf{S} = \int \Omega \rho(x) dV$$

Compute Ω, the surface area of the unit sphere $S^{n-1} \subset H^n$, R^n, or S^n. (Hint: use $\int e^{-x^2} dx = \sqrt{\pi}$, carry the n-fold integral out in Cartesian and radial coordinates, and show $\Omega = 2\pi^{n/2}/\Gamma(n/2)$.)

c. Carry out the integral for a charge q at the origin to show

$$|E| a^{n-1} = q$$

d. Show that the distance d from the origin to the sphere of radius a is

$$d(a) = \int_0^a \frac{dr}{\sqrt{1-kr^2}} \longrightarrow \begin{cases} \sinh^{-1} a & k = -1 \\ a & k = 0 \\ \sin^{-1} a & k = +1 \end{cases}$$

e. Express the electric field strength as

$$|\mathbf{E}| = \frac{q(d)}{d^{n-1}} \qquad q(d) = q \times \begin{cases} \left(\dfrac{d/R}{\sinh(d/R)}\right)^{n-1} \\ 1 \\ \left(\dfrac{d/R}{\sin(d/R)}\right)^{n-1} \end{cases}$$

Here R is some characteristic size scale for the spaces H^n, S^n.

f. Show that in the two curved spaces the observed charge is renormalized upward in S^n, downward in H^n, with lookback time. Give a physical interpretation involving compression or rarefaction of field lines. How does this renormalization depend on R, c, t?

10. The special theory of relativity is based on two assumptions that have been raised to the status of axioms:
    1. The speed of light is the same in all inertial frames.
    2. Physical laws have the same form in all inertial frames.

    The second axiom has been rephrased in the spirit of thermodynamics: "It is impossible, by any experiment, to determine the absolute motion of an inertial frame of reference." This form is motivated by the failure of the Michelson–Morley experiment to detect the motion of the Earth through the "aether." In this form the second axiom is false: This has been shown by measurements of the microwave background radiation, which contains a nonzero dipole moment. This shows that the Solar System of galaxies is moving through the microwave background at a speed of ~370 km/s in the direction with galactic coordinates $(l, b) = (263°, 48°)$.

    a. What effect does the ability to determine an absolute frame of reference have on the special theory of relativity?

    b. Assume the temperature distribution of the microwave background is $T(\theta, \phi; t) = \sum_{l,m} A_m^l(t) Y_m^l(\theta, \phi)$. How do you use this information to determine a frame that is: not translating? not rotating?

    c. Since an absolute rest frame (nontranslating, non rotating) is defined by thermodynamic measurements, argue that this special reference frame is statistically determined.

    d. Show that the determination of this special frame of reference is uncertain due to the uncertainty relations of statistical mechanics: $\Delta U \, \Delta(1/T) \geq k$ in the entropy representation (Gilmore, 1985).

    e. If thermodynamic background fields of spin $\frac{1}{2}$ (neutrinos) and spin 2 (gravitons) also exist, show that they also can be used to determine special rest frames. Argue why, or why not, the special frames defined by $j = \frac{1}{2}, 1, 2$ are the same. What happens if they are different?

f. Assume (for simplicity) that there is only one massive object in the universe and that it moves through the microwave background radiation with a velocity $v(t)$. Show that its velocity decays to zero according to $v(t) \simeq v(t_0)e^{-(t-t_0)/\tau}$ because it is moving through a viscous medium. Estimate τ and present your answer in the form τ/T_p, where T_p is the present age of the universe ($T_p \simeq 13.7$ BY). To carry out this estimate you may assume the massive object is a black body – in fact, assume it is a black hole with mass M, radius R at temperature T_{BH}. Use the standard relations for a neutral nonrotating black hole $R = 2GM/c^2$, $T_{BH} = \hbar c^3/8\pi kGM$. You can assume that the mass M is sufficiently large that the temperature T_{BH} can be neglected (set to zero). Assume that the absorption (geometric) cross section for radiation on a black hole is $\gamma \pi R^2$, where $\gamma = 3^3/2^2$. Note that the problem of slowing down in a viscous medium was discused by Einstein in another of the papers from his "annus mirabilis," the precursor of the fluctuation–dissipation theorem.

16

Lie groups and differential equations

Lie group theory was initially developed to facilitate the solution of differential equations. In this guise its many powerful tools and results are not extensively known in the physics community. This chapter is designed as an antidote to this anemia. Lie's methods are an extension of Galois' methods for algebraic equations to the study of differential equations. The extension is in the spirit of Galois' work: the technical details are not similar. The principle observation – Lie's great insight – is that the simple constant that can by added to any indefinite integral of $dy/dx = g(x)$ is in fact an element of a continuous symmetry group – the group that maps solutions of the differential equation into other solutions. This observation was used – exploited – by Lie to develop an algorithm for determining when a differential equation had an invariance group. If such a group exists, then a first order ordinary differential equation can be integrated by quadratures, or the order of a higher order ordinary differential equation can be reduced.

Galois inspired Lie. If the discrete invariance group of an algebraic equation could be exploited to generate algorithms to solve the algebraic equation "by radicals," might it be possible that the continuous invariance group of a differential equation could be exploited to solve the differential equation "by quadratures"? Lie showed emphatically in 1874 that the answer is YES!, and work has hardly slowed down in the field that he pioneered from that time to the present.

But what is the group that leaves the solutions of a differential equation invariant – or maps solutions into solutions? It turns out to be none other than the trivial constant that can be added to any indefinite integral. The additive constant is an element in a translation group.

We outline Lie's methods for first order ordinary differential equations. First, we study the simplest first order equation in one independent variable x and one dependent variable y: $dy/dx = g(x)$. This is treated in Section 16.1. In that section we set up the general formulation in terms of a constraint equation $dy/dx = p$ and a surface equation $F(x, y, p) = 0$. The special forms of the surface and constraint equations are exploited to write down the solution by quadratures.

Lie's methods are presented in Section 16.2 in a number of simple, easy to digest steps. Taken altogether, these provide an algorithm for determining whether an ordinary differential equation possesses a symmetry and, if so, what that symmetry is. Transformation to a set of canonical variables R, S, T is algorithmic. The canonical variable $R(x, y)$ is the new independent variable (like x), $S(x, y)$ is the new dependent variable (like y), and $T(x, y, p)$ is the new constraint between S and R (like dy/dx). In this new coordinate system the surface and constraint equations assume the desired forms $F(R, -, T) = 0$ and $dS/dR = f(R, -, T)$. The system has been reduced to quadratures, and integration follows immediately.

Despite the simplicity of the algorithm, it is not easy to understand these steps without a roadmap. Such is provided in Section 16.3, where a simple example is discussed in detail.

Lie's methods extend in many different directions. Several of these are indicated in Section 16.4.

16.1 The simplest case

The simplest first order ordinary differential equation to deal with has the form

$$\frac{dy}{dx} = g(x) \tag{16.1}$$

Here x is the independent variable and y is the dependent variable. The solution of this equation is (almost) trivially

$$y = G(x) = \int g(x)\, dx \quad (+\text{ additive constant}) = G(x) + c \tag{16.2}$$

If we write the solution in the form $y - G(x) = 0$, then the surface $y + c - G(x) = 0$ is also a solution of the original equation (16.1). There is a one-parameter group of displacements that maps one solution into another. These displacements can be represented by the Taylor series displacement operator $e^{c\partial/\partial y}$, for

$$e^{c\partial/\partial y}[y - G(x) = 0] = y + c - G(x) = 0 \tag{16.3}$$

In short, the "trivial" additive constant is in fact a one-parameter group of translations that maps solutions (16.2) of (16.1) into other solutions of the original simple equation (16.1). This translation group plays the same role for first order ordinary differential equations that the symmetric group S_n plays for nth degree algebraic equations.

For convenience, we express the derivative dy/dx as a coordinate p. The first order differential equation (16.1) can be written in the form $F(x, y, p) = 0$, where $F(x, y, p) = p - g(x)$ for the particular case at hand. There are two relations among the three variables x, y, p. They are given by the surface equation and the constraint equation:

$$\begin{array}{ll} \text{surface equation} & F(x, y, p) = 0 \\ \text{constraint equation} & p = dy/dx \quad \text{when } F(x, y, p) = 0 \end{array} \quad (16.4)$$

It is useful to express the action of the three partial derivatives $\partial/\partial x, \partial/\partial y, \partial/\partial p$ on the surface $F(x, y, p)$ defining the ordinary differential equation. It is also useful to express the action of the generator of infinitesimal displacements that maps solutions of this equation into other solutions of this equation, on the three coordinates. These two relations are summarized as follows:

$$\begin{bmatrix} \dfrac{\partial}{\partial x} \\ \dfrac{\partial}{\partial y} \\ \dfrac{\partial}{\partial p} \end{bmatrix} [p - g(x)] = \begin{bmatrix} * \\ 0 \\ * \end{bmatrix} \qquad \dfrac{\partial}{\partial y} \begin{bmatrix} x \\ y \\ p \end{bmatrix} = \begin{bmatrix} 0 \\ 1 \\ 0 \end{bmatrix} \quad (16.5)$$

These two equations will be generalized to the determining equation for the infinitesimal generator of the invariance group and the determining equations for the canonical coordinates.

16.2 First order equations

In this section we will summarize Lie's approach to the study of differential equations (Blumen and Cole, 1969; Estabrook and Wahlquist, 1975; Wahlquist and Estabrook, 1976). We do this for equations of first order ($d^n y/dx^n$, $n = 1$) and first degree (depends on $p^m = (dy/dx)^m$, $m = 1$). The results are independent of degree.

If the equation that defines the first order ordinary differential equation, $F(x, y, p) = 0$, is not of the form $p - g(x)$, so that $\frac{\partial}{\partial y} F(x, y, p) \neq 0$, then we can attempt to find the following.

(i) A one-parameter group that leaves $F(x, y, p) = 0$ unchanged.
(ii) A new "canonical" coordinate system (R, S, T). In this coordinate system $R = R(x, y)$ is the independent variable, $S = S(x, y)$ is the dependent variable, and $T = T(x, y, p)$ is the new constraint variable. In this canonical coordinate system the surface equation $F(x, y, p) = 0$ is not a function of the new dependent variable: $F(R, -, T) = 0$.

16.2 First order equations

In this new coordinate system the source term for the constraint equation is also independent of the dependent variable: $dS/dR = f(R, -, T)$.

16.2.1 One-parameter group

We search for a one-parameter group of transformations that leaves the surface equation invariant by changing variables in the (x, y) plane according to

$$\begin{aligned} x &\to \bar{x}(\epsilon) = x + \epsilon\xi(x, y) + \mathcal{O}(\epsilon^2) & \bar{x}(\epsilon = 0) &= x \\ y &\to \bar{y}(\epsilon) = y + \epsilon\eta(x, y) + \mathcal{O}(\epsilon^2) & \bar{y}(\epsilon = 0) &= y \\ p &\to \bar{p}(\epsilon) = p + \epsilon\zeta(x, y, p) + \mathcal{O}(\epsilon^2) & \bar{p}(\epsilon = 0) &= p \end{aligned} \quad (16.6)$$

In the simplest case Eq. (16.1), this one-parameter group is $x \to x$ and $y \to y + \epsilon$, so that $\xi = 0$, $\eta = 1$, and $\zeta = 0$.

16.2.2 First prolongation

The function $\zeta(x, y, p)$ is not independent of the functions $\xi(x, y)$ and $\eta(x, y)$. The former is related to the latter pair by the **first prolongation** formula. Specifically,

$$\bar{p} = \frac{d\bar{y}}{d\bar{x}} = \frac{d\bar{y}/dx}{d\bar{x}/dx} = \frac{p + \epsilon(\eta_x + \eta_y p)}{1 + \epsilon(\xi_x + \xi_y p)} \longrightarrow p + \epsilon[\eta_x + (\eta_y - \xi_x)p - \xi_y p^2] \quad (16.7)$$

to first order in ϵ, where $\eta_x = \partial\eta/\partial x$, etc. As a result

$$\zeta(x, y, p) = \eta^{(1)}(x, y, y^{(1)}) = \eta_x + (\eta_y - \xi_x)p - \xi_y p^2 \quad (16.8)$$

16.2.3 Determining equation

The surface equation must be unchanged under the one-parameter group of transformations, so that

$$F(x, y, p) = 0 \to F(\bar{x}(\epsilon), \bar{y}(\epsilon), \bar{p}(\epsilon)) \xrightarrow{\epsilon \text{ small}} F(x + \epsilon\xi, y + \epsilon\eta, p + \epsilon\zeta)$$

$$= F(x, y, p) + \epsilon\left(\xi\frac{\partial}{\partial x} + \eta\frac{\partial}{\partial y} + \zeta\frac{\partial}{\partial p}\right)F(x, y, p) + \text{h.o.t.} \quad (16.9)$$

These are the leading two terms in the Taylor series expansion

$$F(\bar{x}(\epsilon), \bar{y}(\epsilon), \bar{p}(\epsilon)) = e^{\epsilon X} F(x, y, p) = 0 \quad (16.10)$$

where the generator of infinitesimal displacements for the one-parameter group that leaves the surface equation invariant is

$$X = \xi \frac{\partial}{\partial x} + \eta \frac{\partial}{\partial y} + \zeta \frac{\partial}{\partial p} \qquad (16.11)$$

The first two terms in Eq. (16.9) and (16.10) are

$$F(x, y, p) = 0 \quad \text{and} \quad XF(x, y, p) = 0 \qquad (16.12)$$

These are called the **determining equations**. The determining equations (16.12) are generalizations of equations (16.5).

Specifically, these equations are used to determine the functions $\xi(x, y)$, $\eta(x, y)$, and $\zeta(x, y, p)$ that define the infinitesimal generator X. These functions are determined by an algorithm based on linear algebra. There are recent versions depending on sophisticated methods of algebraic topology. These methods are elegant improvements of a conceptually simple brute strength procedure that we summarize briefly. The surface equation $F(x, y, p) = 0$ is solved for p as a function of x and y: $p = p(x, y)$. This expression is substituted into the determining equation $XF(x, y, p(x, y)) = 0$, so that this equation depends only on two independent variables x and y. The generators of the infinitesimal displacements, $\xi(x, y)$ and $\eta(x, y)$, are represented by Laurent expansions, or Taylor series expansions if convergent solutions are sought:

$$\xi(x, y) = \sum_{i,j} \xi_{ij} x^i y^j \qquad 0 \leq i, j, \ i + j \leq d_\xi \qquad (16.13)$$

and similarly for η. These representations are truncated at finite degrees d_ξ, d_η. The determining equation $XF = 0$ is expanded into the form $\sum C_{ij} x^i y^j = 0$. Each coefficient C_{ij} must vanish separately, by standard linear independence arguments. This gives a set of simultaneous *linear* equations in the expansion amplitudes ξ_{ij}, η_{ij}. In general, there are more equations than unknowns. Since the equations are homogeneous, there are no nontrivial solutions if the rank of this system is equal to the number of unknowns. The number of independent solutions (up to an overall scaling factor) is equal to the corank of this system of equations. This is not larger than one for first order equations but may exceed one for second and higher order equations. This algorithm is effective when $\xi(x, y)$ and $\eta(x, y)$ are polynomials of finite degree.

16.2.4 New coordinates

If an infinitesimal generator X can be constructed from the determining equations, then it is possible to determine a new system of coordinates R, S, T which

16.2 First order equations

"straightens out" the surface equation. This is done by solving the determining equations for canonical coordinates. These are a set of partial differential equations that are analogous to the equations on the right-hand side of Eq. (16.5). For convenience, we summarize the determining equations for the infinitesimal generator and for the canonical coordinates, analogs of the two equations in Eq. (16.5), as follows:

$$XF = 0 \qquad X \begin{bmatrix} R(x, y) \\ S(x, y) \\ T(x, y, p) \end{bmatrix} = \begin{bmatrix} 0 \\ 1 \\ 0 \end{bmatrix} \qquad (16.14)$$

The three linear partial differential equations on the right determine the new canonical coordinates: the independent variable $R(x, y)$, the dependent variable $S(x, y)$, and the new constraint $T(x, y, p)$ between R and S.

16.2.4.1 Dependent coordinate

The dependent coordinate S is determined from the differential equation $X(x, y, p)S(x, y) = 1$. We require S to be independent of p, so the condition defining S reduces to

$$\left(\xi(x, y)\frac{\partial}{\partial x} + \eta(x, y)\frac{\partial}{\partial y} \right) S(x, y) = 1 \qquad (16.15)$$

The solution is not unique: any function of x and y that is annihilated by X can be added to the solution. Further, it is not important that $XS = +1$: we could just as well choose a solution satisfying $XS = -1$ or, for that matter, $XS = k \neq 0$, where k is some constant.

16.2.4.2 Invariant coordinates: independent variable

The two invariant coordinates R and T are unchanged under the one-parameter transformation group. These functions obey $XR = 0$ and $XT = 0$, which are explicitly

$$\left(\xi(x, y)\frac{\partial}{\partial x} + \eta(x, y)\frac{\partial}{\partial y} \right) R(x, y) = 0$$
$$\left(\xi(x, y)\frac{\partial}{\partial x} + \eta(x, y)\frac{\partial}{\partial y} + \zeta(x, y)\frac{\partial}{\partial p} \right) T(x, y, p) = 0 \qquad (16.16)$$

The solutions are most simply found by the method of characteristics. They obey the differential relations

$$\frac{dx}{\xi(x, y)} = \frac{dy}{\eta(x, y)} = \frac{dp}{\zeta(x, y, p)} \qquad (16.17)$$

The first equation is used to construct $R(x, y)$.

16.2.4.3 Invariant coordinates: constraint variable

The second equation in (16.17) is used to construct $T(x, y, p)$. It is often possible to construct T so that it is a function of p to the first power. When this is possible, it is the preferred form of the nonunique expression for the invariant cordinate T.

16.2.5 Surface and constraint equations

In the new coordinate system there is a constraint equation:

$$\frac{dS}{dR} = \frac{dS(x, y)}{dR(x, y)} = \frac{dS/dx}{dR/dx} = \frac{S_x + S_y p}{R_x + R_y p} \qquad (16.18)$$

This derivative is independent of the parameter ϵ of the one-parameter group. Therefore it must be independent of the coordinate S, and depend only on the invariant coordinates R and T. In this new coordinate system the surface and constraint equations are

$$\begin{array}{ll} \text{surface equation} & F(R, -, T) = 0 \\ \text{constraint equation} & dS/dR = f(R, -, T) \end{array} \qquad (16.19)$$

These are directly analogous to Eq. (16.1) and $dy/dx = p$ in Section 16.1.

16.2.6 Solution in new coordinates

To integrate the transformed equation, the surface equation is used to determine T as a function of R: $T = T(R)$. This expression is used in the constraint equation, which can then "easily" be integrated to give

$$S = \int f(R, -, T(R)) \, dR + c \qquad (16.20)$$

The additive parameter c is the image of the parameter ϵ of the one-parameter group of transformations that leaves the original surface equation $F(x, y, p) = 0$ invariant.

16.2.7 Solution in original coordinates

The inverse relation $x = x(R, S)$, $y = y(R, S)$ is used to express the solution Eq. (16.20) of the transformed equation in terms of the original coordinates.

16.3 An example

The algorithm developed in Section 16.2 is, for all practical purposes, impossible to understand without illustrating its workings by a particular example. To illustrate

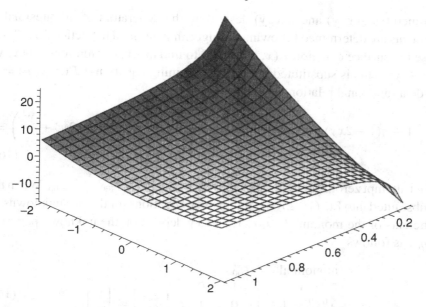

Figure 16.1. The first order ordinary differential equation $xp + y - xy^2 = 0$. Here p (vertical) is plotted over the (x, y) plane for $0.1 \leq x \leq 1.1$ and $-2 \leq y \leq +2$. The shape of the surface depends on both coordinates x and y.

the algorithm, we use it to integrate the equation

$$F(x, y, p) = xp + y - xy^2 = 0 \qquad (16.21)$$

Before setting out on this path, we first attempt the following scaling transformation $y \to \alpha y$ and $x \to \beta x$. Under this transformation the equation transforms to $\alpha(xp + y - (\alpha\beta)xy^2) = 0$. The equation is invariant provided $\alpha\beta = 1$. The one-parameter group that leaves the surface constraint $F(x, y, p) = 0$ invariant is $x \to \lambda x$, $y \to \lambda^{-1} y$, $p \to \lambda^{-2} p$. Since there is a one-parameter invariance group for this differential equation, Lie's methods are guaranteed to work. In fact, it is possible to construct the infinitesimal generator $X(x, y, p)$ from this group directly.

The surface $p = y^2 - y/x$ is shown in Fig. 16.1. The value of p clearly depends on both coordinates x and y. The purpose of the change of variables is to find a new coordinate system in which the surface is independent of the new dependent variable $S(x, y)$.

The determining equation Eq. (16.14) is

$$\xi(p - y^2) + \eta(1 - 2xy) + [\eta_x + (\eta_y - \xi_x)p - \xi_y p^2]x = 0 \qquad (16.22)$$

The functions $\xi(x, y)$ and $\eta(x, y)$ describing the generators of infinitesimal displacements are determined following the algorithm outlined in Section 16.2.3. First, we use the surface equation $F(x, y, p) = 0$ to find an expression for p: $p(x, y) = -y/x + y^2$. This is substituted into the determining equation $XF(x, y, p) = 0$ to provide a functional relation between x and y:

$$\xi\left(-\frac{y}{x}\right) + \eta(1 - 2xy) + \eta_x x + (\eta_y - \xi_x)(xy^2 - y) - \xi_y\left(xy^4 - 2y^3 + \frac{y^2}{x}\right) = 0 \tag{16.23}$$

We first attempt zeroth degree expressions for ξ and η: $\xi = \xi_{00}$, $\eta = \eta_{00}$. When these are substituted into Eq. (16.23) we obtain three equations for the two unknowns. The coefficients of the monomials y/x, 1, and xy depend on the unknown parameters ξ_{00}, η_{00} as follows:

$$\begin{array}{c} \text{monomial} \\ y/x \\ x^0 y^0 = 1 \\ xy \end{array} \quad \begin{array}{cc} \xi_{00} & \eta_{00} \\ \begin{bmatrix} -1 & 0 \\ 0 & 1 \\ 0 & -2 \end{bmatrix} \end{array} \begin{bmatrix} \xi_{00} \\ \eta_{00} \end{bmatrix} = \begin{bmatrix} 0 \\ 0 \end{bmatrix} \tag{16.24}$$

This system of three simultaneous linear equations in two unknowns has rank two, therefore no nontrivial solutions.

We therefore increase the degree of $\xi(x, y)$ and $\eta(x, y)$ to one and repeat the process. The relation Eq. (16.23) between x and y is now

$$(\xi_{00} + \xi_{10}x + \xi_{01}y)\left(-\frac{y}{x}\right) + (\eta_{00} + \eta_{10}x + \eta_{01}y)(1 - 2xy) + \eta_{10}x$$

$$+ (\eta_{01} - \xi_{10})(xy^2 - y) - \xi_{01}\left(xy^4 - 2y^3 + \frac{y^2}{x}\right) = 0 \tag{16.25}$$

This results in the following set of ten equations for six unknowns:

$$\begin{array}{c} \text{monomial} \\ y^2/x \\ y/x \\ 1 \\ x \\ y \\ xy \\ x^2 y \\ xy^2 \\ xy^3 \\ xy^4 \end{array} \begin{array}{cccccc} \xi_{00} & \xi_{10} & \xi_{01} & \eta_{00} & \eta_{10} & \eta_{01} \\ \begin{bmatrix} 0 & 0 & -2 & 0 & 0 & 0 \\ -1 & 0 & 0 & 0 & 0 & 0 \\ 0 & 0 & 0 & +1 & 0 & 0 \\ 0 & 0 & 0 & 0 & +2 & 0 \\ 0 & 0 & 0 & 0 & 0 & 0 \\ 0 & 0 & 0 & -2 & 0 & 0 \\ 0 & 0 & 0 & 0 & -2 & 0 \\ 0 & -1 & 0 & 0 & 0 & -1 \\ 0 & 0 & -2 & 0 & 0 & 0 \\ 0 & 0 & +1 & 0 & 0 & 0 \end{bmatrix} \end{array} \begin{bmatrix} \xi_{00} \\ \xi_{10} \\ \xi_{01} \\ \eta_{00} \\ \eta_{10} \\ \eta_{01} \end{bmatrix} = \begin{bmatrix} 0 \\ 0 \\ 0 \\ 0 \\ 0 \\ 0 \end{bmatrix} \tag{16.26}$$

16.3 An example

This set of equations has rank five, so there is one nontrivial solution. From the first four equations we determine $\xi_{01} = \xi_{00} = \eta_{00} = \eta_{10} = 0$, and from the coefficient of xy^2 we learn $-\xi_{10} - \eta_{01} = 0$ so that, up to some overall scaling factor, we can take $\xi(x, y) = x$ and $\eta(x, y) = -y$. Since we have found one nontrivial solution for an infinitesimal generator of a one-parameter group of a first order equation, we can stop searching for additional solutions to the determining equation (for second order equations there may be additional solutions).

With this solution $\xi(x, y) = x$ and $\eta(x, y) = -y$ the prolongation formula Eq. (16.8) gives $\zeta = -2p$, so that the generator of infinitesimal displacements is

$$X = x\frac{\partial}{\partial x} - y\frac{\partial}{\partial y} - 2p\frac{\partial}{\partial p} \tag{16.27}$$

The infinitesimal generator is now used to determine the new set of coordinates. We first determine the dependent coordinate $S(x, y)$ by attempting to solve

$$\left(x\frac{\partial}{\partial x} - y\frac{\partial}{\partial y}\right) S(x, y) = 1 \tag{16.28}$$

It is useful first to seek a solution $S(x, y)$ depending only on the single variable y. Such a solution can be found if the equation $-y dS(y)/dy = 1$ can be solved. The solution, up to an additive constant, is $-\ln(y)$. We will adopt this solution, neglecting the negative sign: $S(x, y) = \ln(y)$.

The invariant coordinates are determined using the method of characteristics:

$$\frac{dx}{x} = \frac{dy}{-y} = \frac{dp}{-2p} \tag{16.29}$$

The first equation for the new independent variable simplifies to $y dx = -x dy$ or $d(xy) = 0$, from which we conclude that $R(x, y) = xy$ is an invariant coordinate that obeys Eq. (16.14). The invariant coordinate involving p is determined by setting $-dp/2p$ equal to either of the other two differentials. We set it equal to dx/x to avoid having the second invariant coordinate dependent on y. The equation is $dx/x = -dp/2p$ and the solution is $(1/x)d(x^2 p) = 0$, so that $T(x, y, p) = x^2 p$. The forward and backward transformations between the two coordinate systems are

$$\begin{pmatrix} R \\ S \\ T \end{pmatrix} = \begin{pmatrix} xy \\ \ln(y) \\ x^2 p \end{pmatrix} \qquad \begin{pmatrix} x \\ y \\ p \end{pmatrix} = \begin{pmatrix} Re^{-S} \\ e^S \\ Te^{2S}/R^2 \end{pmatrix} \tag{16.30}$$

In the new coordinate system the surface equation transforms to

$$F(x, y, p) = xp + y - xy^2 = 0 \longrightarrow e^S \left[\frac{T}{R} + 1 - R\right] = 0 \tag{16.31}$$

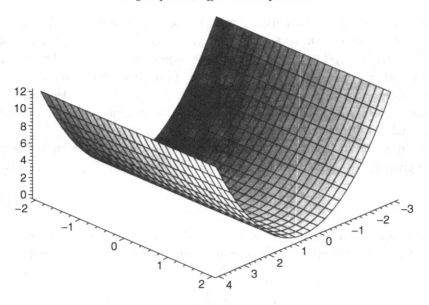

Figure 16.2. The surface $xp + y - xy^2 = 0$ transforms to the surface $T/R + 1 - R = 0$ in canonical coordinates. Here T (vertical) is plotted over the (R, S) plane for $-3 \leq R \leq +4$ and $-2 \leq S \leq +2$. The function is a simple ruled surface, independent of S.

The expression within the brackets is the transformed surface equation. It is independent of S. This surface $T = T(R, S)$ is plotted in Fig. 16.2. It has the desired form: a ruled surface whose shape (height) is independent of the dependent variable S. Such a surface is sometimes called a "cylinder."

The new constraint equation is

$$\frac{dS}{dR} = \frac{d(\ln y)}{d(xy)} = \frac{p/y}{y + xp} = \frac{Te^S/R^2}{e^S + (Re^{-S})(Te^{2S}/R^2)} \quad (16.32)$$

The surface and constraint equations are

$$\begin{aligned} \text{surface equation} &\quad T/R + 1 - R = 0 \\ \text{constraint equation} &\quad dS/dR = (T/R)/(T + R) \end{aligned} \quad (16.33)$$

The surface equation is solved for T as a function of R: $T(R) = R^2 - R$. This expression is substituted into the constraint equation to give a first order differential equation in quadratures:

$$\frac{dS}{dR} = \frac{1}{R} - \frac{1}{R^2} \implies S = \ln(R) + \frac{1}{R} + c \quad (16.34)$$

The parameter c is the parameter of the translation group that leaves invariant the transformed equation.

The inverse transformation, Eq. (16.30), from (R, S) to (x, y) is finally used to rewrite the solution in terms of the original set of variables:

$$y = \frac{-1}{x(c + \ln x)} \qquad (16.35)$$

Remarks The operator $x \, d/dx$ is the infinitesimal generator for scaling transformations, since $e^{\lambda x d/dx} x = e^{\lambda} x$. As a result, the infinitesimal generator X has the following effect on the coordinates (x, y, p):

$$\text{EXP}\left(\lambda \left\{x \frac{\partial}{\partial x} - y \frac{\partial}{\partial y} - 2p \frac{\partial}{\partial p}\right\}\right) \begin{bmatrix} x \\ y \\ p \end{bmatrix} = \begin{bmatrix} e^{\lambda} x \\ e^{-\lambda} y \\ e^{-2\lambda} p \end{bmatrix} \qquad (16.36)$$

From this scaling behavior, it is easy to see that $\ln(y)$ is linear in the Lie translation group parameter: $\ln(e^{-\lambda} y) = \ln(y) - \lambda$. The invariant operators come right out of the scaling transformations: xy and $x^2 p$ are unchanged by the scaling transformation. None of these operators is unique. The operator $\ln(xy^2)$ is linear and $x^3 yp$ is invariant. We have just chosen the most convenient (simplest) solutions to the equations defining the new coordinates.

16.4 Additional insights

Lie's theory of infinitesimal transformation groups has been extended in many different directions, all of which are powerful and beautiful. It is barely possible to scratch the surface here. Instead, we content ourselves by indicating some of the directions in which it can be extended. These directions are simple consequences of the analyses presented in the previous two sections.

16.4.1 Other equations, same symmetry

Many differential equations can share the same invariance group. The most general first order ordinary differential equation invariant under the scaling group Eq. (16.36) has the form $F(R, -, T) = 0$ or more simply $F(xy, x^2 p) = 0$. The most general first order equation of first degree with this symmetry has the form $x^2 p = h(xy)$ or $dy/dx = x^{-2} h(xy)$. For the equation studied in Section 16.3, $h(z) = -z + z^2$. For the Riccati equation $dy/dx + y^2 - 2/x^2 = 0$, $h(z) = z^2 - 2$.

16.4.2 Higher degree equations

These methods work equally well with first order equations of higher degree. For example, the first order, second degree equation $y'^2 + y^4 - x^{-4} = 0$ has canonical form $R^4 + T^2 = 1$. The original equation has two solution branches

$p = \pm\sqrt{x^{-4} - y^4}$, corresponding to the two solution branches in the canonical coordinate system $T = \pm\sqrt{1 - R^4}$.

16.4.3 Other symmetries

The methods described in Section 16.2 and illustrated by example in Section 16.3 apply to any first order ordinary differential equation with a one-parameter group. Table 16.1 provides a list of symmetries that may be encountered for ordinary differential equations. For each symmetry the functions $\xi(x, y)$ and $\eta(x, y)$ are tabulated, as well as the first prolongation $\zeta(x, y, p) = \eta^{(1)}(x, y, p)$. We also present the canonical coordinates (R, S, T). Since the constraint equation dS/dR depends only on the change of variables, it also can be tabulated, and has been. The simplest case, Eq. (16.1), is present in the first line of this table. The equation studied in Section 16.3 is present in the eighth line of this table.

The Lie symmetries leaving the equation invariant can be determined from this table in one of two ways. We can use the generator of infinitesimal displacements to compute them, as in Eq. (16.36). Or we can look at the transformations effected by $S \to S' = S + c$, $R' = R$. In the latter case we find $\ln(y) \to \ln(y) + c = \ln(e^c y) = \ln(\bar{y}(c))$ and since $xy = \bar{x}(c)\bar{y}(c)$, the transformation is $\bar{x}(c) = e^{-c}x$ and $\bar{y}(c) = e^{+c}y$.

16.4.4 Second order equations

Second order equations can be studied by simple extensions of the methods used to study first order equations. The infinitesimal generator for displacements now involves derivatives with respect to $y^{(2)}$ and is given by

$$X = \xi \frac{\partial}{\partial x} + \eta \frac{\partial}{\partial y} + \eta^{(1)} \frac{\partial}{\partial y^{(1)}} + \eta^{(2)} \frac{\partial}{\partial y^{(2)}} \qquad (16.37)$$

The second prolongation can be determined from the first in a straightforward computation

$$\frac{d^2\bar{y}}{d\bar{x}^2} = \frac{d}{d\bar{x}}\left(\frac{d\bar{y}}{d\bar{x}}\right) = \frac{d}{d\bar{x}}(p + \epsilon\eta^{(1)}) = \frac{D^{(1)}(p + \epsilon\eta^{(1)})}{D^{(0)}(x + \epsilon\xi)} = \frac{y^{(2)} + \epsilon D^{(1)}\eta^{(1)}}{1 + \epsilon D^{(0)}\xi}$$

$$= y^{(2)} + \epsilon\left(D^{(1)}\eta^{(1)} - y^{(2)}D^{(0)}\xi\right) \qquad (16.38)$$

As a result,

$$\eta^{(2)}(x, y, y^{(1)}, y^{(2)}) = D^{(1)}\eta^{(1)} - y^{(2)}D^{(0)}\xi \qquad (16.39)$$

16.4 Additional insights

Table 16.1. *Infinitesimal generators ξ, η, ζ, canonical coordinates R, S, T, and constraint equation dS/dR for some Lie symmetries*

Infinitesimal generators			Canonical coordinates			Constraint
$\xi(x,y)$	$\eta(x,y)$	$\zeta(x,y,p)$	$R(x,y)$	$S(x,y)$	$T(x,y,p)$	dS/dR
0	1	0	x	y	p	T
1	0	0	y	x	p	$1/T$
$1/a$	$-1/b$	0	$ax+by$	$bx-ay$	p	$(b-aT)/(a+bT)$
x	0	$-p$	y	$\ln x$	xp	$1/T$
0	y	p	x	$\ln y$	p/y	T
x/a	y/b	$(1/b-1/a)p$	y^b/x^a	$b\ln y$	$p/x^{(a/b-1)}$	$(bT/R)/(bT-aR^{(1/b)})$
x	y	0	y/x	$\ln y$	p	$(T/R)/(T-R)$
x	$-y$	$-2p$	xy	$\ln y$	$x^2 p$	$(T/R)/(T+R)$
$2x$	y	$-p$	y^2/x	$\ln y$	yp	$(T/R)/(2T-R)$
x	$2y$	p	y/x^2	$\ln y$	p/x	$(T/R)/(T-2R)$
y	0	$-p^2$	y	x/y	$x-y/p$	$-T/R^2$
0	x	1	x	y/x	$xp-y$	T/R^2
$-y$	x	$1+p^2$	$\sqrt{x^2+y^2}$	$\tan^{-1}(y/x)$	$(y-xp)/(x+yp)$	$-T/R$
1	y/x	$(px-y)/x^2$	y/x	x	$(xp-y)/x^2$	$1/T$
a	x	1	x^2-2ay	x/a	$x-ap$	$1/(2aT)$
a	y	p	$x-a\ln y$	x/a	p/y	$(1/a)/(1-aT)$
x	b	$-p$	e^y/x^b	y/b	$e^y * p^b$	$(bR)^{-1}/[1-b(R/T)^{(1/b)}]$
y	b	$-p^2$	y^2-2bx	y/b	$y-b/p$	$1/(2bT)$
0	$e^{f(x)}$	$f'e^f$	x	y/e^f	$p-yf'$	$T/e^{f(R)}$
x^2	xy	$y-xp$	y/x	$1/x$	$xp-y$	$1/T$
xy	y^2	$yp-xp^2$	y/x	$1/y$	$y/p-x$	$1/(TR^2)$
xy	0	$-yp-xp^2$	y	$(\ln x)/y$	$y/(xp)-\ln x$	T/R^2
0	xy	$y+xp$	x	$(\ln y)/x$	$xp/y-\ln y$	T/R^2
$g(y)$	0	$-g'p^2$	y	x/g	$1/p-xg'/g$	$T/g(R)$
0	$f(x)$	f'	x	y/f	$fp-f'y$	$T/f^2(R)$
$f(x)$	0	$-f'p$	y	$F\ (F'f=1)$	pf	$1/T$
0	$g(y)$	$g'p$	x	$G\ (G'g=1)$	p/g	T
x^{k+1}	$kx^k y$	$x^k(k^2 y/x-p)$	y/x^k	$1/x^k$	$xp-ky$	$-k/T$
kxy^k	y^{k+1}	$y^k(p-k^2xp^2/y)$	x/y^k	$1/y^k$	$y/p-kx$	$-k/T$

where

$$D^{(n)} = \frac{\partial}{\partial x} + \frac{dy}{dx}\frac{\partial}{\partial y} + \frac{dy^{(1)}}{dx}\frac{\partial}{\partial y^{(1)}} + \cdots + y^{(n+1)}\frac{\partial}{\partial y^{(n)}} \qquad (16.40)$$

It is explicitly

$$\eta^{(2)} = \eta_{xx} + (2\eta_{xy} - \xi_{xx})y' + (\eta_{yy} - 2\xi_{xy})y'^2 - \xi_{yy}y'^3 \\ + (\eta_y - 2\xi_x - 3\xi_{yy'})y'' \qquad (16.41)$$

The determining equations are

$$F(x, y, y^{(1)}, y^{(2)}) = 0 \qquad X(x, y, y^{(1)}, y^{(2)})F(x, y, y^{(1)}, y^{(2)}) = 0 \qquad (16.42)$$

16.4.5 Reduction of order

If a higher order equation has a known one-parameter symmetry group, the order of the equation can be reduced by one. We illustrate as usual by example. The general case can easily be inferred from the example.

Suppose a second order equation $F(x, y, y', y'') = 0$ is invariant under the scaling group (16.36). Then the dependent coordinate is $S = \ln y$ and the surface equation can be expressed in terms of three invariant coordinates as $F(R, -, T, U) = 0$. Here as before R depends only on x and y, $T = T(x, y, y')$, and $U = U(x, y, y', y'')$ is another invariant coordinate. How does one construct such an invariant coordinate? It is simple to see that the derivative dT/dR is invariant under the group. Not only is it invariant, but it is of first degree in the second order term $y^{(2)}$, for

$$\frac{dT}{dR} = \frac{dT/dx}{dR/dx} = \frac{T_x + T_y y^{(1)} + T_{y^{(1)}} y^{(2)}}{R_x + R_y y^{(1)}} \tag{16.43}$$

For the scaling group the new invariant coordinate is

$$\frac{dT}{dR} = \frac{2xy' + x^2 y''}{y + xy'} \tag{16.44}$$

and the most general second order equation invariant under this group is

$$G\left(R, -, T, \frac{dT}{dR}\right) = 0 \tag{16.45}$$

This is a *first* order equation in the invariant coordinate T. The result is that we have used a one-parameter symmetry group to reduce the order of a second order equation by one. If an additional symmetry can be identified, the equation can be reduced to quadratures a second time (i.e., completely integrated).

The most general second order equation invariant under the group of scaling transformations Eq. (16.36) that is of first degree in y'' is

$$\frac{dT}{dR} = \frac{x^2 y'' + 2xy'}{y + xy'} = g(xy, x^2 y') = g(R, T) \tag{16.46}$$

This is a first order equation in T. Certain forms of the function g may admit another Lie symmetry. If such a symmetry can be found, the order of the equation can again be reduced by one.

16.4 Additional insights

16.4.6 Higher order equations

These ideas can be extended to higher order equations. We begin with an nth order equation $F(x, y, \ldots, y^{(n)}) = 0$. As usual, we seek an infinitesimal generator

$$X = \xi \frac{\partial}{\partial x} + \eta^{(0)} \frac{\partial}{\partial y^{(0)}} + \eta^{(1)} \frac{\partial}{\partial y^{(1)}} + \cdots + \eta^{(n)} \frac{\partial}{\partial y^{(n)}} = \xi \frac{\partial}{\partial x} + \sum_{j=0}^{n} \eta^{(j)} \frac{\partial}{\partial y^{(j)}} \tag{16.47}$$

The functions in the prolongation formulas are determined following the procedure demonstrated in Eq. (16.38). They are recursively related:

$$\begin{aligned}
\eta^{(0)}(x, y) &= \eta(x, y) \\
\eta^{(1)}(x, y, y^{(1)}) &= D^{(0)} \eta^{(0)} - y^{(1)} D^{(0)} \xi \\
\eta^{(2)}(x, y, y^{(1)}, y^{(2)}) &= D^{(1)} \eta^{(1)} - y^{(2)} D^{(0)} \xi \\
\eta^{(3)}(x, y, y^{(1)}, y^{(2)}, y^{(3)}) &= D^{(2)} \eta^{(2)} - y^{(3)} D^{(0)} \xi \\
&\vdots \quad \vdots \quad \vdots
\end{aligned} \tag{16.48}$$

The operator X is used as described in Section 16.2 to compute the functions $\xi(x, y)$ and $\eta(x, y)$. There will be as many linearly independent infinitesimal generators as the corank of the set of simultaneous linear equations for the Taylor series coefficients of these functions.

If one or more generators can be constructed, a dependent coordinate S can be computed by solving Eq. (16.15). The remaining invariant coordinates are obtained from the equations

$$\frac{dx}{\xi} = \frac{dy}{\eta^{(0)}} = \frac{dy^{(1)}}{\eta^{(1)}} = \cdots = \frac{dy^{(n)}}{\eta^{(n)}} \tag{16.49}$$

In fact, only the first two invariant coordinates $R(x, y)$ and $T(x, y, y^{(1)})$ need be computed. The remaining invariant coordinates are $dT^{(j)}/dR^{(j)}$, $j = 0$ (for T) and $j = 1, 2, \ldots, n-1$. Each of these latter is of first degree in $y^{(j+1)}$. As a result, the existence of a Lie symmetry can be used to reduce an nth order equation to an $(n-1)$st order equation.

16.4.7 Partial differential equations: Laplace's equation

Lie's methods can be extended to partial differential equations. We illustrate a small part of the theory by treating Laplace's equation in this subsection and the heat equation in the following.

In n dimensions, Laplace's equation with a source term is

$$\nabla^2 u(x^1, x^2, \ldots, x^n) = \delta(x) \tag{16.50}$$

This equation is clearly invariant under rotations, so that the infinitesimal generators of rotations are Lie symmetries. The equation is also invariant under scaling transformations $x^i \to \lambda x^i$, $u \to \alpha u$. Under the scaling transformation $\delta(x) \to \delta(\lambda x) = \lambda^{-n}\delta(x)$, so that

$$\nabla^2 u = \delta(x) \longrightarrow \frac{\alpha}{\lambda^2}\nabla^2 u = \lambda^{-n}\delta(x) \tag{16.51}$$

The equation is invariant provided $\alpha = \lambda^{2-n}$. The infinitesimal generators of symmetries for this equation therefore consist of generators of rotations and scale transformations (Blumen and Cole, 1969):

$$\begin{aligned} X_{ij} &= x^i \partial_j - x^j \partial_i \\ Z &= x^i \partial_i + (2-n)u\frac{\partial}{\partial u} \end{aligned} \tag{16.52}$$

A new independent coordinate $R = R(x, u)$ satisfies $XR = 0$, where X is any linear combination of the generators in Eq. (16.52). A solution is $R \sim u|x|^{n-2}$. As a result, $u \sim |x|^{2-n} = k|x|^{2-n}$. The constant of proportionality can be computed using the divergence theorem. Both sides of Eq. (16.50) are integrated over the interior of a unit sphere in R^n. The volume integral on the right is $+1$. The volume integral on the left is transformed into a surface integral using the divergence theorem:

$$\int_V k\nabla^2 |x|^{2-n} dV = \int_{S=\partial V} k(2-n)\frac{\hat{n} \cdot dS}{|x|^{n-1}} = (2-n)kV(S^n) = 1 \tag{16.53}$$

Here $V(S^{n-1}) = 2\pi^{n/2}/\Gamma(\frac{n}{2})$ is the surface area of a unit sphere in R^n. As a result, the solution of Laplace's equation in R^n ($n \neq 2$) with unit source term at the origin is

$$u(x) = \frac{k}{|x|^{n-2}} \qquad k = \frac{-1}{(n-2)V(S^n)} \tag{16.54}$$

16.4.8 Partial differential equations: heat equation

The heat equation on R^n for $u(x, t)$ with source term

$$u_t - \nabla^2 u = \delta(x, t) \tag{16.55}$$

is treated similarly (Olver, 1993). It is invariant under rotations, so the operators X_{ij} are Lie symmetries. Under the scaling transformation $u \to \alpha u$, $t \to \beta t$, and $x^i \to \lambda x^i$ the equation transforms as follows:

$$u_t - \nabla^2 u = \delta(x, t) \longrightarrow \frac{\alpha}{\beta}u_t - \frac{\alpha}{\lambda^2}\nabla^2 u = \frac{1}{\lambda^n \beta}\delta(x, t) \tag{16.56}$$

Invariance under the scaling transformations places the following two constraints on the three scaling variables (since there is only one equation): $\alpha \lambda^n = 1$ and $\beta/\lambda^2 = 1$.

From these relations it is possible to construct $n+1$ additional Lie symmetries, so that the entire set is

$$X_{ij} = x^i \partial_j - x^j \partial_i$$
$$Y_i = 2t \frac{\partial}{\partial x^i} - x^i u \frac{\partial}{\partial u} \qquad (16.57)$$
$$Z = 2t \frac{\partial}{\partial t} + x^i \frac{\partial}{\partial x^i} - nu \frac{\partial}{\partial u}$$

An invariant coordinate depending on the x^i, t and u is $R = ut^{n/2}e^{|x|^2/4t}$, from which we obtain as before

$$u = kt^{-n/2}e^{-|x|^2/4t} \qquad k = \left(\frac{1}{2\sqrt{\pi}}\right)^n \qquad (16.58)$$

16.4.9 Closing remarks

Galois resolved the problem of determining whether an algebraic equation could be solved by radicals, and if so how, between 1829 and 1832. His manuscripts were lost, rejected, or filed for posterity. His accomplishments were unrecognized at his death in 1832. They were rescued from oblivion, the black hole of French indifference to its greatest mathematician, by Cauchy in 1843.

Lie's discoveries began in 1874. He realized that the hodgepodge of seemingly different techniques for solving differential equations that existed at that time (and still does) were almost all special manifestations of one single principle – the invariance of solutions of ordinary differential equations under a continuous group. Lie was luckier than Galois when it came to recognition during his lifetime.

There are several problems in the implementation of Lie's algorithms that have either been lightly addressed or passed over in our discussion.

1. Under what conditions is it possible to solve the determining equations for the surface? That is, when is it possible – or impossible – to solve the linear partial differential equations for $\xi(x, y)$ and $\eta(x, y)$?
2. Under what conditions is it possible to solve the determining equations for the canonical variables?
3. Under what conditions is it possible to solve the canonical surface equation $F(R, -, T) = 0$ for T as a function of R? When it is possible, what is the algorithm for accomplishing this?
4. Under what conditions is it possible to integrate a function of a single variable: $\int f(R, -, T(R)) dR$?

The final question was resolved for algebraic functions by Risch in (1969). He exploited the tools of Galois theory in a heavy way to provide an algorithm for determining when an algebraic function can be integrated in closed form, and determining the integral when the answer to the first question is positive. We summarize the dates of these accomplishments here:

1830	Galois	solve algebraic equations
1874	Lie	solve differential equations
1969	Risch	integrate in closed form
?	–	solve determining equations for ξ, η
?	–	solve determining equations for R, S, T
?	–	solve $F(R, -, T) = 0$ for R.

It is clear that additional algorithms are possible and desirable.

16.5 Conclusion

Lie set out to extend Galois' treatment of algebraic equations to the field of ordinary differential equations. Galois observed that an algebraic equation has a symmetry group: a set of operations that maps solutions into solutions. If the symmetry group has certain properties, these properties can be used to generate an algorithm for solving the equation.

It was Lie's genius to see that the "trivial" additive constant that occurs in the solution of a differential equation that has been reduced to quadratures is in fact a group operation. The symmetry group in this simplest case is simply the one-parameter group of translations. Armed with this observation, he developed algorithmic methods to attack ordinary differential equations by searching for their symmetry groups. Lie in fact studied local groups of transformations. The even more beautiful study of global Lie groups was a later development.

In Section 16.2 we presented Lie's algorithm for solving first order ordinary differential equations in a number of simple steps. These involve the following.

(i) Introduce a set of point transformations in the x–y plane. These are defined by the functions $\xi(x, y)$ and $\eta(x, y)$.
(ii) Construct the first prolongation $\zeta(x, y, p) = \eta^{(1)}(x, y, y^{(1)})$ from the functions defining the local change of variables.
(iii) Introduce the operator $X = \xi \partial/\partial x + \eta \partial/\partial y + \zeta \partial/\partial p$. This describes a Taylor series expansion of the surface equation $F(x, y, p) = 0$ that defines the first order ordinary differential equation.
(iv) Solve the determining equation $XF = 0$ when $F = 0$ for the functions $\xi(x, y)$ and $\eta(x, y)$.
(v) Solve the determining equations $XR = 0, XS = 1, XT = 0$ for the canonical coordinates. These are the coordinates in which the surface is a "cylinder." The surface equation is independent of the new dependent variable: $F \to F(R, -, T) = 0$.

(vi) Construct the constraint equation $dS/dR = f(R, -, T)$ in this new coordinate system.
(vii) Solve the surface equation for T as a function of R: $T = T(R)$.
(viii) Solve the constraint equation for S: $S = \int f(R, -, T(R)) + c$.
(ix) Backsubstitute the original coordinates for the new coordinates, $x = x(R, S)$ and $y = y(R, S)$, to obtain the solution of the original equation.

The steps in this algorithm have been illustrated by working out a simple example in Section 16.3.

These methods extend in any number of ways. We have indicated a number of useful directions by example in Section 16.4.

16.6 Problems

1. Show that invariance under a one-parameter group of transformations can also be expressed in the form

$$\frac{d^n}{d\epsilon^n} F[\bar{x}(\epsilon), \bar{y}(\epsilon), \bar{p}(\epsilon)]|_{\epsilon=0} = 0 \qquad n = 0, 1, 2, \ldots \qquad (16.59)$$

Show that the first two terms $n = 0, 1$ are exactly the determining equations (16.12).

2. Construct the invariance group for each of the transformations presented in Table 16.1.

3. **Mechanical similarity** The classical Newtonian equation of motion for a particle of mass m in the presence of a potential $V(\mathbf{x})$ is

$$m \frac{d^2 \mathbf{x}}{dt^2} = -\nabla V(\mathbf{x})$$

Assume that under a scaling transformation, the mass scales with a factor α (i.e., $m \to \alpha m$), $\mathbf{x} \to \beta \mathbf{x}$, $t \to \gamma t$. Assume also that the potential is homogeneous of degree k: $V(\beta \mathbf{x}) \to \beta^k V(\mathbf{x})$ (Landau and Lifshitz, 1960). Under this scaling transformation show that the equation of motion transforms to

$$\alpha^1 \beta^1 \gamma^{-2} m \frac{d^2 \mathbf{x}}{dt^2} = -\beta^{k-1} \nabla V(\mathbf{x})$$

a. Show that the scaled equation is identical to the original provided $\alpha^1 \beta^{2-k} \gamma^{-2} = 1$.
b. Set $\alpha = 1$. Show that trajectories are invariant under the scaling transformation with $\gamma^2 = \beta^{2-k}$. Show that in the cases $k = -1, k = 0, k, = +1, k = +2$ the following

scaling results hold:

k	Potential type	Transformation
−1	Coulomb	$\gamma^2 = \beta^3$
0	no force	$\gamma^2 = \beta^2$
+1	local gravitational potential	$\gamma^2 = \beta^1$
+2	harmonic oscillator	$\gamma^2 = \beta^0$

The first line is a statement of Kepler's third law: for closed planetary orbits, the square of the period (γ^2) is proportional to the cube of the semiaxis (β^3). If R' and T' are the semiaxis and period of planet P' and R and T are the semiaxis and period of planet P, and the two planets P and P' have geometrically similar orbits, $\beta^3 \to (R'/R)^3 = (T'/T)^2 \leftarrow \gamma^2$. The second line is a statement of the integral of Newton's second law in the absence of forces in an inertial frame: the distance traveled (β) is proportional to the time elapsed (γ). The third line is a statement that in a local gravitational potential of the form $V = mgz$, the distance fallen increases like the square of the time elapsed. The fourth line is a statement of Hooke's law: in harmonic motion the period (γ) is independent of the size of the orbit.

c. Fix $\gamma = 1$ and construct a table relating the mass and orbital scale under the four forces described in the table above.

d. Fix $\beta = 1$ and show that the period scales like $\sqrt{M}$ for all homogeneous potentials. Reconcile this result with the well-known result that the period of a planet is independent of its mass in lowest order.

e. If the motion is bounded for all times, show

$$2\langle T \rangle = \langle \mathbf{x} \cdot \nabla V(\mathbf{x}) \rangle = \langle kV(\mathbf{x}) \rangle$$

where T is the kinetic energy. This is the virial theorem for homogenoeous potentials.

f. Show that the kinetic energy scales like $\alpha\beta^2\gamma^{-2} = \beta^k$ (use **a**). Since the potential energy scales the same way, the total energy has this scaling property.

4. Assume that the dynamics of a system are derivable from an action principle. For example, the Euler–Lagrange equations are derived from the variation of an action: $\delta \int \mathcal{L}(\mathbf{x}, \dot{\mathbf{x}}) d\mathbf{x} = 0$. Show that if a scaling transformation leaves the Lagrangian invariant up to an overall scaling factor, the trajectories will scale under this transformation.

5. The heat equation in one dimension is

$$\frac{\partial^2 u}{\partial x^2} = \frac{\partial u}{\partial t}$$

Show that the following six differential operators v_i are infinitesimal generators of the invariance group of this equation. Show that $e^{\epsilon v_i} f(x, t)$ has the action shown for each of the six generators (Olver, 1993):

v_i	Infinitesial	$e^{\epsilon v_i} f(x,t) =$
v_1	∂_x	$f(x-\epsilon, t)$
v_2	∂_t	$f(x, t-\epsilon)$
v_3	$u\partial_u$	$e^\epsilon f(x,t)$
v_4	$x\partial_x + 2t\partial_t$	$f(e^{-\epsilon}x, e^{-2\epsilon}t)$
v_5	$2t\partial_x - xu\partial_u$	$e^{-\epsilon x + \epsilon^2 t} f(x - 2\epsilon t, t)$
v_6	$4xt\partial_x + 4t^2\partial_t - (x^2 + 2t)u\partial_u$	$\lambda e^{-\epsilon \lambda^2 x^2} f(\lambda^2 x, \lambda^2 t)$

where $\lambda^2 = 1/(1+4\epsilon t)$

6. The two-dimensional wave equation is

$$\frac{\partial^2 u}{\partial x^2} + \frac{\partial^2 u}{\partial y^2} = \frac{\partial^2 u}{\partial t^2}$$

Show that the following vector fields map solutions into solutions:

displacements $\quad P_i \ \partial_x, \partial_y, \partial_t$
rotations $\quad L_z \ x\partial_y - y\partial_x$
boosts $\quad B_i \ x\partial_t + t\partial_x, y\partial_t + t\partial_y$
dilations $\quad D_i \ x\partial_x + y\partial_y + t\partial_t, u\partial_u$

inversions
$$\begin{bmatrix} i_x \\ i_y \\ i_t \end{bmatrix} = \begin{bmatrix} x^2 - y^2 + t^2 & 2xy & 2xt & -xu \\ 2yx & -x^2 + y^2 + t^2 & 2yt & -yu \\ 2tx & 2ty & x^2 + y^2 + t^2 & -tu \end{bmatrix} \begin{bmatrix} \partial_x \\ \partial_y \\ \partial_t \\ \partial_u \end{bmatrix}$$

Show that $D_2 = u\partial_u$ commutes with all remaining generators. Construct the commutation relations of the remaining ten generators, and show they satisfy the commutation relations of the conformal group in 2+1 dimensions. Show that this group is $SO(2+1, 1+1) = SO(3, 2)$.

7. Construct the invariance group for the wave equation in $3 + 1$ dimensions. This is the Maxwell equation without sources in space-time. There are 16 infinitesimal generators. Show that 15 satisfy the commutation relations for the conformal group $SO(3+1, 1+1) = SO(4, 2)$ (Bateman, 1910). The extra generator commutes with all the rest, and is $u\partial_u$.

8. The heat equation in one dimension is $u_{xx} - u_t = 0$. The infinitesimal generator of symmetries for this equation is $X = \xi^i \frac{\partial}{\partial x^i} + \eta \frac{\partial}{\partial u} + \cdots = \xi^1 \frac{\partial}{\partial x} + \xi^2 \frac{\partial}{\partial t} + \eta \frac{\partial}{\partial u} + \cdots$. Show that (Stewart, 1989)

$$\xi^1 = a_1 + a_2 x + a_3 t + a_4 xt$$
$$\xi^2 = 2a_2 t + a_4 t^2 + a_5$$
$$\eta = -\tfrac{1}{2} a_3 xu - a_4(\tfrac{1}{2}t + \tfrac{1}{4}x^2)u + a_6 u + h(x,t)$$

Here $h(x, t)$ is any function that satisfies the homogeneous heat equation. Construct the infinitesimal generators corresponding to the arbitrary real coordinates a_i and compute their commutation relations. What is the structure of this Lie algebra?

9. Show that the scalar operator

$$S = t^2 \frac{\partial}{\partial t} + t\mathbf{x} \cdot \nabla - \frac{1}{4}(\mathbf{x} \cdot \mathbf{x} + 2nt) u \frac{\partial}{\partial u}$$

is also a Lie symmetry of Eq. (16.55) with source term.

10. **Noether's theorem for physicists** Many dynamical problems can be expressed in an action principle format:

$$I = \int_{t_1}^{t_2} L(t, x, \dot{x}) dt \qquad \delta I = 0$$

Specifically, the action I is stationary on a physically allowed trajectory. The first variation leads to the Euler–Lagrange equations

$$\frac{d}{dt}\left(\frac{\partial L}{\partial \dot{x}_i}\right) - \frac{\partial L}{\partial x_i} = 0$$

Under a one-parameter family of change of variables ($t \to t' = T(t, x, \epsilon) = t + \epsilon \xi(t, x)$, $x_i \to x'_i = X_i(t, x, \epsilon) = x_i + \epsilon \eta_i(t, x)$) the action integral transforms to

$$I = \int_{t'_1}^{t'_2} L(t', x', \dot{x}')dt' = \int_{t_1}^{t_2} L(t', x', \dot{x}')\frac{dt'}{dt}dt$$

where $dt'/dt = \partial T/\partial t + (\partial T/\partial x_i)dx_i/dt$. Show that if you differentiate the action integral with respect to ϵ, then set $\epsilon = 0$ the result is

$$\int_{t_1}^{t_2} \left(\xi \frac{\partial L}{\partial t} + \eta_i \frac{\partial L}{\partial x_i} + \eta_i^{(1)} \frac{\partial L}{\partial \dot{x}_i} + \frac{d\xi}{dt}L\right) dt = 0$$

Show that by standard arguments the integrand must itself be zero. Show that along an allowed trajectory the vanishing of the integrand can be expressed in the form

$$\frac{d}{dt}\left[\xi L + (\eta_i - \xi \dot{x}_i)L_{\dot{x}_i}\right] = 0$$

The expression within the square brackets is a constant of the motion. Apply this theorem to a Lagrangian that is invariant under space displacements, time displacements, and rotations around a space axis to construct the following conserved quantities:

Symmetry	Conserved quantity
Space displacements	momentum
Time displacements	energy
Space-time displacements	four-momentum
Rotations	angular momentum

16.6 Problems

11. **Noether's theorem, more general** We present a more general form of Noether's theorem than is presented above. This form is very powerful and sufficient for most physical applications. It is not the most general form of Noether's theorem. Suppose the dynamics of a system is derivable from an action integral of the form $L[u] = \int \mathcal{L}(x, u) dx$, $x \in R^p$, $u \in R^q$, and suppose the infinitesimal generators that leave the dynamics invariant have the form

$$\mathbf{v} = \sum_{i=1}^{p} \xi^i(x, u) \frac{\partial}{\partial x^i} + \sum_{\alpha=1}^{q} \phi^\alpha(x, u) \frac{\partial}{\partial u^\alpha}$$

Show that the components P_i defined by

$$P^i = \xi^i \mathcal{L} + \sum_{\alpha=1}^{q} \phi^\alpha(x, u) \frac{\partial \mathcal{L}}{\partial u_i^\alpha} - \sum_{\alpha=1}^{q} \sum_{j=1}^{p} \xi^j u_j^\alpha \frac{\partial \mathcal{L}}{\partial u_i^\alpha}$$

satisfy a conservation law of the form

$$\nabla P = \text{div } P = \frac{\partial P^i}{\partial x^i} = 0$$

12. **Representation theory** G is a compact Lie group with invariant measure $d\rho(g)$ and volume $\text{Vol}(G) = \int d\rho(g)$, $\Gamma^\lambda_{\mu\nu}(g)$ are the irreducible representations of G constructed by reduction of tensor products (Wigner–Stone theorem), and $\phi(g)$, $\psi(g)$ are functions defined on the group manifold. The orthogonality and completeness relations are

$$\int \frac{\dim \lambda}{\text{Vol}(G)} \Gamma^{\lambda'*}_{\mu'\nu'}(g) \Gamma^{\lambda}_{\mu\nu}(g) d\rho(g) = \delta^{\lambda'\lambda} \delta_{\mu'\mu} \delta_{\nu'\nu}$$

$$\sum_\lambda \sum_\mu \sum_\nu \frac{\dim \lambda}{\text{Vol}(G)} \Gamma^{\lambda*}_{\mu\nu}(g') \Gamma^{\lambda}_{\mu\nu}(g) = \delta(g', g)$$

Introduce Dirac notation for these matrix elements:

$$\left\langle g \middle| \begin{matrix} \lambda \\ \mu\nu \end{matrix} \right\rangle = \sqrt{\frac{\dim \lambda}{\text{Vol}(G)}} \Gamma^\lambda_{\mu\nu}(g) \qquad \left\langle \begin{matrix} \lambda \\ \mu\nu \end{matrix} \middle| g \right\rangle = \sqrt{\frac{\dim \lambda}{\text{Vol}(G)}} \Gamma^{\lambda*}_{\mu\nu}(g)$$

(a) **a.** Write the orthogonality and completeness relations in Dirac notation and show:

$$\int d\rho(g) \left\langle \begin{matrix} \lambda' \\ \mu'\nu' \end{matrix} \middle| g \right\rangle \left\langle g \middle| \begin{matrix} \lambda \\ \mu\nu \end{matrix} \right\rangle = \left\langle \begin{matrix} \lambda' \\ \mu'\nu' \end{matrix} \middle| \begin{matrix} \lambda \\ \mu\nu \end{matrix} \right\rangle$$

$$\sum_\lambda \sum_\mu \sum_\nu \left\langle g' \middle| \begin{matrix} \lambda \\ \mu\nu \end{matrix} \right\rangle \left\langle \begin{matrix} \lambda \\ \mu\nu \end{matrix} \middle| g \right\rangle = \langle g'|g \rangle$$

b. Show that the orthogonality and completeness relations can be expressed in the form of "resolutions of the identity" in appropriate spaces:

$$|g\rangle\langle g| \;=\; \int |g\rangle d\rho(g)\langle g| \;=\; I \quad \text{in group space}$$

$$\left| \begin{matrix} \lambda \\ \mu\nu \end{matrix} \right\rangle\!\!\left\langle \begin{matrix} \lambda \\ \mu\nu \end{matrix} \right| = \sum_\lambda \sum_\mu \sum_\nu \left| \begin{matrix} \lambda \\ \mu\nu \end{matrix} \right\rangle\!\!\left\langle \begin{matrix} \lambda \\ \mu\nu \end{matrix} \right| = I \text{ in representation space}$$

c. Carry out a Fourier decomposition on the functions $\psi(g) = \langle g|\psi\rangle$ and $\langle {}^{\ \lambda}_{\mu\nu}|\psi\rangle = \int d\rho(g)\langle {}^{\ \lambda}_{\mu\nu}|g\rangle\langle g|\psi\rangle$ (and similarly for $\phi(g) = \langle g|\phi\rangle$) using the Dirac representation. Write down the Parseval equality for the inner product $\int \phi^*(g)\psi(g)d\rho(g)$ expressed in terms of the discrete and continuous basis vectors in this Hilbert space.

Bibliography

F. T. Arecchi, E. Courtens, R. Gilmore, and H. Thomas (1972), Atomic coherent states in quantum optics, *Phys. Rev. A* **6**, 2211–2237.

G. A. Baker, Jr. (1958), Degeneracy of the n-dimensional isotropic harmonic oscillator, *Phys. Rev.* **103**, 1119–1120.

M. Bander and C. Itzykson (1966a), Group theory and the hydrogen atom (I), *Rev. Mod. Phys.* **38**, 330–345.

M. Bander and C. Itzykson (1966b), Group theory and the hydrogen atom (II), *Rev. Mod. Phys.* **38**, 346–358.

V. Bargman and E. P. Wigner (1948), Group theoretical discussion of relativistic wave equations, *Proc. Nati. Acad. Sci. (US)* **34**, 211–223.

A. O. Barut and G. L. Fronsdal (1971), $SO(4,2)$-formulation of symmetry-breaking in relativistic Kepler problems with or without magnetic charges, *J. Math. Phys.* **12**, 841–846.

A. O. Barut and R. Raczka (1977), *Theory of Group Representations and Applications*, Warsaw: PWN Polish Scientific Publications.

A. O. Barut and W. Rasmussen (1971), Non-relativistic and relativistic Coulomb amplitude as the matrix element of a rotation in $SO(4,2)$, *Phys. Rev. D* **3**, 956–959.

H. Bateman (1910), The transformation of the electrodynamical equations, *Proc. London Math. Soc.* **8**, 223–264.

O. Bely (1966), Quantum defect theory III. Electron scattering by He^+, *Proc. Phys. Soc.* **88**, 833–842.

G. Berendt, E. Weimar, and R. Gilmore (1975), Harmonic oscillator Green's function from a BCH formula, *J. Math. Phys.* **16**, 1231–1233.

H. A. Bethe and E. E. Salpeter (1957), *Quantum Mechanics of One- and Two-Electron Atoms*, Berlin: Springer-Verlag.

L. C. Biedenharn (1962), Invariant operators of the Casimir type, *Phys. Lett.* **3**, 69–70.

G. W. Blumen and G. D. Cole (1969), *Symmetries and Differential Equations*, New York: Springer-Verlag.

C. E. Burkhardt and J. J. Leventhal (2004), Lenz vector operators on spherical hydrogen atom eigenfunctions, *Am. J. Phys.* **72**, 1013–1016.

H. D. Doebner and O. Melsheimer (1967), On a class of generalized group contractions, *Nuovo Cimento A* **49**, 306–311.

L. Dresner (1999), *Applications of Lie's Theory of Ordinary and Partial Differential Equations*, Bristol: IOP Publishing.

F. Estabrook and H. Wahlquist (1975), Prolongation structures of nonlinear evolution equations. *J. Math. Phys.* **16**, 1–7.

V. A. Fock (1935), Zur theorie des Wasserstoffatoms, *Z. Phys.* **98**, 145–154.

L. L. Foldy and S. A. Wouthuysen (1950), On the Dirac theory of spin 1/2 particles and its nonrelativistic limit, *Phys. Rev.* **78**, 29–36.

C. Fronsdal (1965), Infinite multiplets and the hydrogen atom, *Phys. Rev.* **156**, 1665–1677.

T. Fulton, F. Rohrlich, and L. Witten (1962), Conformal invariance in physics, *Rev. Mod. Phys.* **34**, 442–457.

G. Gabrielse, D. Hanneke, T. Kinoshita, M. Nio, and B. Odom (2006), New determination of the fine structure constant from the electron g value and QED, *Phys. Rev. Lett.* **97**, 030802.

I. M. Gel'fand and M. L. Tsetlein (1950), Matrix elements for the unitary groups, *Dokl. Akad. Nauk SSSR* **71**, 825–828.

I. M. Gel'fand and M. L. Tsetlein (1950), Matrix elements for the orthogonal groups, *Dokl. Akad. Nauk SSSR* **71**, 1017–1020.

R. Gilmore (1970), Construction of weight spaces for irreducible representations of $A_n; D_n, B_n, C_n$, *J. Math. Phys.* **11**, 513–523.

R. Gilmore (1970), Spin representations of the orthogonal groups, *J. Math. Phys.* **11**, 1853–1854.

R. Gilmore (1970), Spectrum of Casimir invariants for the simple classical Lie groups, *J. Math. Phys.* **11**, 1855–1856.

R. Gilmore (1970), Diagrammatic technique for constructing matrix elements, *J. Math. Phys.* **11**, 3420–3427.

R. Gilmore (1974a), *Lie Groups, Lie Algebras, and Some of Their Applications*, New York: Wiley, 1974; republished New York: Dover.

R. Gilmore (1974b), Baker–Campbell–Hausdorff formulas, *J. Math. Phys.* **15**, 2090–2092.

R. Gilmore (1977), Structural stability of the phase transition in Dicke-like models, *J. Math. Phys.* **18**, 17–22.

R. Gilmore (1985), Uncertainty relations of statistical mechanics, *Phys. Rev. A* **31**, 3237–3239.

R. Gilmore (2004), *Elementary Quantum Mechanics in One Dimension*, Baltimore, MD: Johns Hopkins University Press.

R. Gilmore (2006), Lie groups: general theory. In: J.-P. Francoise, G. Naber, and S. T. Tsu, eds., *Encyclopedia of Mathematical Physics*, Amsterdam: Elsevier, pp. 286–304.

R. Gilmore and C. M. Bowden (1976a), Coupled order-parameter treatment of the Dicke hamiltonian, *Phys. Rev. A* **13**, 1898–1907.

R. Gilmore and C. M. Bowden (1976b), Bifurcation properties of the Dicke hamiltonian, *J. Math. Phys.* **17**, 1617–1625.

R. Gilmore and J. M. Yuan (1987), Group theoretical approach to semiclassical dynamics: single mode case, *J. Chem. Phys.* **86**, 130–139.

R. Gilmore and J. M. Yuan (1989), Group theoretical approach to semiclassical dynamics: multimode case, *J. Chem. Phys.* **91**, 917–923.

R. Gilmore, H. G. Solari, and S. K. Kim (1993), Algebraic description of the quantum defect, *Found. Phys.* **23**, 873–879.

R. J. Glauber (1963), Coherent and incoherent states of the radiation field, *Phys. Rev.* **131**, 2766–2788.

H. Goldstein (1950), *Classical Mechanics*, Reading, MA: Addison-Wesley.

S. HELGASON (1962), *Differential Geometry and Symmetric Spaces*, New York: Academic Press.

S. HELGASON (1978), *Differential Geometry, Lie Groups, and Symmetric Spaces*, New York: Academic Press.

L. K. HUA (1963), *Harmonic Analysis of Functions of Several Complex Variables in the Classical Domains*, Translations of Mathematical Monographs, Vol. 6, Providence, RI: American Mathematical Society.

E. INÖNÜ AND E. P. WIGNER (1953), On the contraction of groups and their representations, *Proc. Natl. Acad. Sci. (US)* **39**, 391–402.

S. KAIS AND S. K. KIM (1986), Unstable bound states of the Dirac equation by an algebraic approach, *Phys. Lett. A* **114**, 47–50.

P. KUSTAANHEIMO AND E. STIEFEL (1965), Perturbation theory of Kepler motion based on spinor regularization, *J. Reine Angew. Math.* **218**, 204.

L. D. LANDAU AND E. M. LIFSHITZ (1960), *Mechanics*, Reading, MA: Addison-Wesley.

S. LANG (1984), *Algebra*, Reading, MA: Addison-Wesley.

I. A. MALKIN AND V. I. MAN'KO (1965), Symmetry of the hydrogen atom, *Sov. Phys. JETP Lett.* **2**, 146–148.

H. V. MCINTOSH, Symmetry and the hydrogen atom, http://delta.cs.cinvestav.mx/~mcintosh/comun/symm/symm.html.

W. MILLER, JR. (1968), *On Lie Algebras and Some Special Functions of Mathematical Physics*, Memoirs of the American Mathematical Society, vil **50**, Providence, RI: American Mathematical Society.

V. I. OGIEVETSKII AND I. V. POLUBARINOV (1960), Wave equations with zero and nonzero rest masses, *Sov. Phys, JETP* **10**, 335–338.

P. OLVER (1993), *Applications of Lie Groups to Differential Equations*, 2nd edn., New York: Springer.

W. PAULI (1926), On the hydrogen spectrum from the standpoint of the new quantum mechanics, *Z. Phys.* **36**, 336–363. English translation in: B. L. VAN DER WAERDEN, ED., *Sources of Quantum Mechanics*, New York: Dover, 1967, pp. 387–415.

J. RAMOS AND R. GILMORE (2006), Derivation of the source-free Maxwell and gravitational radiation equations by group theoretical means, *Int. J. Mod. Phys.* **15**(4), 505–519.

R. H. RISCH (1969), The problem of integration in finite terms, *Trans. Am. Math. Soc.* **139**, 167–189.

D. A. SADOVSKIÍ AND B. I. ŽHILINSKIÍ (1998), Tuning the hydrogen atom in crossed fields between the Zeeman and Stark limits, *Phys. Rev. A* **57**, 2867–2884.

E. J. SALETAN (1961), Contraction of Lie groups, *J. Math. Phys.* **2**, 1–21.

L. I. SCHIFF (1968), *Quantum Mechanics*, 3rd edn., New York: McGraw Hill.

J. SCHWINGER (1965), On angular momentum. In L. C. BIEDENHARN AND H. VAN DAM, EDS., *Quantum Theory of Angular Momentum*, New York: Academic Press, pp. 229–279.

M. J. SEATON (1966a), Quantum defect theory I. General formulation, *Proc. Phys. Soc.* **88**, 801–814.

M. J. SEATON (1966b), Quantum defect theory II. Illustration on one-channel and two-channel problems, *Proc. Phys. Soc.* **88**, 815–832.

H. STEPHANI AND M. MACCALLUM (1989), *Differential Equations, Their Solutions Using Symmetery*, Cambridge: Cambridge University Press.

I. STEWART (1989), *Galois Theory*, London: Chapman and Hall.

E. L. STIEFEL AND G. SCHEIFELE (1971), *Linear and Regular Celestial Mechanics*, Berlin: Springer-Verlag.

J. D. TALMAN (1968), *Special Functions: A Group Theoretic Approach (Based on Lectures by Eugene P. Wigner)*, New York: Benjamin.

N. JA VILENKIN (1968), *Special Functions and the Theory of Group Representations, Translations of Mathematical Monographs*, vol. **22**, Providence, RI: American Mathematical Society.

H. WAHLQUIST AND F. ESTABROOK (1976), Prolongation structures of nonlinear evolution equations. II, *J. Math. Phys.* **17**, 1293–1297.

S. WEINBERG (1964), Feynman rules for any spin. II. Massless particles, *Phys. Rev. B* **134**, 882–896.

G. H. WEISS AND A. A. MARADUDIN (1962), The Baker–Campbell formula and a problem in crystal physics, *J. Math. Phys.* **3**, 771–777.

H. WEYL (1946), *The Classical Groups*, Princeton, NJ: Princeton University Press.

E. P. WIGNER (1939), On unitary representations of the inhomogeneous Lorentz group *Ann. Math.*, **40**, 149–204.

E. P. WIGNER (1954), Conservation laws in classical and quantum physics, *Progr. Theor. Phys.* **11**, 437–440.

E. P. WIGNER (1957), Relativistic invariance and quantum phenomena *Rev. Mod. Phys.* **29**, 255–268.

E. P. WIGNER (1959), *Group Theory and its Application to the Quantum Mechanics of Atomic Spectra*, New York: Academic. Press.

R. M. WILCOX (1967), Exponential operators and parameter differentiation in quantum physics, *J. Math. Phys.* **8**, 962–982.

D. P. ZHELOBENKO (1962), The classical groups, spectral analysis of their finite dimensional representations, *Russ. Math. Surveys* **17**, 1–92.

Index

$A(p\,q)$, 39, 48
A_1, 161
A_2, 151, 160
A_3, 46, 161, 162
A_n, 46, 49, 161, 166
B_1, 161
B_2, 151, 160, 161, 164
B_3, 162
B_n, 161, 166, 168
C_1, 161
C_2, 151, 160, 161, 164
C_3, 162
C_n, 161, 166, 168
D_2, 151, 160, 162
D_3, 161, 162
D_n, 161, 166, 168
$E(2)$, 91, 207
$E(3)$, 42
E_6, 162, 168
E_7, 162, 168
E_8, 162, 168
$F(n)$, 45, 49
F_4, 162, 168
$GL(1;\mathbb{Q})$, 40, 47
$GL(2;\mathbb{C})$, 47
$GL(2;\mathbb{R})$, 43
$GL(2;\mathbb{Z})$, 45, 49
$GL(3;\mathbb{Z})$, 45
$GL(n;\mathbb{C})$, 47
$GL(n;\mathbb{F})$, 34, 36, 74
$GL(n;\mathbb{Q})$, 47
$GL(n;\mathbb{R})$, 47, 104
$GL(n;\mathbb{Z})$, 44, 45, 49, 81
G_2, 151, 160, 162, 165
$HT(p,q)$, 37, 48
H_1^2, 103, 189
H_2^2, 102, 189, 191
H_4, 211
$ISO(2)$, 91, 206, 207, 268
$ISO(2)$, little group, 267
$ISO(3)$, 208, 209
$Nil(n)$, 38, 48

$O(3\,1)$, 261
$O(3)$, 40, 78
$O(3;\mathbb{Z})$, 46
$O(n)$, 40, 43, 145
$O(n;\mathbb{G})$, 41
$O(n;\mathbb{Z})$, 45, 49
$O(p,q)$, 43
$OU(2n)$, 43
P_n, 45
$SL(2;\mathbb{C})$, 43
$SL(2;\mathbb{R})$, 26, 28, 29, 30, 41, 43, 56, 58, 62, 100, 102, 189
$SL(n;\mathbb{R})$, 30
$SL(n;\mathbb{C})$, 43, 47
$SL(n;\mathbb{Q})$, 43
$SL(n;\mathbb{R})$, 43, 47, 164
$SL(n;\mathbb{Z})$, 45
$SO(2,1)$, 105
$SO(2,1)$, little group, 267
$SO(2,1)/SO(2)$, 106
$SO(2)$, 48, 164
$SO(2n)$, 164
$SO(2n+1)$, 164
$SO(3,1)$, 263
$SO(3,1)$, little group, 267, 269
$SO(3,2)$, 210
$SO(3)$, 49, 90, 106
$SO(3)$, little group, 267
$SO(3)/SO(2)$, 107
$SO(4,1)$, 210
$SO(5)$, 164
$SO(n)$, 43, 145
$SO(p,q)$, 43, 164
$SU(1;\mathbb{Q})$, 40, 48
$SU(1,1)$, 38, 43, 48, 105
$SU(1,1)/U(1)$, 106
$SU(2)$, 48, 106
$SU(2)/U(1)$, 107
$SU(n)$, 43, 90, 164
$SU(p,q)$, 43, 164
S^2, 189, 191
S_3, 5, 46

S_n, 45, 49
$Sol(n)$, 38, 48
$Sp(1)$, 40
$Sp(2; \mathbb{R})$, 41
$Sp(2n; \mathbb{R})$, 41
$Sp(n)$, 40, 164
$Sp(n; \mathbb{C})$, 41
$Sp(n; \mathbb{G})$, 41
$Sp(n; \mathbb{R})$, 41
$Sp(p, q)$, 164
$U(1, 1)$, 43
$U(2)$, 40, 78
$U(2)$, contraction of, 211
$U(2; \mathbb{Q})$, 164
$U(n)$, 40, 43, 90
$U(n)$, representations of, 90
$U(n; \mathbb{G})$, 41
$U(p, q)$, 43
$USp(2n)$, 44
$UT(1, 1)$, 83
$UT(p, q, r)$, 37, 48
$UT(p, q)$, 48
V_4, 15
$\mathbb{Z}$, integers, 44

abelian group, 39
active interpretation, of group action, 93
aether, 282
affine transformations, 37
algebraic constraints, 29
algebraic equations, 3
algebraic manifold, 29, 104
algebras, contraction of, 211
alternating group, 5, 46
amplitudes, external, 54
    internal, 54
analytic, continuation, 40, 86, 142, 143, 176
    reparameterization, 113
angular momentum, matrix elements, 217
    operators, 258
    states, 213
annihilation operators, 84, 88
    bosons, 88
    fermions, 89
    two photon, 77
anticommutation relations, 89
anticommutator, 89
anticommute, 47
antihermitian matrices, 78
antipodal points, 106
Araki–Satake root diagram, 192
associativity, 4, 24, 25
Automorphism, involutive, 177
auxiliary equation, 11
    for cubic, 14, 20
    for quartic, 15, 18

Baker–Campbell–Hausdorff formulas, 108
basis, 61
basis functions, 9

basis states, contraction of, 214
BCH formulas, 108
    contraction of, 215
Bessel functions, 217
bilinear constraints, 39
block diagonal, 64
block matrix decomposition, 178
Bohr radius, 253
Boltzmann constant, 116
boost, 31
Bose–Einstein counting problem, 95, 96
Bose–Einstein statistic, 256
boson operator algebras, 88
boson operators, 88
bounded, 27
building up principle, 159
building up process, 161

c-number, 127
canonical commutation relations, 151, 159, 172
canonical coordinates, 286, 302
Cartan, covering theorem, 107
    decomposition, 84
Cartan–Killing form, 65
Cartan–Killing inner product, 65, 82, 102, 139, 147
Casimir covariants, 157
Casimir invariants, 143, 148
Casimir operators, 153, 159, 192, 201, 217
    contraction of, 207, 212
    higher order, 146
Cauchy, 301
Cayley–Hamilton theorem, 58, 157
character table, 9
    of S_2, 10
    of S_3, 12
    of S_4, 16
character, of real form, 175
characteristics, method of, 289
Christoffel symbol, 200
classical functions, 2
classical problems, double a cube, 2
    square a circle, 2
    trisect an angle, 2
Clebsch–Gordan series, 264
closed, 27
closure, 4, 24, 25
Columbus, 25
Commutation, 59
commutation relations, 89
    C_2, 153
commutative, 3, 133
commutative group, 39
commutator, 59
    in algebra, 59
    in group, 59
commuting operators, 192
compact, 26
    and metric, 65
compass, 22
complementary series, of representations, 187

completeness relations, 307
  special functions, 216
complex extension, 164
complex numbers, 34, 35
conformal condition, 235
conformal group, 201, 305
conformal map, 203
conjugate subgroups, 6
connectivity matrix, 54
conservation, of momentum, 51
constraint equation, 285, 294, 303
constraints, 35
constructable numbers, 22
contraction, 205
Contraction, of $U(2)$, 211
  of algebras, 211
  of basis states, 214
  of BCH formulas, 215
  of Casimir operators, 212
  of Dynkin diagram, 167
  of groups, 205
  of matrix elements, 214
  of parameter space, 213
  of representations, 213
  of special functions, 215
coordinate representation, 273
coordinate, dependent, 289
  independent, 289
coset, 8, 103, 104
Coset representative, 104, 267
cover, open, 25
covering group, 105, 107
  $\overline{SO(2,1)/SO(2)}$, 108
  $\overline{SU(1,1)/U(1)}$, 108
  universal, 107
covering problem, 100
creation operators, 84, 88
  bosons, 88
  fermions, 89
  two photon, 77
crossing symmetry, 52
cubic equation, 1, 11, 22
  Galois group, 12
cylinder, 294, 302

defining matrix representation, 131
degeneracy, and symmetry, 230
dependent coordinate, 289
DeSitter symmetry, 235
determining equation, 286, 287, 302
Dicke model, 126
diffeomorphism, 109
differential equations, 284
  and Lie groups, 284
differential operators, first order, 90
dimension, 61
  of manifold, 26
  of root space, 153
direct product group, 8
discrete invariant subgroup, 107

discrete series, of representations, 187
discriminant, 11
dispersion relation, 223
double the cube, 22
dynamical symmetry, 230
Dynkin diagram, 159, 165, 166
  contraction of, 167

eigenoperator, commutation relations, 140
  decomposition, 139
electromagnetic field, 259
embedded groups, 43
entropy representation, 282
equation, constraint, 285, 294
  determining, 286, 287
  surface, 285, 294
equilibrium, thermodynamic, 116
equivalence principle, 93, 223, 250
Euclidean, group, 42
  motions, 207
  submanifold, 192
  transformations, 79
EXP, 57
EXPonential, 55, 58
  operation, 59
EXPonentiation, 99

factor group, 8
faithful, 7
  representation, 5, 122
fermion operator algebras, 89
fermion operators, 89
Fibonacci number, 45, 49
Fibonacci-type series, 49
field, 259
  equations, 262
  theory, 3
fine structure constant, 225
first order equations, 286
first prolongation, 287
fluctuation–dissipation theorem, 283
Fock space, 213
four-group, 15
Frobenius method, 225
fully reducible, 63, 134
fundamental roots, 166

Galilei group, 42, 80, 86
Galilean transformation, 48
Galois, 1, 284, 301
Galois group, 4, 21
  for quartic, 15
Galois theory, 3
Galois' theorem, 9
general linear, algebras, 74
  groups, 36
generating function, 217
geometric symmetry, 227
globally symmetric spaces, 190
gravitons, 283

group theory, 3
group, "infinite", 1
  abelian, 6
  axioms, 3, 24
  commutative, 6
  composition function, 28
  composition map, 28
  elements, 24
  generators, 6
  inversion map, 28
  multiplication, 3, 5, 24
  operations, 3, 24
group-subgroup chain, 12, 15
group-subgroup diagram, 7
Groups, intersections of, 80

Hamilton's equations, 39, 41, 180
harmonic oscillator wavefunctions, 215
harmonic oscillator, isotropic, 96
heat equation, 300
Heisenberg, algebra, 89
Heisenberg, commutation relations, 77
  group, 38
  identity, 110
helicity, of photon, 259
  state, 259
Hermite polynomials, 97, 215
higher order equations, 299
Hilbert–Schmidt inner product, 64
homogeneous Lorentz group, 261
homogeneous Lorentz transformation, 263
homogeneous polynomials, 140, 256
homomorphic image, 7
homomorphism, 7
Hooke's law, 304
hyperbolic plane, 202
hyperboloid, 27, 29
  single-sheeted, 102, 103, 189
  two-sheeted, 102, 189

identity, 4, 24, 25
Inönü–Wigner contraction, 205, 206
indefinite metric, 40, 197
independent coordinate, 289
independent functions, 192
independent roots, 192
index, of real form, 175
inertial frame, 282
infinitesimal generator, 286, 295
inhomogeneous Lorentz group, 210, 261
inner product, 61, 64
integrability condition, 61
interpretations of group action, active, 93
  passive, 93
intersections, of groups, 43
invariance algebra, 96
invariant, measure, 66, 193
  metric, 66, 193
  operators, 143, 148, 159
  subalgebra, 134

subgroup, 6, 8
subspace, 36
inverse, 4, 24, 25
  image, 7
inversion mapping, 30
involutive automorphism, 177
irreducible, 63, 134
  representations, 10
isomorphism, 7
  problem, 105
isotropic, 191

Jacobi identity, 59, 60, 149
Jacobi polynomials, 215, 217

Kepler's third law, 304
Klein four-group, 15
Klein group, 15
Klein–Gordon equation, 224
Kustaanheimo–Stiefel transformation, 240

Laplace equation, 299
Laplace–Beltrami operators, 192, 200
Laplace–Runge–Lenz vector, 230
Laplacian operators, 208
laziness, principle of maximum, 211
Legendre polynomials, 215, 217
Levi–Civita skew tensor, 156
Levi–Civita symbol, 143
Lie, 1, 284, 307
Lie algebra, $\mathfrak{a}(p,q)$
  $\mathfrak{a}(p,q)$, 77, 129
  $\mathfrak{gl}(n;\mathbb{F})$, 74, 83
  $\mathfrak{ht}(p,q)$, 75
  $\mathfrak{nil}(n)$, 77, 130
  $\mathfrak{ou}(2n)$, 179
  $\mathfrak{o}(n;G)$, 79
  $\mathfrak{o}(p,q)$, 78
  $\mathfrak{sl}(2;\mathbb{R})$, 100, 102, 154, 173
  $\mathfrak{sl}(n)$, 80
  $\mathfrak{sl}(n;\mathbb{C})$, 80, 85, 86
  $\mathfrak{sl}(n;\mathbb{Q})$, 80, 86
  $\mathfrak{sl}(n;\mathbb{R})$, 80, 85, 178, 180
  $\mathfrak{sol}(n)$, 77, 130
  $\mathfrak{so}(2,1)$, 78
  $\mathfrak{so}(2n)$, 180
  $\mathfrak{so}(3,1)$, 79
  $\mathfrak{so}(3,2)$, 86
  $\mathfrak{so}(3)$, 86, 90, 154
  $\mathfrak{so}(4,1)$, 86
  $\mathfrak{so}(4)$, 132
  $\mathfrak{so}(5)$, 86, 146
  $\mathfrak{so}(n)$, 132, 145, 178
  $\mathfrak{so}(p,q)$, 84, 178
  $\mathfrak{so}^*(2n)$, 180
  $\mathfrak{sp}(2n;\mathbb{R})$, 178, 179, 180
  $\mathfrak{sp}(G;\mathbb{C})$, 79
  $\mathfrak{sp}(G;\mathbb{R})$, 79
  $\mathfrak{sp}(n)$, 132, 178
  $\mathfrak{sp}(n;G)$, 79
  $\mathfrak{sp}(p,q)$, 78, 178

$\mathfrak{su}(1,1)$, 140, 141, 143, 173
$\mathfrak{su}(2)$, 111, 140, 141, 143, 173
$\mathfrak{su}(2n)$, 180
$\mathfrak{su}(n)$, 80, 132, 178
$\mathfrak{su}(p,q)$, 85, 178
$\mathfrak{su}^*(2n)$, 180
$\mathfrak{usp}(2n)$, 179
$\mathfrak{ut}(1,1)$, 83
$\mathfrak{ut}(p,q,r)$, 76
$\mathfrak{ut}(p,q)$, 75, 131
$\mathfrak{u}(n)$, 80
$\mathfrak{u}(n; \mathbb{F})$, 178
$\mathfrak{u}(n; G)$, 79
$\mathfrak{u}(p,q)$, 78
$\mathfrak{u}(p,q; \mathbb{F})$, 178
$\mathfrak{sl}(2; \mathbb{C})$, 154
Lie algebra, $\mathfrak{sl}(2; \mathbb{R})$, 62
Lie algebras, 55, 56
Lie algebras, properties of, 59
Lie groups, 2, 21, 28
   and differential equations, 284
   global properties, 57
   local properties, 57
Lie symmetries, 296, 300
light cone, 101
limit points, 27
linear constraints, 36
little group, 267
local groups, 302
local Lie groups, 302
loops, none in Dynkin diagrams, 167
Lorentz group, 31, 40, 79, 260
   homogeneous, 261
   in a plane, 78
   inhomogeneous, 261
Lorentz transformations, 31, 42, 210, 259
   homogeneous, 263
lowering operators, 228

Manifestly covariant, 259
   representations, 264
manifold, 25, 55
matrix elements, 2
Matrix elements
   angular momentum, 217
   contraction of, 214
matrix groups, 29
Matrix groups, 34
matrix inversion, 29
matrix multiplication, 5, 29
matrix representations, 2, 5, 7
Maxwell's Equations, 259, 260, 305
measure, 66, 193
   invariant, 66, 193
mechanical similarity, 303
method of characteristics, 289
metric, 66, 193
metric preserving groups, antisymmetric, 79
metric-preserving groups, antisymmetric metric, 41
   compact, 39, 78
   general metric, 41
   noncompact, 40, 78
   singular, 79
metric preserving groups, antisymmetric, 79
metric tensor, 193, 197
metric, invariant, 66, 193
Michelson–Morely experiment, 282
microwave background radiation, 282
minimal electromagnetic coupling, 223
Minkowski, transformation, 176
   trick, 177
modular groups, 44, 81
momentum conservation, 51
momentum representation, 273
multilinear constraints, 42, 80
multiplication table, 9
multiply connected, 197
Mutually commuting operators, 153, 159

network, 54
network topology, 54
neutrinos, 283
nilpotent, 65, 130, 133
   algebras, 77, 141
   groups, 38
Noether's theorem, 307
noncompact, 26
nonsemisimple, 63, 134
   group, 2
normally ordered, 112

one-parameter group, 287
operator algebras, 88
operators, momentum, 38
   position, 38
order, normal, 112
   of a group, 8
orthogonal groups, 40, 78
orthogonality relations, 307
   special functions, 216

parameter space, contraction of, 213
parameterization problem, 108
Parseval inequality, 308
partial differential equations, 299
partition function, 116
Pascal triangle, 257
Passive interpretation, of group action, 93
Pauli spin matrices, 31, 78
Periodic table, Mendelyeev, 50
permutation, group
   group, 4
   matrix, 4, 5
   representation, 45
   transformation, 141
phase shift, 248
photon, 259, 275
   number states, 213
   operators, 38, 77, 84, 110, 130, 136, 140, 146, 211
Poincaré plane, 202
Poincaré group, 42, 80, 86, 210

318 Index

point transformations, 302
polarization, 259
  and inner products, 69
polynomial equation, 4
principal series, 187
  of representations, 187
principle of equivalence, 223, 250
principle of relativity, 223, 250
problems, of antiquity, 22
projective transformation, 234
prolongations, first
  first, 287, 302
  higher order, 299
  second, 296
pseudo–Riemannian symmetric space, 190, 197

quadratic constraints, 39
quadratic equation, 1, 10
  Galois group, 10
quadratic resolvent, 20
quadrature, 2, 284
quadrupole tensor operators, 258
quantum number, principle, 50
quartic equation, 1, 15
quaternions, 34, 35, 47
quintic equation, 1, 17
  Galois group, 2
quotient, 8, 103, 104
quotient, space, 8

radial quantum number, 225
radicals, 1, 284, 301
raising operators, 228
rank, 143, 148, 153
  for symmetric space, 192
real form, 172
  character of, 175
  classical algebras, 181
  classical equivalences, 181
  compact, 174
  exceptional algebras, 182
  index of, 175
  least compact, 174
real numbers, 34, 35
recursion relation, root chain, 149
reducible, 63, 134
reduction of order, 298
regular elements, 146
regular representation, 62, 129, 139
relativity, principle of, 250, 223
reparameterization, local, 113
representation, 4
  contraction of, 213
  coordinate, 273
  faithful, 122
  irreducible, 187
  manifestly covariant, 264
  momentum, 273
  reducible, 187
  unitary, 187
  unitary irreducible, 262, 264, 266

representations, of $SU(2)$, 187
  of $SU(1, 1)$, 187
resolvent equation, 13
Riccati equation, 295
Riemannian globally symmetric space, 192
Riemannian space, 191
Riemannian symmetric space, 189, 190
Risch, 302
Rodriguez formula, 97
root chain, 150
  recursion relation, 149
root reflections, 150
root space, 148, 159
  decomposition, 160
  diagram, 147, 151, 153, 159, 160, 172
roots, 148, 153
  of secular equation, 159
  properties of, 159
ruler, 22
Rydberg electron, 248

scaling transformation, 291, 295, 300
scattering matrix, 52
scattering phase shift, 248
Schrödinger equation, 52, 223, 224
Schrödinger prescription, 224
Schur's Lemma, 107
Schwarz inequality, 160, 167
Schwinger representation, 94, 232, 238
second order equations, 296
second prolongation, 296
secular equation, 58, 139, 140, 148, 159, 192
  independent coefficients, 148, 153
  independent functions, 159
  roots of, 159
self-conjugate, 6
semidirect sum, 206
semisimple, 63, 134
  group, 2
  Lie algebras, 147
sheets, 49
similarity transformations, 62
simple, 63, 134
  group, 2
simply connected, 107
single-sheeted hyperboloid, 102, 103
solution surface, cylinder, 294
solvable, 133
  algebras, 77
  group, 2, 38
space-time, 176
  coordinates, 31
special functions, 215
  completeness relations, 216
  contraction of, 215
  orthogonality relations, 216
special linear groups, 43, 80
special relativity, 282
spectrum generating, algebra, 96, 258
  group, 245
speed of light, c, 282

spherical harmonics, 215, 217, 225
spin groups, and $SO(n)$, 183
spin states, 40, 259
spinor, of $SO(3)$, 164
  of $SO(5)$, 164
splitting map, 177
splitting transformation, 177
square the circle, 22
squeezed states, 38
stability subgroup, of a vector, 267
structure constants, 61, 151, 153, 160
Structure factor, 122
structure theory, for lie algebras, 129
  for simple lie algebras, 139
subalgebra, 65
  invariant, 134
subfield restriction, 178
subgroup, 5
  invariant, 6
  normal, 6
surface equation, 285, 294, 302
symmetric, group
  group, 4
  matrix, 27
  polynomials, 9
  spaces, 189
symmetry, and degeneracy, 230
  crossing, 52
symplectic group, 40, 78
symplectic transformations, 180

tensor, 259
thermal expectation values, 116
Thomas precession, 31
time-ordered product, 114
time-reversal operator, 267
topological space, 25
topology, 25

transfer matrix, 51, 52
transformation, scaling, 295
translation group, 39
trisect an angle, 23
Tschirnhaus transformation, 11, 20
  for cubic, 13
  for quartic, 18
Tschirnhaus transformation, for quartic, 15
two-photon algebra, 77, 146
two-sheeted hyperboloid, 102

uncertainty relations, of statistical mechanics, 282
unimodular groups, 43
unit disk, 203
unit sphere, 25
unitary groups, 40, 78, 90
unitary irreducible representations, 262, 264, 266
unitary representation, 38
Universal covering group, 107
upper half-plane, 202
upper triangular, 130
  algebras, 75
  and photon operators, 109
  groups, 36

Van der Monde matrix, 158
variables, dependent, 285
  independent, 285
velocity addition law, 31
vierergruppe, 15, 199
viscous medium, 283

wave equation, 224
Weyl group, 156
  of reflections, 155
Weyl symmetry, 150
Wick rotation, 114
Wigner–Stone theorem, 216, 307

Printed in the United States
By Bookmasters